# Communications and Control Engineering



For further volumes:
www.springer.com/series/61

Arben Çela • Mongi Ben Gaid • Xu-Guang Li •
Silviu-Iulian Niculescu

# Optimal Design of Distributed Control and Embedded Systems

Springer

Arben Çela
Department of Computer Science
and Telecommunication
Université Paris-Est, ESIEE Paris
Noisy-le-Grand, France

Mongi Ben Gaid
Electronic and Real-Time Systems
Department
IFP New Energy
Rueil-Malmaison, France

Xu-Guang Li
School of Information Science
and Engineering
Northeastern University
Shenyang, People's Republic of China

Silviu-Iulian Niculescu
L2S—Laboratoire des signaux et systèmes
Supélec
Gif-sur-Yvette, France

ISSN 0178-5354 ISSN 2197-7119 (electronic)
Communications and Control Engineering
ISBN 978-3-319-37663-9 ISBN 978-3-319-02729-6 (eBook)
DOI 10.1007/978-3-319-02729-6
Springer Cham Heidelberg New York Dordrecht London



Printed on acid-free paper

Springer is part of Springer Science+Business Media (www.springer.com)

*To our families*

# Preface

*Distributed control and embedded systems* (*DCESs*) incorporating real-time control and communication functions implemented on some appropriate distributed *Hardwareare/Software dedicated architecture* are ubiquitous in areas of avionics, automotive, production of energy, space exploration and many others. It would be impossible to run the existing nuclear plants with the current level of safety requirements without their presence. Furthermore, *DCESs* appear to be essential to send robots exploring space, shuttles and satellites orbiting the earth. Our cars are more fun to drive, safer and cleaner, thanks to small embedded computers that perform functions such as cruise control, navigation system, *ABS*, *ESP*, airbag, optimal control of injection and many others and which exchange information between them in real-time. The quality of their design is directly related to the quality, performance and security of the final products (planes, cars, ...) and they directly influence the results of the related companies. This can be explained by the fact that the strong constraints of mass production imply a greater sensitivity on the prices of embedded computers and communication components. Requirements in terms of quality and performance of new products not only involve new methodologies and tools for designing of *DCESs*, but also imply a better use of resources they offer. Finally, the competition in which companies are involved forces engineers to design more efficient products that are, more and more complex, less expensive and with shorter design and production times than the competing firms.

To address these scientific and technical challenges, methodologies and design tools for *DCESs* are proposed by different teams of researchers and engineers according to their own scientific cultures. The proposed methods as well as the modeling and implementation tools are mainly based on the model of computation, discrete or continuous, and often involve computer science or control systems researchers and engineers. The results obtained so far have justified this scientific dichotomy which is nothing more than a view on the same object with discrete and continuous dynamics, commonly known as Hybrid Dynamic Systems (*HDS*). In the 1990s, several projects offering modeling, design and verification tools of *HDS* have emerged (*HyTech* [230], *Ptolemy* [229], UPPAAL [239], *KRONOS* [127], *HYSDEL* [228] ...). Experiments on these tools showed that the way to handle the *model complexity* is essential in the design and verification of *DCESs*. The pragmatic approach involving a thorough knowledge of the model is essential in managing the complexity and therefore in the design and verification of *DCESs*.

Other tools for *DCESs* design, specially related to calculation models, have been developed up to now. In the field of real-time systems, modeling and code generation tools such as *SCADE* [128], *SynDEx* [129] ... continue to be developed. They are used to model the temporal aspect of control tasks according to the dynamical characteristics of the objects to check, to generate real-time code on various targets and to elaborate the verification of different properties. This diversity of tools is explained not only by the existing competition but also by the importance that each particular view of a *HDS* has in the design. It does not only highlight a problem of training and scientific culture of our engineers and designers but also the importance that a particular view has in the life cycle of an *HDS*. At the same time, this fact reveals that current paradigm of computer science does not necessarily apply to embedded systems, as underlined by *Henzinger and Sifakis* in [116], and even more to *DCESs* design which has to be completed by a control system and signal processing view. In our opinion, a holistic approach and methodology for the design of *DCESs* which integrates consistently the essential paradigm from control theory, computer science and signal processing is of great actuality operating a necessary convergence toward the study of *DCESs*.

From the system control point of view, a fundamental question that determines the balance is whether the *DCESs* can be more efficient than their counterparts based on a centralized architecture and if this is the case, the price to be paid needs to be evaluated. There are at least three criteria which help answering to such a question. The *first* criterion relates to the nature of time-delays generated by each architecture. Surprisingly, the delay, can be more problematic in the case of centralized architecture. The *second* criterion concerns the existence of appropriate design tools. Concerning the existing design tools, in our opinion, we are at the beginning of a long road. Finally, the *third* criterion is related to the reliability and scalability. Naturally, distributed architectures are more reliable and scalable due to the distribution of their computing resources. Increasing or modifying the number of networks nodes and the number of real-time tasks to be executed on each of them may be done also dynamically. Moreover, if we observe the deployment of real-time networks operated since the arrival of the *Control Area Networks* (*CAN*) buses (in its standard version, by the middle of 1986) in various products such as cars, aircrafts, trains, ... and combined it with the actual scientific and technology development in the field allows thinking that we are at the beginning of a movement towards a generalization of distributed *Hardware/Software* (*HW/SW*) architecture in which different network nodes share their information and computing resources. Naturally, this requires a thorough understanding of real-time communication and computing phenomena in order to construct relevant and reduced complexity/order models. These models should represent accurately the temporal properties of states and control signals to enable their integration in the design of control laws and thus ensure the desired performance of *DCESs*.

In this monograph, several results on joint design of control laws and scheduling algorithms as well as stability analysis of some special cases of *DCESs* including, among others, the study of the effects induced by the delays are presented. This study is addressed by considering different aspects of the limitations imposed by the use of communication channels as well as embedded node processors composing the *HW/SW* architecture of *DCESs*. We specially focused on limitations in terms of network communication bandwidth and processor calculation power inducing sampling and period jitter, communication delay and signal quantization limitation. In our opinion, this approach of *DCESs* control allows to better emphasize on optimal use of its computing and communication resources. The algorithms used to solve the joint design of optimal control laws and message communication and/or real-time task scheduling have *NP-complete complexity*[1] [37, 93]. Reducing their complexity requests a deeper study of their models and their stability with respect to the delays induced by both signal communication between nodes and those induced by the real-time task scheduler at each node. We observed two interesting results in the stability study of *DCESs*. *First*, their stability, in some special cases, represents some unexpected behaviors with respect to the standard intuition. An increase in the input/output delay does not necessarily imply a potential destabilization of the system. This is in line with the design goal of reducing the computational and communication resources by guaranteeing the same level of *DCESs* performance. *Second*, it is not necessary to satisfy the stability conditions for each sampling period. Thus, we can handle and control the temporal expression of input/output delays allowing the design and implementation of optimal controllers for *DCESs* satisfying communication and calculation constraints.

The objectives and the structure of this monograph are different with respect to those proposed in two excellent books [124] and [17]. We try to construct a unified approach of the analysis and design of *DCESs*. The approach of the networked control systems design presented in [100] appears to be quite close to our methodology of design of *DCESs*. However, the major difference lies in the modeling and the switch control design to handle the induced delay and especially in the way it is taking into account concurrent calculation of control signals and their time scheduling.

As mentioned before, we focus on the optimal design of *DCESs* with respect to the communication and calculation resources constraints as well as the design of special control algorithms based on the analysis of induced time-delay system. In this context, a particular emphasis is put on the optimal control signals scheduling based on the systems state. In order to render this complex optimization problem feasible in real-time, a time decomposition is operated based on periodicity induced by the static scheduling. It is natural that our approach in the design and analysis of *DCESs* can not cover all the classes of *DCESs* which appear to be extremely rich and various. We do not claim either to give the best methods and tools in the optimal design and analysis of *DCESs* whose solution depends on the particular nature of each of them. We believe that the co-design approaches which consist in the synthesis of the optimal control laws and the generation of an optimal scheduling of control signals on the real-time network based on a thorough analysis of the

[1] NP-complete is the set of all decision problems whose solutions can be verified in polynomial time.

induced time-delay system have the best chances to render this problem feasible and to find optimal or some sub-optimal solution.

## Book Outline and Content

This book is organized in three parts. In *Part* I, composed of the first three chapters, a general overview, the state of art as well as the description of an abstract model of *DCESs* are given. In *Part* II, the problem of optimal co-design of *DCESs* is addressed under calculation and communication constraints focusing more on the scheduling of control signals on the networks as well as on the scheduling of control tasks on the *DCES* nodes. Finally, in *Part* III composed of four chapters, a particular attention is paid to various control configuration strategies as well as to the effects induced by delays (constant or time-varying) on the overall system's stability.

The main contributions are summarized as follows:

- *Optimal integrated control and scheduling of resource-constrained systems:* We start by adopting a model introduced by [122], where control and communication resource allocation aspects are strongly dependent. By interpreting such a model as a hybrid dynamical system *HDS* [21] with two types of inputs: *control inputs* and *scheduling inputs*, we formalize and solve the problem of the joint optimization of control and scheduling, by using an appropriate quadratic cost function as a performance criterion. The study of the properties of the optimal schedule, through some selected illustrative examples, shows that it is strongly dependent on the *DCES* state and dynamics. This dependence offers some ideas for improving *DCESs* performances using on-line scheduling algorithms based on their state information. However, this dependence shows that it is necessary to find appropriate performance metrics for the synthesis of optimal off-line schedules.
- *Optimal integrated control and off-line scheduling in the sense of the $\mathcal{H}_2$ norm:* In the context above, we motivate the use of the $\mathcal{H}_2$ norm [14] as a design criterion for obtaining optimal off-line schedules that only depend on the intrinsic characteristics of the system. We propose a method for the joint control and off-line scheduling in the sense of the $\mathcal{H}_2$ criterion. We show that this problem can be decomposed into two sub-problems, that can be solved separately. The first sub-problem aims at determining the optimal off-line scheduling in the sense of the $\mathcal{H}_2$ criterion and can be solved by applying the so-called *branch and bound* method [85, 86, 146]. The second sub-problem aims at computing the optimal control gains and can be solved by adopting tools from optimal control theory of periodic systems [45].
- *The use of the model predictive control as a means for the joint optimal control and on-line scheduling:* We propose an approach that allows on-line calculation of the optimal values of both control signals and scheduling decisions, in the sense defined by an appropriate quadratic cost function. This approach relies on the use of the model predictive control (*MPC*) technique [50], which was applied successfully in the past for the control of *HDS* [21]. We illustrate the performance

improvements, in terms of quality of control, which are brought by this approach, compared to static approaches, where the used scheduling is pre-computed off-line. We also state some appropriate stability conditions of the corresponding predictive controller.

- *Optimal integrated control and on-line scheduling of resource-constrained systems:* The major drawback of the model predictive control technique is that it requires solving an optimization problem which is of *NP-complete* [37] type. For that reason, an on-line sub-optimal scheduling algorithm, called *OPP* for *optimal pointer placement* is proposed. While being based on an off-line pre-computed optimal schedule, *OPP* makes possible to allocate on-line the communication resources, by taking into account the state of the controlled dynamical systems. It is shown that, under mild conditions, *OPP* ensures the asymptotic stability of the controlled systems and enables in all the situations the improvement of the control performance compared to the basic static scheduling. Furthermore, *OPP* is applied to two typical examples of a distributed control and embedded systems: the car active suspension controller [27] and the control of a quadrotor.
- *Optimal relation between quantization precision and sampling rates:* We extend the model that was first considered in order to take into account quantization related aspects [72]. Consequently, the communication constrains are modeled at the bit level, in bits per second. In general, increasing the sampling frequency improves the disturbance rejection abilities whereas increasing the quantization precision improves the steady state precision. However, when the bandwidth is limited, increasing the sampling frequency involves the reduction of the quantization precision. As the opposite, augmenting the quantization precision requires the lowering of the sampling frequency. Based on these observations, we propose an approach allowing the dynamical on-line assignment of sampling frequencies and control inputs quantization [24]. This approach based on the model predictive control technique enables to choose the sampling frequency and the quantization levels of control signals from a predefined set, in order to optimize the control performance.
- *Optimal state-feedback resource allocation:* A new approach for the co-design of control and real-time scheduling is proposed. This approach decomposes the problem into two sub-problems solved separately. The first sub-problem amounts to find the optimal non-preemptive off-line schedule, and can be solved by using the *branch and bound* method [85, 86, 146]. The second sub-problem resolution makes use of the lifting technique [45] to determine the optimal control gains, based on the solution of the first sub-problem. In the second part, a plant state feedback scheduling algorithm, called reactive pointer placement (*RPP*) scheduling algorithm is proposed. Its objective is to improve the control performance by reacting fastly to unexpected disturbances. Performance improvements as well as stability guarantees using the *RPP* algorithm are formally proven and then illustrated on a comprehensive implementation model, which was simulated using the tool TRUETIME [5]. Finally, the *RPP* algorithm is implemented on an embedded processor in order to achieve the concurrent real-time speed regulation of two DC motors.

- *Insight in delay system modeling of* DCESs*:* The *DCESs* stability and performance robustness depend on the model of communication and calculation applied to the associated *Hardware/Software* application architecture and related scheduling policy of state/control communication messages as well as of control tasks based on system state. The scheduling policy organizes the distribution of communication and calculation resources between competing nodes and tasks and handles contention situation. As it will be seen throughout this document, the sampling period of sensor and actuation signals update as well as execution of related control task will vary with time. This phenomenon induces variable delays in the *DCES* whose time characteristics, strongly conditions their performance. In this context, we propose and briefly discuss various delay models that can be used for representing *DCESs*.
- *Stability analysis of* DCESs *subject to induced delays:* We analyze some possible scenarios or time-delay models based on the off-line periodic and on-line aperiodic scheduling. We have to point out the fact that the control signal scheduling on the network as well as control tasks scheduling on each *DCES* node are non preemptive ones. The objectives of this analysis are twofold: *first,* to obtain less conservative stability domain with respect to the network delay and sampling period variation, and *second*, to shed some light on the interplay between resource allocation and system stability and performances. In some special cases, we observe a contradiction with the generally accepted intuition which consists in the fact that more computation or communication resources will easily stabilize a given *DCES*.
- *Design of the hyper-sampling sequence of* DCESs*:* Optimal scheduling of a number of control tasks on a processor or sensors and actuators signals on a real-time communication network depends mainly on the relative dynamics of the related subsystems composing a *DCES*. More a given subsystem dynamics is important, more it needs calculation and/or communication resources. As it will be seen in Chap. 2, the scheduling chronograms are periodic and their period depends on relative dynamics of the sub-systems sharing the given resource. Generally, such a period is called *hyperperiod* or *hyper-sampling period*. In Chap. 11, we propose a new method to optimally design the *hyper-sampling period* (including the standard single-sampling period as a special case) for *DCESs*. Furthermore, we will develop an analytic relation between the dynamic performance index and the *hyper-sampling period*. Thus, given an *average sampling frequency*, we will be able to design optimally the *hyper-sampling period* corresponding to the minimum value of dynamic performance index.
- *An optimal control strategy for distributed control and embedded systems:* Enhancing the stability as well as dynamic performances of some special class of *DCESs* using switched sampled-data (*SD*) control strategy is also studied. Regarding the stability issue, we will show that the use of the switched *SD* control strategy allows to enlarge the stability bound on the sampling period. In order to take into account the inter-sample behavior, we choose a continuous-time cost function to evaluate the system performances. It will be seen that the performance index can be explicitly calculated as a function of the switching time-parameter

and the optimal value of the performance index can be analytically found. Another advantage of the applied approach consists in analyzing standard control systems problems and more specifically in the definition of their optimal sampling period. The results obtained clearly show that increasing the sampling period does not necessarily reduce the system's performances. This phenomenon is quite interesting since it allows us, in some cases, increasing the sampling period or reducing the computational resources and achieving simultaneously better system's performances.

- *Optimal design of switched hold-zero compensation strategy for* DCESs *subject to control missing:* The admissible control input missing rate (*ACMR*) of some sampled-data *DCESs* is an important index which reflects the stability robustness with respect to control input missing induced by packets dropout or induced delays. We propose a simple switched hold-zero (*HZ*) control law as a control-missing compensator. More precisely, the switched *HZ* control has two control modes: the hold-control and the zero-control, respectively. The switching between two control modes is determined by an appropriate switching parameter. To obtain an *ACMR* as large as possible, we present a method to optimally design the corresponding switching parameter. It will be seen that the switched *HZ* control leads to better results than both the zero-control and the hold-control strategies acting separately and independently. In addition, the *ACMR* index can be used to calculate an hyper-period of messages scheduling optimizing the network bandwidth.

This book is mainly addressed to post-graduated students willing to operate research studies on *DCESs* and especially on the optimization of their performances with respect to communication and calculation resources. In our opinion, the modeling and analysis tools given in this monograph may be useful for research engineers in modeling and analyzing of *DCESs* for industrial applications purposes. Finally, the design methodology combined with complexity reduction objectives may help them to consistently formulate the corresponding *DCESs* design problem. The principal concepts and the methods are introduced and explained via some concrete illustrations and examples borrowed from robotics, automotive application and unmanned aerial vehicle (*UAV*).

## How to Read the Book?

This monograph can be read in different ways depending on the proximity of the reader with the subject. The first part is necessary if the associated calculation and communication model of *DCESs* are, in large part, unknown to them. The proposed models are simple and general enough allowing to integrate them easily in the design of control and scheduling algorithms of *DCESs*. A structural reading of the second part allows to understand the general idea of the proposed approaches. Naturally, a more informative reading where the methodological aspect is complemented by implementation of concrete examples is also possible, and even recommended, paving the way for real-world and/or industrial applications. Concerning the third part of this document, except the ninth chapter, where we give a number of induced input delayed models of *DCESs*, the other chapters can be read independently.

Noisy-le-Grand, France — Arben Çela
Rueil-Malmaison, France — Mongi Ben Gaid
Shenyang, People's Republic of China — Xu-Guang Li
Gif-Sur-Yvette, France — Silviu-Iulian Niculescu

# Acknowledgements

We greatly acknowledge our institutions; *ESIEE Paris* (*Noisy-Le-Grand France*), *IFP New Energy* (*Rueil Malmaison, France*), *Northeastern University* (*Shenyang, PR China*) and the *French CNRS* for their support as well as for the existing collaborative programs which helped us during the whole process.

Parts of the book have been presented at the *International Master Program on Computer Science and Embedded Systems* of *ESIEE Paris* as well as at *ESIEE Paris* graduate program. The students' feedback was constructive, and helped us to reorganize some material, and to present results from a different point of view.

We are grateful to OLIVER JACKSON and CHARLOTTE CROSS from Springer, for their help and patience during the preparation of the manuscript.

We would like to thank our friends, past and present collaborators, who implicitly or explicitly made a significant contribution to the research results presented in this volume. Among them, we mention ABDELLLATIF REAMA (ESIEE Paris, France), CESAR-FERNANDO MENDEZ-BARRIOS (Universidad Autonoma de San Luis Potosi, Mexico), COSMIN IONETE (Universitatea Din Craiova, Romania), LU ZHANG (Northeastern University, China), REDHA HAMOUCHE (ESIEE Paris, France), REMY KOCIK (ESIEE Paris, France), SHI-GUANG WEN (Northeastern University, China), TIAO-YANG CAI (Northeastern University, China), XU LI (Northeastern University, China), YSKANDAR HAMAM (ESIEE Paris, France), WIM MICHIELS (KU Leuven, Belgium).

Finally, we thank our families for their immense patience, love, and ongoing support.

UPE, ESIEE Paris, Noisy-le-Grand, France — Arben Çela
IFP New Energy, Rueil-Malmaison, France — Mongi Ben Gaid
Northeastern University, Shenyang, P.R. China — Xu-Guang Li
CNRS/SUPELEC, Gif-Sur-Yvette, France — Silviu-Iulian Niculescu

# Contents

# Acronyms

| | |
|---|---|
| $\mathbb{N}$ | Set of positive integers |
| $\mathbb{N}_0$ | Set of non-negative integers |
| $\mathbb{R}$ | Set of real numbers |
| $\mathbb{R}_0^+$ | Set of non-negative real numbers |
| $\mathbb{R}^-$ | Set of negative real numbers |
| $\mathbb{R}^n$ | Real vector space of order $n$ |
| $\mathbb{C}$ | Set of complex numbers |
| $\mathbf{A}^{-1}$ | Inverse of matrix $\mathbf{A}$ |
| $\mathbf{A}^T$ | Transpose of matrix $\mathbf{A}$ |
| $\mathbf{A} > 0$ | Imply that matrix $\mathbf{A} \in \mathbb{R}^{n \times n}$ positive definite |
| $\mathbf{A} \geq$ | Imply that matrix $\mathbf{A} \in \mathbb{R}^{n \times n}$ is positive semidefinite |
| $\mathrm{tr}(\mathbf{A})$ | Trace of matrix $\mathbf{A}$ |
| $\det(\mathbf{A})$ | Determinant of matrix $\mathbf{A}$ |
| $\lambda_{\min(\mathbf{A})}$ | Minimum eigenvalue of $\mathbf{A}$ |
| $\lambda_{\max(\mathbf{A})}$ | Maximum eigenvalue of $\mathbf{A}$ |
| $\mathrm{diag}(\mathbf{A}_1, \ldots, \mathbf{A}_n)$ | Block-diagonal matrix which entries are $\mathbf{A}_1, \ldots, \mathbf{A}_n$: $\mathrm{diag}(A_1, \ldots, A_n) = \begin{bmatrix} A_1 & & \\ & \ddots & \\ & & A_n \end{bmatrix}$ |
| $\|X\|_2$ | Euclidian norm: $\|X\|_2 = \sqrt{\sum_{i=1}^{n} X_i^2}$ |
| $\|X\|_\infty$ | The $\infty$-*norm* of $X$: $\|X\|_\infty$ |
| OPP | Optimal Pointer Placement |
| MLD | Mixed Logical Dynamical |
| MPC | Model Predictive Control |
| StS | Static Scheduling |
| RPP | Reactive Pointer Placement |
| WCET | Worst Case Execution Times |
| FBR | Feedback Scheduler |
| CAN | Control Area Network |
| TTP | Time-Triggered Protocol |
| ECU | Electronic Control Units |

| | |
|---|---|
| RM | Rate Monotonic |
| FP | Fixed Priority |
| EDF | Earliest Deadline First |
| MAC | Medium Access Control |
| TDMA | Time Division Multiple Accesses |
| GSM | Global System for Mobile Communications |
| CSMA/CA | Carrier Sense Multiple Access with Collision Avoidance |
| CSMA/CD | Carrier Sense Multiple Access with Collision Detection |
| SISO | Single Input Single Output |
| LTI | Linear Time Invariant |
| LQR | Linear Quadratic Regulator |
| LQG | Linear Quadratic Gaussian |
| CRC | Cyclic Redundancy Check |
| MATI | Maximum Allowable Transfer Interval |
| MEF-TOD | Maximum Error First—Try Once Discard |
| GLPC | Generalized Linear Complementarity Problem |
| QP | Quadratic Programming |
| MIQP | Mixed-Integer Quadratic Program |
| ZOH | Zero Order Holder |
| SPL | Sampler |
| CAD | Computer Aided Design |

# Part I
# Abstract Model of a Distributed Control and Embedded Systems

# Chapter 1
# Introduction to Distributed Control and Embedded Systems (*DCES*)

A *Distributed Control and Embedded System* (*DCES*) *is a control system having one or more control loops that are closed via a communication network associated to a* Hardware/Software (HW/SW) architecture *responsible for calculation and communication tasks scheduling and execution*. In this class of systems, the sensors and the actuators are situated at distant locations. This distribution is primarily related to the physical location of the system's components, which may be either sources of information (sensors) or means of action (actuators). In order to control such systems, the information that is provided by the sensors is transmitted to the controllers. Based on the received information, the controllers determine the corrective actions and the control commands thus computed are sent to the actuators.

Nowadays, we observe a substantial increase in the use of *networks* in *DCES*. Several factors explain such a choice. The price of the hardware components (network nodes and cables) has continuously decreased over last last years to make them more affordable. Consequently, the use of networks has become an economically conceivable solution in many application fields. Furthermore, the use of networks offers many technological advantages, such as the reduction of the obstruction (compared to the point-to-point wires), better flexibility and modularity, easier diagnosis and maintenance and finally, an improved reliability. As a consequence, networks are largely used in *DCES* that belong to many application fields such as automotive industry, aeronautics, robotics or manufacturing. An example of the distributed architecture of a *DCES* is given in Fig. 1.1. In this figure, abbreviations **S**, **C** and **A** denote sensor, controller and actuator nodes, respectively.

Roughly speaking, networks are generally classified into two main categories: *control networks* and *data networks*. The first category was developed in order to be employed in *DCES*. The networks belonging to this category were designed to support frequent and regular exchanges of small-size data (measurements or control commands for example) that must satisfy some strict temporal constraints. The network must guarantee an upper bound on the transmission delay of a given message. Various architectures are supported but the bus topology remains the most used one. *Flex-Ray*, *CAN*, *TTP/C*, or *Profinet-DP* are typical examples of control networks. The second category was designed in order to support an infrequent and bursty

A. Çela et al., *Optimal Design of Distributed Control and Embedded Systems*, Communications and Control Engineering, DOI 10.1007/978-3-319-02729-6_1,

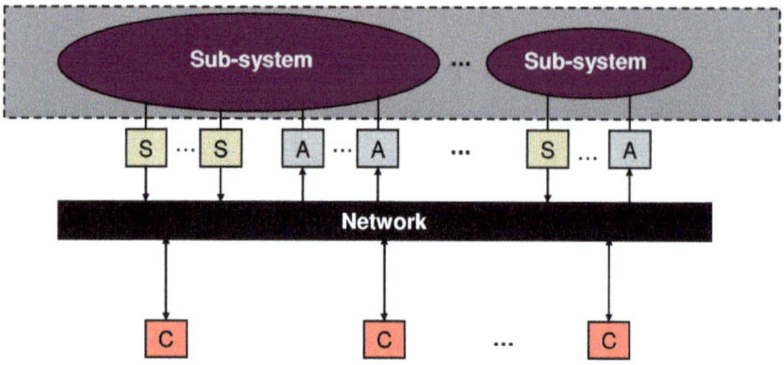


**Fig. 1.1** Architecture of a distributed control systems

transfer of a large quantity of data, encapsulated in packets, and without any criticality. Due to these design constraints, the data-rate that is offered by data networks is much more important than the one supported by control networks. In general, data networks, like *Ethernet* can cover a wider geographical area than control networks.

The fast development of Internet and wireless networks, which have become inexpensive, ubiquitous and pervasive means of communication, represents an opportunity for the development of new distributed control applications. By allowing the number of theoretically assignable IP addresses to be equal to $3.4 \times 10^{38}$ addresses (which makes approximately $6.5 \times 10^{23}$ addresses per square meter of the earth surface, including the oceans), the designers of version 6 of the IP protocol, have probably anticipated this development.

However, the introduction of networks throws down new challenges. In fact, depending on the nature of the employed network, problems like bandwidth limitations, delays or information loss may appear and need to be taken into account. The controllers may be implemented on limited *CPU* and memory target platforms, which may be shared between several concurrent tasks. In these situations, the basic assumption, consisting on separating control, communication and computation problems, deserves a careful reexamination. In fact, the presence of the network considerably affects the behavior of the controlled system. The control system performance becomes strongly dependent on the network and computer nodes behaviors. For that reason, the comprehensive study of the distributed control system requires the integrated study of the control, the communication and the computation problems.

As a first example, consider pursuit evasion games given by *Sinopoli et al.* in [213], where a team of pursuing robots must take the control of a team of evaded robots, and minimize a given criterion, for example the evasion time. The communication between the different robots is a fundamental aspect of the operation of the system. Any disturbance affecting the communication deteriorates the global behavior of the system and degrades the performances expressed by the criterion. Taking into account the communication constraints becomes crucial in understanding the global operation of the system.

The second example refers to the problem of the distributed control of the automobile suspension described by *Chalasani* in [60]. The objective is to maximize passengers' comfort as well as road handling. This amounts to minimize the accelerations of the car body, and to maximize the tire-ground contact effort. The correct operation of this system requires the coordinated operation of the four actuators, which are located at the four corners of the vehicle. It is clear that without the communication between these distributed components, the achievement of control objectives such as the minimization the roll acceleration is not possible any more. For that reason, the anti-roll bars are maintained when local control strategies are employed.

## 1.1 Motivations

With the new prospects offered by the diffusion of the communication means on the one hand, and the abilities of the control engineering and science to model and handle different categories of systems belonging to a broad range of application fields on the other hand, in our opinion, it appears easy to predict the development of new applications in the future, where the *dynamics* and the *information* will be strongly dependent. For that reason, the integrated study of control, communication and computation has involved many research teams and people all round the world, defining it as a major research direction challenge [130, 181].

Furthermore, as underlined by *Årzèn et al.* in [13], such a research area (interplay between control, communication and computation) becomes more important when the communication bandwidth or the processing power is limited. Communication and computing resource limitations are generally presented as being a common characteristic to an important range of *embedded systems.* The term embedded system refers to an electronic system, which is in close relationship to a physical one. Embedded systems, as defined by *Buttazzo* in [48], are reactive system: *they must correctly respond to the stimuli of their physical environment. They are characterized by a given degree of autonomy, which often appears in the autonomy of the computational resources, and less frequently in the autonomy of the energy supply.*

### *1.1.1 Communication Resources Limitations*

Communication resources limitations may have several reasons. For instance, these limitations may be caused by the signal's propagation medium. As clearly explained by *Sozer et al.* in [218], the underwater medium represents an example of signal's propagation limitation. In this communication medium, as given by *Stojanovic* in [220], the acoustic communications, which are preferred to the electromagnetic communications (for more details, please see the paper of *Quazi and Konrad* [195]), offer data-rates varying between a few kbps for long-range communications (over several tens of kilometers), to a few hundred kbps for short-range communications (over several tens of meters). Some other wireless networks present bandwidth limitations. In the *Bluetooth* networks, for example, only one node can have access to the network every 1.25 ms.

In other situations, network nodes may be autonomous and battery-powered. The transmission of the control information through the network interface requires an important energetic power, particularly in wireless networks. In order to satisfy the lifetime requirements of the batteries, the power consumption that is due to the transmissions, must be limited. Consequently, the data-rate is limited.

To guarantee of a deterministic data transmission introduces technological constraints limiting the maximal possible data-rate. For example, consider the characteristics of *CAN* networks as given by *Rachid and Collet* in [196]. In these networks, the maximal available data-rate depends on the length of the network cable. This data-rate is equal to 1 Mbps for cables whose length is less than 40 m. It goes down to 125 kbps when the cable length becomes equal to 130 m. This limitation is due to the relationship between the bit length on the network and the data-rate on the one hand, and to the used collision resolution mechanisms on the other hand (which require that all the nodes observe the same information during a predefined amount of time). In fact, collision resolution mechanisms require that the length of a bit must be greater than twice the distance between the two most distant network stations. Let $R$ be the data rate, $\mathscr{D}_x$ the distance between the two most distant stations, $t_{\text{bit}}$ the duration of a bit on the bus, $c_{\text{bit}}$ its propagation velocity and $l_{\text{bit}}$ its length. Knowing that $t_{\text{bit}} = \frac{1}{R}$, $l_{\text{bit}} = c_{\text{bit}} t_{\text{bit}}$, $l_{\text{bit}}$ must verify $l_{\text{bit}} > 2\mathscr{D}_x$, which imposes that $R < \frac{c_{\text{bit}}}{2\mathscr{D}_x}$. Bandwidth limitations affect many areas such as embedded systems (see *Arzén and Cervin* [12]), formations control (*Belanger et al.* [20]), underwater robotics (see *Speranzon et al.* [219]), haptics (see *Kuschel et al.* [143]) or large arrays of micro-electro-mechanical systems (*MEMS*) (see *Berlin and Gabriel* [33]).

### *1.1.2 Computational Resources Limitations*

Embedded systems are generally found in products that are designed for mass production. Their cost may represent a significant part of the cost of these products. Due to market requirements, and to a strong competition between the different manufacturers, their cost is subject to very strict constraints. This may be easily understood since economizing a few centimes in the cost of a hardware component that will be produced in around one million copies represents a very significant economy. Nowadays, a modern car contains more than 80 embedded microcomputers [183], called *Electronic Control Units* (*ECUs*). Control laws are still being computed using fixed point operations, on limited *CPU* power and memory micro-controllers. It is just recently that some automotive suppliers have begun to implement the engine controller, which represents the "most" complex controller of a modern car, on floating point micro-controllers.

## 1.2 Goals and Contributions

When computation or communication resources are limited, they have to be used as efficiently as possible. In (*DCES*), the *scheduler* is the entity that is responsible for the allocation of communication or/and computation resources. Consequently, the efficient use of these resources amounts to the design of appropriate scheduling algorithms. The induced delays may deteriorate the system performances or, even worst, make them unstable.

The objective of this book is to propose appropriate methods and algorithms allowing a more effective exploitation of communication or computation resources in *DCES* as well as the analysis, in some cases, of the induced delays influences on their stability and performances. The proposed approach relies on three complementary ideas: *first*, to refine the dynamical model of the plant by taking into account the available communication and computation resources; *second*, to use performance criteria of controlled system for the joint synthesis of the control and scheduling, and *finally*, to analyze the induced delays influence on the stability and reallocate the communication and computation resources based on this analysis as well as propose new *ad hoc* control algorithms to enhance the *DCES* performances.

## 1.3 Methodology

The *DCES* control approaches, which consists in emphasizing the need for optimal use of computing and communication resources, is adopted by more and more researchers. One of the first question can be asked is related to the means we have to influence on *DCES* performances. For a control engineer is natural to think on the design of new control laws based on model refinement which imply to take into account patterns of communication between computation nodes with sensor and actuator nodes.[1] These communication patterns are highly dependent on the type of networks used and the priorities, if any, assigned to *HW/SW* nodes and messages. Their influence in the control loop performances is, first, related to the messages order and, second, to the induce delay in uplink and downlink channels and to the sensibility of *DCES* to them. Finally, it is related to the given state of *DCES*. In short, the *DCES* performances are closely related to time pattern of messages in uplink and downlink channels and the given state of *DCES*.

In order to obtain a relevant model of network, we can not ignore the constraints on bandwidth otherwise say quantization, in a broad sense, of measurement and control signals. These constraints imply a quantization and coding of transmitted signals and necessarily induce a delay, bounded from below, between two successive samples of the same signal. So, in general, we have to deal with time variable quantized and delayed signals. We can rephrase the above question as follows: *"Can*

[1]These two communication links are commonly called by uplink and downlink communication channels.

*we control the induced delays by a real-time communication network and if so what are the means at our disposal?"*. In order to answer correctly to this question, we have to look particularly at two techniques that we thought are natural: the *first* one is to use system dynamical model to predict the evolution of its state and the *second* one is based on analysis of networks messages induced delays. With respect to this last idea, the contribution of *Lian et al.* [153] showed that messages scheduling makes *networks deterministic* and *minimize* their *induced delays*. Based on this study and the need to master the delays, we are particularly concerned by the following question: *What should be the physical basis on which we should rely in order to achieve an optimal ordering of messages based on the DCES application, that is, between different control systems sharing the same communication network*? Naturally, the most relevant information that we have to take into account in handling such problem is the dynamics of systems sharing the same network, node calculation and network bandwidth availability and the systems current state. Then we focus on the methodology allowing us to define a scientific framework reflecting our physical insights.

We set two goals: *First*, the obtained tasks and messages scheduling must be optimal with respect to the dynamical systems sharing the real-time network and *second*, it must be real-time relative to the varying states of dynamical systems composing *DCES*. Proceeding in such a temporal decomposition, we can handle the induced *NP-complete complexity*[2] [93] of mixed optimization problem involving continuous and discrete variables and solve it in real-time.

In the sequel, the basic concepts of the real-time scheduling theory are first reviewed in the forthcoming Chap. 2. Next, an overview of the state of the art of the integrated approaches for control and resource allocation in distributed embedded control systems is given. Chapter 3 describes the abstract model of a distributed embedded control system, with communication resources constraints that was adopted throughout the book. The main theoretical results, established previously in the literature, are reviewed and discussed.

[2]NP-complete is the set of all decision problems whose solutions can be verified in polynomial time.

# Chapter 2
# Resource Allocation in Distributed Control and Embedded Systems

Traditionally, control design problems are decoupled from software design and implementation considerations. Such a separation allowed the control and computer science communities to focus on specific problems independently, and led to the development that we are familiar with nowadays. However, as explained in *Årzèn et al.* [13], this separation relies on the fact that these two fields use very simplified models of their interface with each other. In fact, control designers disregard the characteristics of the implementation and the available computational and communication resources. On the other hand, real-time designers see the control loop as a periodic task with a hard deadline, that have sometimes to fulfill data dependency constraints (especially when the sensing and actuation are distant). Recently, researchers from these two communities have shown that if more elaborate models are used, significant improvements in terms of implementation efficiency and quality of control may be achieved. This *integrative co-design approach* is central to this book.

This chapter is organized into two parts. The first part presents the state of the art of the real-time scheduling theory, focusing on the most used results in distributed control and embedded system (*DCES*) applications. We start by presenting real-time single-processor scheduling problems. Next, we focus on the problem of ensuring real-time networked communications. We emphasize different methods for managing the access concurrency having a determinant impact on the guarantee of deterministic real-time communications. Finally, an overview of the problem of guaranteeing end-to-end real-time constraints in distributed systems is given. In the second part, we present a state of the art of the new approaches, that are based on more elaborate models, and that take into account both the dynamic nature of the controlled systems as well as some characteristics of their implementation. Various problems and models were discussed in the literature. We propose a classification of these different approaches and illustrate the different problems and models that were addressed.

## 2.1 Real-Time Scheduling Theory

As mentioned above, the objectives of this section concern different tools and mechanisms which handle the concurrency in real-time systems at the processor level as

A. Çela et al., *Optimal Design of Distributed Control and Embedded Systems*,
Communications and Control Engineering, DOI 10.1007/978-3-319-02729-6_2,

well as at the distributed communication architecture. We focus on the models and tools related to control applications without claiming to be exhaustive. Throughout this section, we give some references that may help for a deeper and better understanding of the models to be used in the particular applications we may be faced with.

## 2.1.1 Real-Time Single-Processor Scheduling

This subsection is devoted to real-time scheduling of multiple task on a single processor. The objective is to obtain a calculation model related to concurrent execution of a given number of tasks. For more details, please refer to two excellent books of *Buttazzo* [48] and *Cottet* [67]. In real-time processing systems [48, 67], the processor is a resource that is shared between various concurrent *tasks*. A *task* represents a sequence of instructions that are intended to be executed by the processor. The service that is delivered by a task may be performed several times during the lifetime of the application. For such reasons, a task may be *"instantiated"* several times in the form of *jobs* or *task instances*. Jobs or task instances represent the execution flow that corresponds to the effective execution of the task code.

### 2.1.1.1 Events Characterizing the Lifetime of a Job

A job or task instance is characterized by the following temporal parameters:

- *release time*: the time instant at which the scheduler is requested to execute the job, which have just become ready to run;
- *start time*: the time instant at which the job starts its execution;
- *preemption times*: time instants when the scheduler suspends the execution of the job on behalf of other jobs having a more important priority;
- *resumption times*: time instants at which the execution of the job is resumed after a preceding preemption;
- *completion time*: time instant at which the job finishes its execution;
- its *absolute deadline*: time instant before which the job should have terminated.

Figure 2.1 illustrates the single processor scheduling of three jobs $j_1$, $j_2$ and $j_3$. More precisely, in this figure:

- instants $t_1$, $t_2$ and $t_3$ represent the respective release time instants (depicted by up arrows) of jobs $j_1$, $j_2$ and $j_3$,
- instants $t_1$, $t_2$ and $t_6$ represent the respective start times of jobs $j_1$, $j_2$ and $j_3$,
- instants $t_2$ and $t_4$ represent respectively preemption and resumption times of job $j_1$,
- instants $t_6$, $t_4$ and $t_7$ represent the respective completion time instants of jobs $j_1$, $j_2$ and $j_3$,
- instants $t_5$ and $t_8$ represent the respective absolute deadlines (depicted by down arrows) of jobs $j_1$ and of $j_2$ and $j_3$.

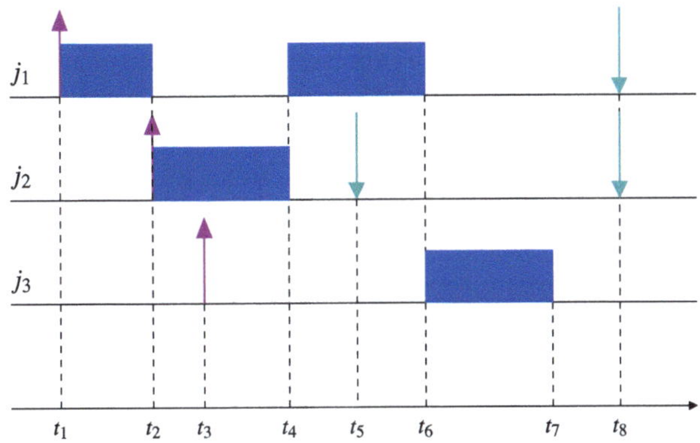


**Fig. 2.1** Single-processor scheduling of three tasks

#### 2.1.1.2 Task Model

A real-time task $\tau^{(i)}$ is characterized by:

- worst case execution time (*WCET*) $c^{(i)}$;
- activation law of its jobs: the jobs of a given task may be activated periodically with a period $T^{(i)}$, sporadically with a minimum inter-arrival time or aperiodically if no temporal constraints are imposed on the activation of its jobs;
- relative deadline $D^{(i)}$: the time interval between the release time of the job and its absolute deadline.

A real-time task is called *periodic*, *sporadic* or *aperiodic* according to the activation law of its jobs [48, 67].

#### 2.1.1.3 Scheduling Algorithms Classification

This paragraph describes the commonly used terminology for classifying the existing scheduling algorithms [48, 67].

- *Preemptive/non-preemptive*: A scheduler is called *preemptive* if it is able to suspend a running task on behalf of other tasks that have more important priorities. It is called *non-preemptive* in the opposite case. Preemption is supported by the majority of real-time operating systems. In the opposite, the scheduling of packets in networks is always non-preemptive.
- *Off-line/on-line*: In *off-line* scheduling algorithms, the sequencing of the tasks to be executed is described at design time, as a *schedule* or *execution plan*. Consequently, the scheduler is simply a sequencer, that executes the different tasks according to the schedule that was pre-computed off-line. The execution order of the different tasks is then identical to that specified in the execution plan. In practice, the schedule is executed in a repetitive way. The period of repetition is called *major cycle* or *hyperperiod*. In general, the schedule describes the start

time instants of the different tasks instances, and possibly, their preemption and resumption time instants, which are expressed as a function of an elementary time unit, called *minor cycle* of the schedule. In the opposite, in *on-line* scheduling algorithms, the choice of what task to execute is determined at run-time by the scheduler. When activated, the scheduler performs a given processing in order to determine the next tasks to execute. In most cases, this processing amounts to the comparison of the priorities of the ready tasks. These priorities may be fixed in the case of *fixed-priority* scheduling algorithms or dynamic (i.e., adjustable at run-time) in the case of *dynamic scheduling algorithms*.

#### 2.1.1.4 Schedulability Analysis

Real-time scheduling theory [48, 67] aims at providing the sufficient conditions (and preferably the necessary and sufficient conditions) guaranteeing that a task set (which is defined by a given model) will respect its real-time constraints (which are defined by the assigned deadlines). An important theoretical tool that it provides is the *schedulability analysis*. Schedulability analysis is performed off-line, in order to ensure that the scheduling of a task set (which satisfies a given model), using a given scheduling algorithm, ensures the respect of the tasks deadlines. In the sequel, we present the fundamental results that are related to the preemptive real-time scheduling of periodic tasks whose relative deadlines are equal to their periods, by fixed-priority and dynamic-priority scheduling algorithms.

Consider a task set containing $\mathscr{N}$ independent tasks $\tau = (\tau^{(1)}, \ldots, \tau^{(\mathscr{N})})$. Each task $\tau^{(i)}$ is characterized by its *worst-case execution time* (*WCET*) $c^{(i)}$, its period $T^{(i)}$ and its relative deadline $D^{(i)} = T^{(i)}$. The task set $\tau$ is said to be *schedulable* by a given scheduling algorithm if all the jobs of all the tasks forming $\tau$ finish their execution before their absolute deadlines.

- *Fixed-priority scheduling*: In this scheduling policy, a fixed priority $p^{(i)}$ is assigned to each task $\tau^{(i)}$. At each instant, the scheduler executes the ready task whose priority is the most important. Since preemption is authorized, if during the execution of a given task $\tau^{(i)}$, another task $\tau^{(j)}$ that has a more important priority becomes ready, then task $\tau^{(i)}$ is preempted and the processor is allocated to task $\tau^{(j)}$. In 1973, *Liu and Layland* [156] have shown that the optimal priority assignment is given by the *rate monotonic* (*RM*) rule (i.e., the smaller the period is, the higher is the assigned priority). Thus, the rate monotonic scheduling algorithm is a fixed-priority scheduling algorithm where priorities are assigned according to the rate monotonic rule. The optimality of a real-time scheduling algorithm was defined by *Liu and Layland* by its ability to schedule a given task set, which verifies a given task model, such that the predefined real-time constraints are met. A fixed-priority (resp. dynamic-priority) scheduling algorithm is optimal (for a given task model) in the sense that *any task set* (*satisfying the task model*) *that is not schedulable under this algorithm will not be schedulable under any other fixed-priority* (*resp. dynamic-priority*) *scheduling algorithm*. A sufficient schedulability condition using *RM* is

$$\mathcal{U} = \sum_{i=1}^{\mathcal{N}} \frac{c^{(i)}}{T^{(i)}} \leq \mathcal{N}\left(2^{1/\mathcal{N}} - 1\right).$$

The necessary and sufficient schedulability condition by *RM* requires the analysis of the maximum response time $R^{(i)}$ of each task (i.e., the maximum among the response times of its jobs) (see, for instance, *Joseph and Pandya* [135]). The response time of a job is defined by the duration between its release time and its completion time. $R^{(i)}$ is given by:

$$R^{(i)} = c^{(i)} + \sum_{j \in hp(i)} \left\lceil \frac{R^{(i)}}{T^{(j)}} \right\rceil c^{(j)}.$$

where $hp(i)$ is the set of tasks which have priority over $\tau^{(i)}$. The task set $\tau$ is schedulable by *RM* if and only if $R^{(i)} \leq D^{(i)}$, for all $i \in \{1, \ldots, \mathcal{N}\}$.

- *Dynamic-priority scheduling*: In this scheduling policy, the priority $p^{(i)}$ that is assigned to the task $\tau^{(i)}$ may vary over time. *Liu and Layland* [156] proved that the optimal dynamic priority assignment policy consists in assigning the most important priority to the task that is the closest to its deadline. This priority assignment rule is called *Earliest Deadline First* (*EDF*). The necessary and sufficient schedulability condition under *EDF* is simpler than that established for *RM* and is given by:

$$\mathcal{U} = \sum_{i=1}^{\mathcal{N}} \frac{c^{(i)}}{T^{(i)}} \leq 1.$$

Real-time scheduling theory progressed substantially since the fundamental article of *Liu and Layland* [156], to take into account the problems involving the scheduling of sporadic and aperiodic tasks, the scheduling of tasks whose deadlines are lower or higher than their periods, the protected access to shared resources, the precedence constraints and the non-preemptive scheduling. A detailed description of some fundamental results concerning these problems can be found in *Shin* [142], *Buttazzo* [48], *Liu* [157], *Burns* [46] and *Decotigny* [71].

### *2.1.2 Real-Time Medium Access Control in Communication Networks*

Ensuring real-time communications is the responsibility of the entire communication stack. Nevertheless, as explained by *Zimmermann* in [267], the crucial role falls on the *MAC* (medium access control) sub-layer of the layer 2 of the *open systems interconnection* (*OSI*) model. The *MAC* sub-layer has the responsibility of managing the access to the communication medium, which may be shared between several

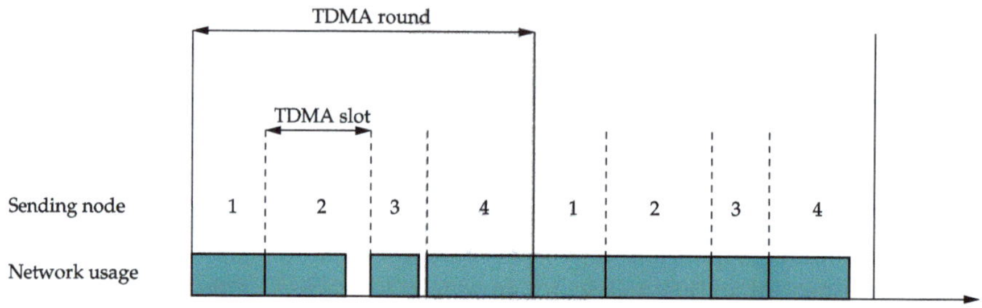


**Fig. 2.2** Bandwidth sharing using the *TDMA* protocol and used terminology

nodes of the network. In real-time networks, there are several access protocols. The most deployed ones are described in the sequel:

#### 2.1.2.1 Time Division Multiple Accesses (*TDMA*)

The *TDMA* [153, 183] protocol makes it possible to statically divide, the available bandwidth, in the temporal domain, between several competing nodes. In this protocol, each node knows exactly the moments when it is allowed to transmit over the network. As a consequence, each node has to transmit during the *time slot* which is allocated to it, called *TDMA slot*. In this way, collisions are avoided. The time slots are predetermined off-line. Their sequencing has a periodic structure. The minimal sequence of time slots allowing to describe the sequencing of the time slots of the various nodes is called *TDMA round* (Fig. 2.2).

The *TDMA* protocol may be implemented in a centralized or a distributed way. In centralized implementations, a master node has the responsibility of triggering the communications of the other slave nodes by transmitting a synchronization signal. The major inconvenient of this approach is that a breakdown of the master node leads to a total breakdown of the network. The distributed implementations require the establishment of a sufficiently precise global time in all the nodes of the network, requiring thus the use of some appropriate clock synchronization algorithms. The *TDMA* protocol is the cornerstone of the mobile communications *GSM* protocol. The *TTP/C* communication protocol [238] manages the concurrent access to the communication medium using a distributed implementation of the *TDMA* access protocol. In order to ensure a global clock, the *FTA* (*Fault-Tolerant Average*) algorithm proposed by *Kopetz and Ochsenreiter* in [141] is used to ensure the clock synchronization of the different network nodes.

#### 2.1.2.2 Token-Bus

The *Token-Bus* access protocol was specified by the *IEEE* (*IEEE standard 802.4*) and by the *ISO* (*ISO standard 8802.4*). It represents the cornerstone of many communications protocols that are employed in industrial field busses [231], like *Profibus* [18], *ControlNet* [66], *MAP* [164] and *ProfiNet* [39].

In this protocol, the network nodes are logically organized in a *ring topology*: each node knows its logical predecessor and its logical successor. The access arbitration is performed by the circulation of the *token* between the nodes. At any given moment, only one node has the token. The possession of the token gives to the possessing node the right to transmit over the network. The node that takes possession of the token can start transmitting over the network. It must pass the token to its successor if the time it has held the token reaches a limit or if it finishes transmitting before the expiry of this duration. If a node that does not have any information to transmit, receives the token, then it transmits it directly to its successor node. Concerning the real-time properties of this protocol, the analysis of the worst-case response times of *Profibus* and *ControlNet* messages are respectively given by *Tovar and Vasques* in [236] and by *Lian et al.* in [153]. The worst-case response time of a message is defined as the duration between the moment the transmission is requested and the moment when the message is received by the destination process.

#### 2.1.2.3 Carrier Sense Multiple Access with Collision Avoidance (*CSMA/CA*)

The carrier sense multiple access with collision avoidance access protocol (*CSMA/CA* is used in many networks such as *CAN*, *DeviceNet* or IEEE 802.11 wireless networks. In this access model, each message is characterized by a unique priority. Since the shared communication medium can transmit only one message at a time, each node that wishes to transmit a message must initially check whether the network is free (by sensing the network to find whether or not a carrier signal is being transmitted). If the network is free, then the node can start transmitting. However, it is possible that other nodes start transmitting at the same time, because they have detected at the same time that the communication medium has become free. In this situation, the transmission of the highest priority message is continued; the other messages with lowest priorities are discarded. By this way, collisions are explicitly avoided.

*CAN* networks [132, 196] use the *CSMA/CA* protocol in order to manage the concurrent access to the shared bus. Their implementation of *CSMA/CA* relies on a bit synchronization mechanism at the bus level. In *CAN* networks, each message is characterized by a unique identifier. Furthermore, no node is particularly addressed: all the sent messages are broadcasted to all the other network nodes. When several nodes are emitting and if at least one node sends one a '0' (called dominant level in *CAN* terminology), then all listening network nodes will detect a '0' at the same time, even if there are other nodes which have transmitted a '*1*'. Reciprocally, when all the transmitting nodes send a '*1*' (called recessive level), all the listening nodes will detect a '*1*'. The CAN bus behaves like a logical AND gate. Since the identifier field is located at the beginning of the frame (the most significant bit is first coded, the highest priority is *0*), and the binary synchronization is implemented on the bus, when a collision occurs, the different nodes directly compare the identifiers of

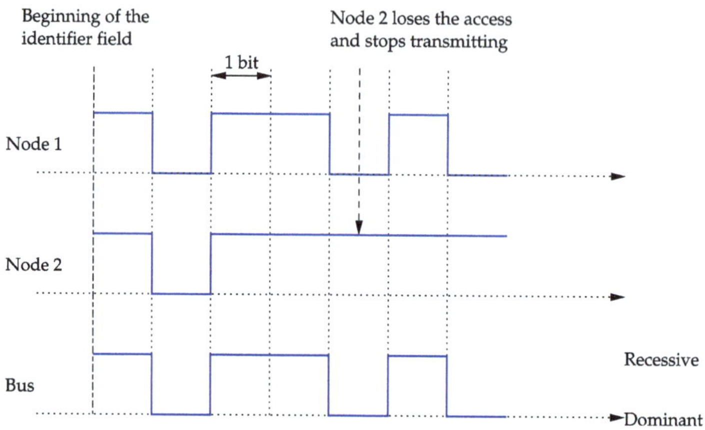


**Fig. 2.3** Illustration of the bus arbitration mechanism in CAN networks

their messages to the resulting logical level at the bus. A node that detects that the resulting level is dominant (*'0'*) whereas it has sent a *'1'*, knows that it has tried to send a message whose priority is lower than another message and must consequently stop transmitting. Figure 2.3 describes a collision between two messages, which were sent at the same moment by node 1 and node 2. Node 2 stops transmitting when it detects that it has a lower priority than node 1 (because it has sent a *'1'* and the resulting level on the bus was a *'0'*).

*CAN* protocol is a deterministic protocol. Consequently, it is possible to compute an upper bound over messages response times. A *CAN* bus may be seen as a non-preemptive scheduler. The computation of the worst-case response time of fixed-priority messages have been tackled in *Tindell and Burns et al.* [232, 233]. The evaluation of a *RM* priority assignment policy was also addressed in these references. The use of *EDF* scheduling in *CAN* networks was studied by *Di Natale* in [74].

The binary synchronization over the bus, which is necessary to collisions arbitration, introduces a relationship between the maximum data-rate and the length of the network cable. In fact, to use a data-rate of 1 Mbps, the maximum length of the cable must be less than 40 m. If it is necessary to use a cable whose length is greater than 620 m, then the maximum data-rate falls down to 100 kbps.

Note that in data networks, like Ethernet or the Internet, the *CSMA/CD* (carrier sense multiple access with collision detection) protocol is the most used medium access protocol. The fundamental difference between *CSMA/CD* and *CSMA/CA* resides in the collision arbitration mechanism. In fact, in the *CSMA/CD* protocol, the nodes that generate a collision are able to detect this collision and to stop their transmission during a random duration. For that reason, it is impossible to bound messages response times in networks employing the *CSMA/CD* access protocol. In wireless networks, such as *IEEE 802.11* networks, the *CSMA/CA* protocol is employed because it is not possible to detect the collisions (and thus to deploy the *CSMA/CD* protocol).

### *2.1.3 Real-Time Scheduling of Distributed Systems*

Real-time multiprocessor scheduling problems are still not as well understood as real-time single-processor scheduling problems; the most obscure points, as underlined by *Sha et al.* in [211], are related to the schedulability analysis. In this field, the impact of the *Dhall and Liu's* paper [73] on real-time multiprocessor scheduling theory was equivalent to the impact of the *Liu and Layland's* paper on real-time single-processor scheduling theory [156].

Multiprocessor scheduling algorithms may be classified into two categories [211]:

- *partitioned scheduling*, where each task is assigned to only one processor,
- *global scheduling*, where all the tasks compete for the use of all the processors.

The problem of the optimal partitioning of tasks among processors, as underlined by *Garey and Johnson* [93], is *NP-complete complexity.*[1] For that reason, simulated annealing or branch and bound-based heuristics were proposed to tackle this problem. The use of these heuristics relies on the modeling the scheduling problem as an optimization problem. A detailed presentation of these approaches is given in [16]. An outline of the most important results concerning multiprocessor schedulability analysis may be found in *Sha et al.* [211].

The tasks, that may be located on the different processors, may have data-dependencies and thus exchange messages via a communication medium. The communication medium may be seen as "a processor" that only supports the non-preemptive scheduling. Among the tools for off-line partitioned scheduling generation for tasks with precedence, on distributed architectures, one may cite [139, 217], which is based on the so-called *Adéquation Algorithme Architecture* (for efficient matching of the algorithm on the architecture) approach [102]. In *Syndex*, the architecture and the algorithm are described by two graphs. The algorithm graph is a direct acyclic graph, where the vertices represent the operations to be performed and where the edges represent the data-dependencies between the operations. The architecture graph describes the available parallelism as well as the communication possibilities between the various processors. Based on these two graphs, and possibly on placement constraints which may be specified by the user, *Syndex* uses the greedy list scheduling algorithm given by *Yang and Gerasoulis* and *Kwok and Ahmad* [144, 257] to synthesize the scheduling of the operations on the different architecture elements.

## 2.2 Integrated Approaches for Control and Resource Allocation

In the previous subsection, we have described some important results of the real-time scheduling theory. Disregarding the nature of the considered applications, these

---

[1]NP-complete is the set of all decision problems whose solutions can be verified in polynomial time.

results mainly addressed the problem of communication or computation resource allocation, in order to respect strict temporal constraints. In this subsection, we outline other approaches, which jointly consider the problems of control and communication or computation resources allocation. First, we review some problems and methods for the adaptive sampling. Second, we give an outline of the methods allowing to jointly considering the problems of control and communication resource allocation. Finally, we review the state of the art of the methods for the joint control and computational resource allocation in embedded systems.

## 2.2.1 Adaptive Sampling of Control Systems

The analysis of asynchronously sampled systems was undertaken at the end of the fifties. The research works, which were performed during the sixties and at the beginning of the seventies mainly focused on *single-input single-output* (*SISO*) systems. To the best of our knowledge, the first adaptive sampling method was proposed by *Dorf et al.* [77]. The proposed adaptive sampler changes the sampling frequency according to the absolute value of the first derivative of the error signal. Using analog simulations, the authors have shown that this method reduces between 25 % and 50 % of the number of required samples, with respect to the periodic sampling, given the same response characteristics.

Other approaches were proposed thereafter, in particular those of *Gupta* [108, 109], *Tomovic and Bekey* [234, 235] and *Mitchell and McDaniel* [176]. The evaluation of these approaches and their comparison to the periodic sampling was performed by *Smith* in [216]. Simulations have shown that these adaptive sampling methods are not always better than periodic sampling, especially where the input is subject to unknown disturbances. Remarking that the methods of *Dorf et al.* [77] and [176] are closely related to [125], *Hsia* proposed a generic approach allowing to derive adaptive sampling laws [126].

More recently, an important direction of research on the *DCES* performance enhancement through adaptive sampling is that of *Event Driven Controllers* (*EDC*). The aim of this class of controller is to reduce the calculation and communication resource utilization in order to provide in priority those tasks which need more. This results in a non periodic sampling depending on the state of all the subsystems composing a *DCES* and the "*policy*" implemented to handle their resources. Naturally, the hight level system specification such as stability needs to be assured which means that each new control signal has to be calculated, at least, with a minimum sampling frequency. In Årzèn [11], the *EDC* adapts its task period with respect to required system performances expressed as an event condition on the system state. These events are usually generated when the system state crosses an hyper-surface in state space. It is clear that the difficulties reside in the detection of this crossing and the event generator structure.

A second class of event based approaches called self-triggered one, consists in emulating event-triggered control by computing for each sampling instants a lower bound of the next sampling interval or the next candidate sampling instant.

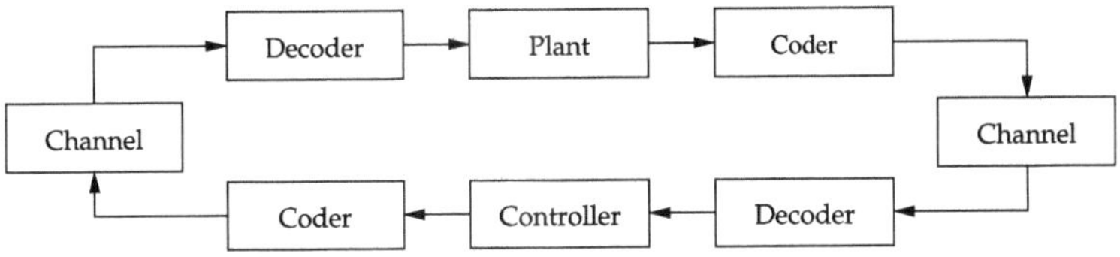


**Fig. 2.4** General model of the information flow in a control system whose control loop is closed through finite bandwidth communication channels

As stated in *Årzèn* [11], event-driven control is closer in nature to the way a human behaves as a controller. Indeed, when a human performs manual control his behavior is event-driven rather than time-driven. This fact conjugated with the need to optimally handle calculation and communication resources partially explains the rationale behind the important research activity in designing event-based controllers.

The *EDC* controller are triggered by *external* events or they are self-triggered. The works of *Heemels et al.* [110], *Åström and Bernhardsson* [15], *Tabuada and Wang* [226], *Suh et al.* [223], *Heemels et al.* [111], *Henningsson et al.* [113], *Lunze* [163], *Wang et al.* [246], *Marchand et al.* [165], and many others, fall in the event-triggered class whereas the works of *Velasco et al.* [242], *Anta and Tabuada* [6–9], *Wang and Lemmon* [247–250], *Mazo and Tabbuada* [170, 172], *Araujo* [10] are some of the contributions in the latter class of self-triggered controllers.

In a general distributed control architecture related to *DCES*, reduction of sampling frequency is not always sufficient to enhance system performances. It is also needed to synchronize the decisions between subsystems sharing given calculation or communication resources. The recent works of *Tabuada* [225], *Seyboth* [210], *Donkers and Heemels* [75], *Mazo and Cao* [171]. *Wang and Lemmon* [251], *de Persis* [70] go in this direction with the objective to coordinate the subsystems' local decision.

## 2.2.2 Allocation of Communication Resources: The "Per Symbol" Paradigm

In this paradigm, the information exchange is modeled at the symbol level. The quantization of measurements and control commands is thus implicitly taken into account. The general model is given in Fig. 2.4.

In this model, the communication channel can transmit at most $R$ bits per time unit. Because of these resource limitations, measurements and control commands must be encoded (as a flow of symbols) before their transmission and decoded at their reception. Various coding techniques may be employed. A fundamental question is to determine the necessary and/or sufficient data-rate allowing the existence of a coder, a decoder and a controller that achieve the stabilization of the system. Many contributions have tried to bring more insight into this fundamental question, by treating various models of resource limitations.

For instance, in [72], *Delchamps* has shown that it is impossible to asymptotically stabilize a discrete-time unstable *linear time invariant* (*LTI*) system, whose output passes through a quantizer having a finite number of quantization levels. In this setting, it is necessary to introduce and use other stability concepts, like, for example, practical stability.

The problem of state estimation, in the presence of state and measurement noise, was studied in *Wong and Brockett* [253]. In the considered model, the state observer is situated at the same location as the plant. However, the controller is located at a distant place. Consequently, the observations must be sent to the controller through a finite bandwidth communication channel. It was shown that this problem is quite different from the classic estimation and vector quantization problems. The concept of *finitely recursive coder-estimator sequence* was then introduced. Necessary conditions as well as sufficient conditions, which are related to the stability and the convergence of various coding and estimation algorithms, were stated. These conditions relate the network data-rate to the dynamical characteristics of the plant.

Next, in [254], *Wong and Brockett* introduced the concept of *containability*, as a weaker stability condition, to tackle the problem of the stabilization of networked systems through limited capacity communication networks, where the values of the measurements (which are received by the controller) and the controls (which are sent to the plant) belong to a finite set of values $\mathscr{J}$ (because of the quantization which is induced by the limited data-rate communication channel). Considering continuous-time *LTI* systems, which are impulsively controlled, they proved that a necessary condition to ensure the containability is $e^{\frac{2}{R}\mathrm{tr}(A_c)} \leq |\mathscr{J}|$ where $|\mathscr{J}|$ is the size of the alphabet, $\frac{1}{R}$ is the transmission duration of one bit and $A_c$ is the state matrix. They also proved that if the initial condition of the system lies in a bounded set, then a memoryless coding and control is sufficient to ensure the containability, if some conditions affecting the data-rate are met.

In [182], *Nair and Evens* considered a class of discrete-time, linear, time-varying and infinite-dimensional plants. No process or measurement noise affects the considered plants. The initial state is the realization of a random variable. Communications constraints only affect the sensors-to-controller link. The controller is directly connected to the actuators. The considered problem is the synthesis of a coder (on the sensors-to-controller link) and of a controller that minimize a cost function of the state over finite and infinite horizons. The cost function over the finite horizon is the $m$th output moment. A coder/controller scheme was proposed. Under some technical assumptions, which are related to the probability density function of the initial state, to some conditions depending on the size of the alphabet, to the data-rate and to the plant dynamics, a necessary and sufficient optimality condition of the proposed coder/controller was established. A necessary and sufficient condition for the existence of a coder/controller that asymptotically stabilizes the system (in the sense that the $m$th output moment converges to zero over an infinite horizon) was stated. In the special case where the plant is invariant, unstable and finite-dimensional, this last condition simplifies to $R > \log_2 |\lambda|$ where $\lambda$ is the unstable open-loop pole with the largest magnitude.

Next, in [42], *Brockett and Liberzon* proposed the idea of "zooming" as a means for ensuring the asymptotic stability of continuous-time and discrete time system whose control loops are closed through a finite bandwidth communication network. The zooming technique consists on changing the sensitivity of the quantizer over the time, based on the available quantized measurements. The relationship between performance and complexity of the quantized stabilization using the zooming technique was further studied by *Fagnani* in [82].

In [80], *Elia and Mitter* addressed the problem of the stabilization of single-input linear systems whose measurements and control commands are quantized. By first considering quantizers with a countable number of levels, and supposing the exact knowledge of the state, they proved that the coarsest quantizer allowing the quadratic stabilization of a discrete-time single-input *LTI* system, is logarithmic, and may be computed by solving a special *linear quadratic regulator* (*LQR*) problem. The state-feedback control problem as well as the state observation problem were solved within this theoretical framework. These results were thereafter extended to the continuous-time single-input and periodically sampled linear systems. The expression of the optimal sampling period (for the suggested quantizers) was established. It only depends on the sum of the unstable eigenvalues of the continuous system. This approach was finally extended to address quantizers with a finite number of levels.

Next, in [227], *Tatikonda and Mitter* considered discrete-time *LTI* systems. The control loop is closed through a limited capacity communication channel. Consequently, before their transmission, the measurements are quantized and encoded in symbols by a coder. At their reception, a decoder reconstructs a state estimate that will be used by the controller, which is directly connected to the plant. Two types of coders were studied:

- class 1 coders, which know past measurements, past controls and past transmitted symbols that were sent over the channel,
- class 2 coders, which only know past measurements.

Stabilization and asymptotic observability properties were particularly studied. It was shown that a necessary conditions for the existence of coders and decoders making it possible to guarantee these two properties is given by $R > \sum_{\lambda(A)} \max\{0, \log|\lambda(A)|\}$, the sum is over the eigenvalues of the state matrix $A$. This necessary condition is independent from coder classes and becomes sufficient if the whole state is measured and if class 1 coders are used.

### 2.2.3 *Allocation of Communication Resources: The "Per Message" Paradigm*

The different approaches that may be related to the "per message" paradigm are motivated by the fact that in all communication networks, protocol frames contain fields with fixed and incompressible length. These fields include, for example, the identifier field, the *CRC* (Cyclic Redundancy Check) field, which is used by error detection algorithms. For example, in *CAN* networks, the minimal length of the fixed protocol fields is 47 bits (the length can be more important because of the bit-stuffing mechanism as given in *Rachid and Collet* [196]). A measure that is encoded in 12 bits and sent in a *CAN* message only represents 20 % of the size of the message. In *Bluetooth* networks, the minimal size of the data field is 368 bits. If an information whose size is less than 368 bits is to be transmitted, then padding by '0' bits must be carried out.

The various approaches, which are related to the "per message" paradigm, may be classified into two categories:

- The decentralized minimization of the network bandwidth usage. The basic idea is that each node tries to locally minimize its bandwidth consumption.
- The scheduling of the concurrent access to the network. In these approaches, a more global view of the application, of its distribution and of the network access mechanism is considered. The information transmission over the network is managed by taking into account static characteristics (the model) or instantaneous information (the state) of the controlled dynamic system.

#### 2.2.3.1 Minimization of the Network Bandwidth Usage

A research direction related to the "per message" paradigm is the *model based control*, which was studied in *Yook et al.* [259], *Montestruque and Antsaklis* [177, 178], *Hespanha and Xu* [117, 118], *Li et al.* in [149, 150]. The basic idea of this approach is to use local open-loop observers in order to reduce the required communications between the sensors and the controller.

In [259], a method allowing the minimization of the required communications in some particular distributed control applications was proposed. This method addresses multivariable discrete-time *LTI* systems that are implemented according to a specific distributed architecture and controlled by using output-feedback. The global distributed system has $n$ states, $m$ inputs and $m$ outputs, such that the $i$th input $u_i(k)$ and the $i$th output $y_i(k)$ are co-located at the same network node, which is also equipped with a computer (Fig. 2.5). The measurements, which are provided by the $m$ sensors, are corrupted by a measurement noise $\nu(k)$. The $m$ nodes are connected through a perfect network (without delays, nor losses of information). Each node contains an estimator that estimates at the same time its own output as well as the outputs of the other nodes. The estimated outputs $\hat{y}_j(k)$, $j \neq i$ are then used by the $i$th node, for the computation of the control $u_i(k)$, instead of the measured outputs $y_j(k)+\nu_j(k)$, $j \neq i$, whose use would require a transmission of their values over the network. By this way, an important communication saving is achieved, at the price of a more important computational load at the computer nodes. This approach relies on the fact that all the estimators compute identical values of the estimated outputs $\hat{y}_j(k)$. If the estimator of the $i$th node realizes that $|y_i(k) + \nu_i(k) - \hat{y}_i(k)| \geq g_i$, where $g_i$ is a fixed threshold, then it broadcasts to the other nodes the value of its measurement $y_i(k) + \nu_i(k)$, enabling them to update the state of their estimators.

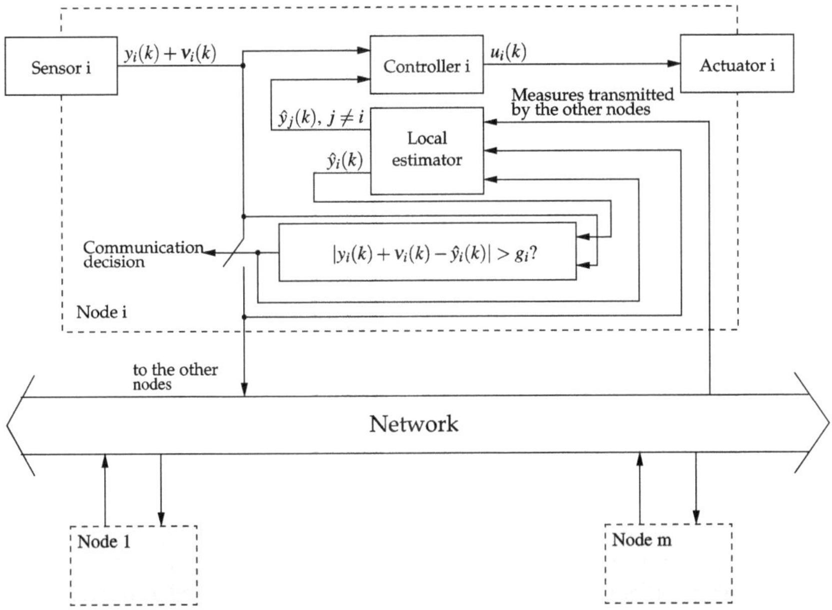


**Fig. 2.5** Global architecture and model of the $i$th node according to the approach of [259]

By this way, $|y(k) + v(k) - \hat{y}(k)|$ is limited by $g = [g_1, \ldots, g_m]^T$. Finally, a result allowing the choice of $H$ to guarantee a maximum degradation of $\varepsilon$ % compared to the ideal system (without networked communication) was stated. The experimental validation of the method was carried out on a two axis contouring system. The experimental results have shown that only 12 % of the communication is necessary in order to guarantee a maximum degradation of 1 %, assuming a model uncertainty of 20 %.

A similar approach was studied in *Montestruque and Antsaklis* [177]. In the considered architecture, the controller is directly connected to the plant. A perfect network connects the sensors to the controller. The sensors periodically transmit (each $T_s$ time instants) the measurements to the controller, which is provided with an open-loop state-observer. The estimated state is then used for the computation of the control commands. At the reception of a message from the sensors, the state of the open-loop observer is updated. The necessary and sufficient stability conditions of this particular model of networked control systems were stated, and generalized to take into account output feedback. Sufficient stability conditions when the sampling period, $T_s$, is time-varying but bounded were presented in *Montestruque and Antsaklis* [178] and *Li et al.* in [149, 150].

Next, in *Hespanha and Xu* [117], a similar architecture was studied. In the considered model, the plant is disturbed by a zero-mean Gaussian white noise. Instead of sending the measurements (of the full state) when the prediction error exceeds a threshold [259] or periodically [177], the transmission of measurements is performed using predefined communication logics. The study and the evaluation of different communications logics (stochastic and deterministic) was also performed. By modeling the problem as an appropriate jump-diffusion process, sufficient conditions for the boundedness of the finite moments of the estimation error were established for the considered logics. Considering a long term average cost, penalizing at the same time the estimation error and the transmission rate, the expression of the optimal communication logics was explicitly given in *Hespanha and Xu* [118].

In *Gommans et al.* [97, 98] the main rationale behind the novel dropout compensators remains the same. They act as model-based, closed-loop observers if information is received and as open-loop predictors if a dropout occurs. These compensators were considered for two dropout models, using either worst-case bounds on the number of subsequent dropouts or stochastic information on the dropout probabilities. For the worst-case bound dropout model, sufficient conditions for global asymptotic stability of the closed-loop networked control systems (*NCS*) with the compensation based strategy are derived. For the stochastic dropout models, necessary and sufficient conditions for (exponential) mean square stability of the closed-loop *NCS* are given. In addition, for both dropout models they developed linear matrix inequality (*LMI*) based conditions for the synthesis of the compensator gains.

In Chap. 12 of this book, we propose a design methodology combining zero and hold strategies in order to optimize the system performance as well as to increase its stability domain in presence of packets dropouts. This static switching strategy may be adapted to operate state dependent one.

#### 2.2.3.2 Medium Access Scheduling

The experimental study of communication networks characteristics was performed in *Nilsson* [187] and *Lian* in [153]. Studying the main characteristic of *ControlNet*, *DeviceNet* and *EtherNet* networks, *Lian et al.* [153] have shown that the transmission time of a message (i.e., the time the message spends on the physical link from the source to the destination) in the most used networks may be neglected. The delays occurring in networked control loops are mainly due to the *contention* between the different messages which are sent by the nodes of the network. The most efficient way of reduction of these delays is the design and use of appropriate message scheduling strategies.

These results show the practical importance of the study of the medium access control as well as scheduling algorithms. The problems of the concurrent access to shared communication resources were studied these last years within various theoretical frameworks and with various modeling assumptions. We present thereafter a brief summary of the approaches taking into account explicitly the concurrent access to the communication network. These contributions were classified into three categories, according to the class of the used scheduling algorithms: off-line scheduling, on-line scheduling of the sensors-to-controller link and the on-line scheduling of the controller-to-actuators link.

- *Off-line scheduling*: The problem of optimal control and off-line scheduling of the controller-to-actuators link was studied in *Rehbinder and Sanfridson* [200].

In the proposed model, the control commands are sent to the actuators through a shared *TDMA* bus. At each slot, only one control command can be sent, the remaining commands for the other actuators are held constant. The choice of which actuator to update at each slot was handled using the notion of communication sequence introduced by *Brockett* [43]. Only periodic communication sequences were considered. A quadratic cost function is associated to each communication sequence, corresponding to the worst-case initial condition and worst-case sequence permutation. Control commands and periodic communication sequences are obtained through the solving of a complex combinatorial optimization problem. The problem of the optimal control and scheduling in the sense of *LQG* was introduced and developed in *Lincoln and Bernhardsson* [155]. The relaxed dynamic programming method was applied for its resolution leading to a more efficient search heuristics. A heuristic approach for the problem of the optimal control and off-line scheduling of the sensors-to-controller link in the sense of the $\mathcal{H}_\infty$ performance index was proposed in *Lu* [161]. The application of branch and bound algorithm to this problem was performed in *Ben Gaid et al.* in [22]. The application of genetic algorithms and particle swarm optimization to this problem was undertaken by *Longo et al.* in [159]. A generalization of this approach to tackle uncertainties of the plant model was undertaken by *Al-Areqi et al.* in [2], and to cope with nonlinearities in *Su et al.* [221].

- *On-line scheduling of the sensors-to-controller link*: The scheduling of sensor measures was studied in *Walsh and Ye* [244]. The addressed configuration consists in a continuous-time plant where the controller is directly connected to the actuators. The network only connects the sensors to the controller. The notion of *MATI* (maximum allowable transfer interval) was introduced, and represents the upper bound on the time between two consecutive sensor messages transmissions that guaranties the stability of the plant. The *MATI* is defined for a given scheduling algorithm. A new on-line scheduling algorithm, called *MEF-TOD* (maximum error first—try once discard), was introduced. In this dynamic priority on-line scheduling algorithm, the priority of a sensor message depends on the error of the measure that it carries; smaller the error is, lower is the assigned priority. The error is defined as the weighted absolute value of the difference between the value of the current measure and the value of the last transmitted measure. If a node fails to send a message, then this message is discarded (dropped from the queue). The authors stated sufficient stability conditions, involving the value of the *MATI*, which ensure the stability of the system, when the *MEF-TOD* algorithm and a round robin like static scheduling algorithm are used. These results are based on the perturbation theory and are very conservative. This approach was generalized to nonlinear systems in *Walsh et al.* [245]. The practical implementation of the *MEF-TOD* algorithm was considered in [243]. This implementation, which was performed on *CAN* networks, is mainly based on the nondestructive bitwise arbitration of *CAN* technology to dynamically encode the dynamic priorities. The effects of the quantization of priorities were experimentally studied. Sufficient input/output $\mathcal{L}_p$-stability results for a class of network scheduling protocols, including *MEF-TOD* and static scheduling algorithms were stated and illustrated in *Nesic and Teel* [185]. These results considerably reduces the conservativeness of the results that were initially stated in *Walsh et al.* [244]. The application of the *Rate Monotonic* scheduling algorithm to the networked control systems was investigated in *Branicky et al.* [41].

- *On-line scheduling of the controller-to-actuators link*: On-line scheduling of control commands to the actuators was studied in *Palopoli et al.* [191]. In the proposed model, it is assumed that every slot, only one command vector can be sent to an actuator group, the other control vectors are set to zero. The stabilization is achieved using a model predictive controller, which calculates on-line the appropriate control law and the allocation of the shared bus. The cost function used by the *Model Predictive Control* (*MPC*) calculates a weighted sum of the infinity norms of the states and the control commands over a specified horizon. The optimization problem solved at each step by the *MPC* algorithm was proven to be equivalent to the generalized linear complementarity problem (*GLCP*) [258]. The same architecture is considered in *Goodwin et al.* [99]. The considered model assumes that it is possible to send only one message during one sampling period. The actuators, which do not receive their control inputs, maintain constant the last received ones. The control commands are quantized with a fixed precision. The expression of the optimal model predictive controller, in the sense of a quadratic cost function, was established. The optimal solution as well as computationally efficient approach (*OPP*) to the problem of joint control and network scheduling was proposed in *Ben Gaid et al.* [27]. A solution expressed as piecewise linear feedback law was proposed in *Görges et al.* [101]. Robustness issues were investigated in *Al-Areqi et al.* [2], using the *OPP* algorithm for complexity reduction. The impact of network induced delays and packet dropouts to this problem were investigated in *Guo and Jin* [107].

## *2.2.4 Allocation of Computational Resources*

### 2.2.4.1 Optimal Control and Mono-Processor Scheduling

The problem of the optimal selection of control tasks periods subject to schedulability constraints was addressed in *Seto et al.* [208]. Assuming that the discrete-time control laws are designed in the continuous-time domain and then discretized, the notion of performance index was introduced. The performance index quantifies the performance of the digitalized control law at a given sampling frequency. In most control applications, the performance index is minimal when the continuous-time control law is used and increases (i.e., degrades) as the sampling frequency is decreased (note that for some control systems this relationship is more complicated, as illustrated in *Eker* [79]). Considering this important class of control applications, the problem of the optimal sampling frequency assignment for a set of control tasks consists on minimizing a weighted sum of the performance indices of the considered control tasks subject to schedulability constraints. In *Rehbinder et al.* [199],

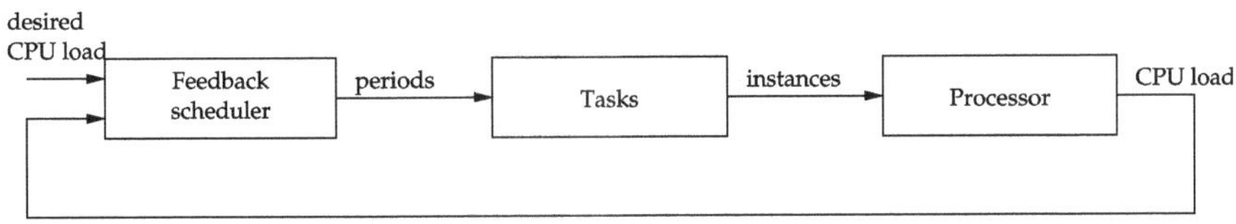


**Fig. 2.6** General model of a feedback scheduler

the optimal off-line scheduling of control tasks in the sense of *LQG* was considered, assuming that all the control tasks have the same constant execution time. The resolution of this problem was performed using the exhaustive search method, which limits the application of this approach to applications with a limited number of tasks. Similarly, the problem of the optimal mono processor scheduling of control tasks in order to optimize a robustness metric (i.e., the stability radius) was treated in *Palopoli et al.* [192, 193].

#### 2.2.4.2 Scheduling of Control Tasks in Environments with Variable Computing Workload

It is well known that worst-case analysis techniques given in *Sha et al.* [211] may be used in order to guarantee the deadlines of tasks with variable but bounded execution times. However, when the average execution time is smaller than the worst-case execution time (*WCET*), these techniques lead to an oversized design. Recently, new approaches were proposed in order to handle variations in tasks execution times and system overload more efficiently than worst-case analysis techniques, among them the feedback scheduling given in *Lu et al.* [160], *Cervin et al.* [55], *Robert et al.* [203], *Xia and Sun* [255] and the elastic task model given in *Buttazzo et al.* [49].

Feedback scheduling (*FSB*) is a control theoretical approach to real-time scheduling of systems with variable workload. The feedback scheduler whose general model is schematically given in Fig. 2.6, may be seen as a *"scheduling controller"* that receives filtered measures of tasks execution times and acts on tasks periods in order to minimize deadline misses. The application of feedback scheduling to robot control was experimentally evaluated in *Simon* [212].

In the elastic task model given in *Buttazzo et al.* [49], a periodic task set containing $\mathcal{N}$ tasks may be seen as a sequence of $\mathcal{N}$ linear springs. In this model, the utilization factor of a task is analogous to the spring's length. Tasks may change their utilization rate in order to handle overload conditions, which may occur, for example, if a new task is admitted to the computing system. In order to ensure the schedulability of the task set, tasks are compressed or decompressed. In *Liu et al.* [158], the elastic task model was applied to the scheduling of control tasks with variable execution times. The use of this method allows the application of the approach of *Seto et al.* [208] in order to find the optimal tasks periods based on tasks average execution times (instead of their worst-case execution times), leading to an improvement of the control performance. *Buttazzo et al.* [47] generalized this approach to take into account the degradations that may occur to the control system if its control task that was designed to work at a given rate runs at another rate. The analytical expressions of the performance degradations were given and a compensation method was proposed. This method allows to trade-off the performance degradations and the required memory space (which is needed to store the parameters of the pre-computed control laws that will be used by the compensation algorithm).

However, all these described approaches are mainly based on the assumption that control performance is a *convex function* of the sampling period. In reality, as illustrated in *Martì* [167] (and further in the forthcoming Chap. 4), the quality of control is also dependent on the dynamical state of the controlled system.

The work of *Bini and Cervin* [34], *Samii et al.* [205, 206], although different in the formulation of the optimization problem in terms of objective functions and constraints, can be included in a subset of works that share in common an off-line approach where sampling periods are derived before run-time and kept constant during execution.

Inside the class of *FBS* methods, in our opinion, there are two main trends. The first one concerns those methods relaying on the instantaneous plant state information and metric in order to adapt the sampling period. In this class, we may classify the work of *Martì et al.* [169]. The second class includes those methods calculating the future sampling periods based on finite or infinite horizon metric as, for example, the methods developed in *Henriksson and Cervin* [114], *Castane et al.* [52], *Cervin et al.* [57]. This type of approach was also adopted in *Ben Gaid et al.* [28, 29] where the sampling period is calculated as a function of the states of the controlled plants, a given static scheduling and a quadratic metric over an periodic infinite horizon.

## 2.3 Notes and Comments

This chapter introduced the basic concepts, the terminology as well as the state of the art of communication and computation resource allocation approaches in *Distributed Control and Embedded Systems* (*DCES*). To this end, the basic concepts of the real-time scheduling theory are outlined, focusing primarily on the hard real-time scheduling of tasks on processors and messages on deterministic networks. An outline of the state of the art of the approaches for the integrated control and communication/computation resource allocation was given, providing an overview of the tackled problems and the provided solutions.

This particular interest on the hard real-time systems is motivated by their industrial realism and utility. Proposing simple and relevant integrated model of *DCESs* are of prime importance for their design as well as for their wide spread in industrial applications. We have clearly seen that different problems posed and treated in literature are related to their stability and performance robustness which are, practically, of different nature. The stability analysis of *DCES* is related to methods and tools used in the field of control and information sciences (*Mitter et al.* [80, 227])

whereas their performance optimization calls for methods and approaches related traditionally to control system domain and in parallel to those of computer science. The necessity of double-breasted view of the *DCES* design problem not only facilitates its analysis but allows optimizing its performances. Application of simple and consistent scheduling policy given respectively in *Yook et al.* [259], *Hristu* [123], *Longo et al.* [159], *Ben Gaid et al.* [27, 28] allows reducing their model complexity by inducing structure properties such as periodicity of control and/or sensing signals/messages. This fact, as it will be seen in the forthcoming chapters of Part II, simplify substantially the design of *DCESs* projecting it in a class of well-known problems in the area of control systems.

Naturally, the *DCES* area encompasses a larger class of applications including those for which a predefined or fixed scheduling policy is impossible. This is mainly due to non-determinism induced by the computation model related to the network nodes as well as to the model of communication related to the network. In this case, we observe variable induced delays on signals/messages sending and reception as well as modification of their reception order. This fact faced us with the problem of system information reduction in general and particularly the way to handle related variable induced delays. As it will be seen in the forthcoming Chaps. 10, 11, 12, 13 and from the literature review done in Sect. 2.2.1, two main trends are observed. The first one concerns the techniques related to system state information enhancement based on the system model as a complementary information source. The second one proposes methods and approaches whose objectives are to use the communication and calculation resources offered by *DCES* with respect to the system state or system performances. These two main trends have the same objective that is injecting complementary information in the system in order to overcome the lack of resources and/or to reduce the resources reserving them in priority to subsystems needing more. The topological structure and the size of *DCES* are different and function of the considered application. The decisions taken at the level of each subsystem composing it are effective if they are coordinated between them. The different approaches and techniques originally borrowed from the computer science such as synchronization help to increase the system performances. In the same time, they induce new models of communication between sub-systems that have to be considered and merged with the dynamic model of *DCES*. In fine, the main problem to handle is related to the structure and the quantity of information communicated between sub-systems in order to reduce the information delay of the critical system. Mastering this delay represents one of the main challenges in the design of *DCES*.

# Chapter 3
# Modeling and Analysis of Resource-Constrained Systems

In this chapter, we present our abstract view of a distributed control and embedded systems (*DCES*) operating under communication constraints. This abstract view is described by the class of *computer-controlled systems*, which was introduced by *Hristu* in [122]. This class allows modeling, in a finely-grained and abstract way, the impact of the resource limitations on the behavior of the controlled system. In this book, we will rather use the term of *resource-constrained systems* to refer to this class of systems. After the introduction of the notation we are using, we present the framework of the *mixed logical dynamical* (*MLD*) systems, which represents a modeling framework for hybrid systems. Such a framework was introduced by *Bemporad and Morari* in [21]. We show that resource-constrained systems may be modeled in the *MLD* framework. Furthermore, we propose a systematic approach allowing establishing the *MLD* model of a resource-constrained system. Finally, we review the main theoretical results that are related to resource-constrained systems, and encountered in the literature. These results include the problems of stabilization, tracking, reachability and observability.

## 3.1 Mixed-Logical Dynamical (*MLD*) Systems

Mixed logical dynamical (*MLD*) systems, represent a class of hybrid systems, that was introduced by *Bemporad and Morari* in [21]. This framework was proposed in order to allow the modeling and the control of a class of systems where dynamics interact in tightly coupled way with logic and heuristic rules. The *MLD* framework provides a general model generalizing various existing models such as linear hybrid systems, linear constrained systems, finite-state machines ... The logical aspects may be formulated in the propositional logic. The fundamental idea of the *MLD* framework is to represent these propositional formulas by equivalent linear constraints, involving continuous and/or Boolean variables. In this way, it is possible to represent many hybrid systems by linear dynamic equations whose variables, which may be continuous and/or Boolean, are subjected to linear inequality constraints.

A. Çela et al., *Optimal Design of Distributed Control and Embedded Systems*,
Communications and Control Engineering, DOI 10.1007/978-3-319-02729-6_3,

Note that the idea of representing propositional formulas by equivalent linear inequalities was developed in the past in the field of artificial intelligence, and more specifically for inference engines. Using this correspondence, the proof of a theorem is reduced to the search of the existence of a feasible solution of an appropriate optimization problem. In fact, by associating a Boolean variable to a propositional formula, such that the truth-value of the formula is equal to the value of the variable, the propositional calculus operations like conjunction, disjunction, negation, implication or equivalence may be translated into equivalent linear programs. Let $X_1$ and $X_2$ be two propositional formulas and $\alpha_1$, $\alpha_2$ two associated Boolean variables verifying

$$X_1 \text{ (resp. } X_2) \text{ is True} \quad \text{if and only if} \quad \alpha_1 = 1 \text{ (resp. } \alpha_2 = 1),$$

and equivalently

$$X_1 \text{ (resp. } X_2) \text{ is False} \quad \text{if and only if} \quad \alpha_1 = 0 \text{ (resp. } \alpha_2 = 0).$$

It is easy to verify the correspondence between the propositional formulas and the conjunction of linear inequalities that are described below.

$$X_1 \vee X_2 \quad \text{is equivalent to} \quad \alpha_1 + \alpha_2 \geq 1,$$

$$X_1 \wedge X_2 \quad \text{is equivalent to} \quad \alpha_1 = 1, \alpha_2 = 1,$$

$$\neg X_1 \quad \text{is equivalent to} \quad \alpha_1 = 0,$$

$$X_1 \quad \Longrightarrow \quad X_2 \quad \text{is equivalent to} \quad \alpha_1 - \alpha_2 \leq 0,$$

$$X_1 \quad \Longleftrightarrow \quad X_2 \quad \text{is equivalent to} \quad \alpha_1 - \alpha_2 = 0.$$

Following the same reasoning, it is also possible to translate terms involving at the same time continuous and Boolean variables into linear inequalities. Let $f : \mathbb{R}^m \longrightarrow \mathbb{R}$ be a linear function. Let $\mathscr{X}$ be a bounded set. Assume that

$$U = \max_{x \in \mathscr{X}} f(x),$$

and

$$L = \min_{x \in \mathscr{X}} f(x).$$

*Morari and Bemporad* have shown that the product $\alpha f(x)$, for $x \in \mathscr{X}$, where $\alpha$ is a Boolean, may be replaced by a conjunction of linear constraints, if the auxiliary variable $y = \alpha f(x)$ is introduced. It is easy to verify that the equality $y = \alpha f(x)$ is equivalent to

$$y \leq U\alpha,$$

$$y \geq L\alpha,$$

$$y \leq f(x) - L(1 - \alpha),$$

$$y \geq f(x) - U(1 - \alpha).$$

This translation will be of great importance when dealing with the problem of the joint optimization of control and scheduling.

In [21], many other useful translations are reviewed. Using these transformations, many hybrid systems may be handled in the *MLD* framework.

The general state-model of a *MLD* system is given by the following equations

$$x(k+1) = A_k x(k) + B_{1_k} u(k) + B_{2_k} \alpha(k) + B_{3_k} \xi(k), \tag{3.1a}$$

$$y(k) = C_k x(k) + D_{1_k} u(k) + D_{2_k} \alpha(k) + B_{3_k} \xi(k), \tag{3.1b}$$

$$E_{2_k} \alpha(k) + E_{3_k} \xi(k) \leq E_{1_k} u(k) + E_{4_k} x(k) + E_{5_k}, \tag{3.1c}$$

where

$$x = \begin{bmatrix} x_c \\ x_l \end{bmatrix}, \quad x_c \in \mathbb{R}^{n_c}, \; x_l \in \{0, 1\}^{n_l}, \; n = n_c + n_l$$

is the state, which contains $n_c$ real-valued components and $n_l$ Boolean-valued components,

$$y = \begin{bmatrix} y_c \\ y_l \end{bmatrix}, \quad y_c \in \mathbb{R}^{p_c}, \; y_l \in \{0, 1\}^{p_l}, \; p = p_c + p_l,$$

and

$$u = \begin{bmatrix} u_c \\ u_l \end{bmatrix}, \quad u_c \in \mathbb{R}^{m_c}, \; u_l \in \{0, 1\}^{m_l}, \; m = m_c + m_l$$

are respectively, the output and the command input of the system, also containing both continuous and Boolean components. Here, $\alpha \in \mathbb{R}^{q_l}$ and $\xi \in \mathbb{R}^{q_c}$ are respectively continuous and Boolean auxiliary variables.

It may be remarked that the inequality (3.1c) might be satisfied for many values of $\alpha(k)$ and $\xi(k)$. For that reason, the determination of $x(k+1)$ and $y(k)$ may be non-unique. A condition that is generally imposed on *MLD* systems is to be *well posed*. This condition guarantees that once $x(k)$ and $u(k)$ are assigned, then $x(k+1)$ and $y(k)$ are uniquely determined authorizing thus the definition of trajectories in the state or output spaces. In general, *MLD* models that are obtained from a real physical system are well posed. The formal definition of a *well posed MLD* system may be found in [21].

## 3.2 *MLD* Modeling of Resource-Constrained Systems

Consider the continuous-time *LTI* plant described by

$$\dot{x}_c(t) = A_c x_c(t) + B_c u_c(t) + W_c w_c(t), \tag{3.2a}$$

$$y_c(t) = C_c x_c(t) \tag{3.2b}$$

where $x_c(t) \in \mathbb{R}^n$, $u_c(t) \in \mathbb{R}^m$, $y_c(t) \in \mathbb{R}^p$ and $w_c(t) \in \mathbb{R}^r$ represent respectively, the state, the command input, the output and the disturbance input. The plant is controlled by a discrete-time controller, with sampling period $T_s$. The plant (3.2a), (3.2b) and the controller are connected through a limited bandwidth communication bus. At each sampling instant $t = kT_s$ ($k \in \mathbb{N}$), the bus can carry at most $b_r$ measures and $b_w$ control commands, with $b_r \leq p$ and $b_w \leq m$. The input to the plant is preceded by a zero-order holder, which maintains the last received control commands constant until new control values are received. Let $u(k)$ be the input of the zero-order holder at instant $kT_s$, then its output is given by

$$u_c(t) = u(k) \quad \text{if } kT_s \leq t < (k+1)T_s. \tag{3.3}$$

First, we assume that $W_c = 0$. Let $x(k) = x_c(kT_s)$ and $y(k) = y_c(kT_s)$ be respectively, the sampled values of the state and the output. A discrete-time representation of the plant (3.2a), (3.2b) at the sampling period $T_s$ is given by

$$x(k+1) = Ax(k) + Bu(k), \tag{3.4a}$$

$$y(k) = Cx(k) \tag{3.4b}$$

where $A = e^{A_c T_s}$, $B = \int_0^{T_s} e^{A_c \tau} B_c \, d\tau$ and $C = C_c$.

We will assume, throughout this monograph, that the pairs $(A, B)$ and $(A, C)$ are respectively reachable and observable. These assumptions are systematically satisfied if the pair $(A_c, B_c)$ is reachable, the pair $(A_c, C_c)$ is observable and the sampling period $T_s$ is *non-pathological*. A sampling period is said pathological if it causes the loss, for the sampled-data model, of the reachability and observability properties, which were verified by the continuous model before its discretization. In [136], *Kalman et al.* have proved that the set of pathological sampling periods is countable, and uniquely depends on the eigenvalues of the state matrix $A_c$. Consequently, in order to avoid the loss of reachability and observability, which may be caused by the sampling, it is sufficient to choose $T_s$ outside this set.

Communication constraints may be formally described by introducing two vectors of Booleans $\sigma(k) \in \{0, 1\}^{b_r}$ and $\delta(k) \in \{0, 1\}^{b_w}$, defined for each sampling instant $k$.

**Definition 3.1** The vector $\sigma(k)$ defined by

$$\begin{cases} \sigma_i(k) = 1 & \text{if } y_i(k) \text{ is read by the controller at instant } k, \\ \sigma_i(k) = 0 & \text{otherwise} \end{cases}$$

is called sensors-to-controller scheduling vector at instant $k$.

**Definition 3.2** The vector $\delta(k)$ defined by

$$\begin{cases} \delta_i(k) = 1 & \text{is } u_i(k) \text{ updated at instant } k, \\ \delta_i(k) = 0 & \text{otherwise} \end{cases}$$

is called controller-to-actuators scheduling vector at instant $k$.

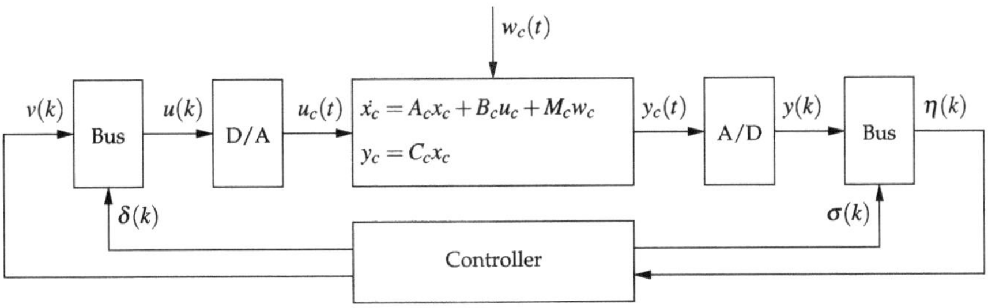


**Fig. 3.1** Schematic representation of a resource-constrained system

The vector $\sigma(k)$ indicates the measures that the controller may read at instant $k$. In a similar way, the $\delta(k)$ indicates the control inputs to the plant that the controller may update at instant $k$. The introduction of the scheduling vectors allows modeling in a simple way the communication constraints. The limitations that affect the transmission of the measures to the controller may be described by the following inequality:

$$\sum_{i=1}^{p} \sigma_i(k) \leq b_r. \tag{3.5}$$

In a similar way, the limitations concerning the sending of the control commands to the actuators may be modeled by

$$\sum_{i=1}^{m} \delta_i(k) \leq b_w. \tag{3.6}$$

The last received control inputs (through the communication bus) are kept constant. Consequently, if a control input is not updated at the $k$th sampling period, then it is maintained constant. This assertion may be modeled by the logic formula:

$$\delta_i(k) = 0 \quad \Longrightarrow \quad u_i(k) = u_i(k-1). \tag{3.7}$$

The plant, the analog-to-digital and digital-to-analog converters, the communication bus and the controller are schematically depicted in Fig. 3.1. In this figure, $\eta(k) \in \mathbb{R}^{b_r}$ represents the vector of partial measurements that the controller receives (through the communication bus) at the sampling period $k$. In a similar way, vector $v(k) \in \mathbb{R}^{b_w}$ represents the vector of partial control commands that the controller may send to the actuators (through the limited bandwidth communication bus) at the sampling period $k$. Blocks D/A and A/D respectively represent the digital-to-analog and analog-to-digital converters. The controller may also assign the values of the sensors-to-controller scheduling vector ($\sigma(k)$) as well as the controller-to-actuators scheduling vector ($\delta(k)$).

Knowing $v(k)$ and the relation (3.7), $u(k)$ is given by

$$\begin{cases} u_i(k) = v_j(k) & \text{if } \delta_i(k) = 1 \text{ and } \sum_{l=1}^{i} \delta_l(k) = j, \\ u_i(k) = u_i(k-1) & \text{otherwise.} \end{cases} \tag{3.8}$$

It may be easily verified that if $v_{j_1}$ and $v_{j_2}$ are mapped to respectively $u_{i_1}$ and $u_{i_2}$, than $(j_1 < j_2)$ implies that $(i_1 < i_2)$.

The mapping (3.8) may be written in an appropriate matrix form. Let $[D_\delta(k)]_{1 \le i \le m, 1 \le j \le b_w}$ the matrix defined by

$$\begin{cases} \left[D_\delta(k)\right]_{ij} = 1 & \text{if } \delta_i(k) = 1 \text{ and } \sum_{l=1}^{i} \delta_l(k) = j, \\ \left[D_\delta(k)\right]_{ij} = 0 & \text{otherwise,} \end{cases}$$

and

$$E_\delta(k) = \begin{bmatrix} 1-\delta_1(k) & & \\ & \ddots & \\ & & 1-\delta_m(k) \end{bmatrix},$$

then

$$u(k) = D_\delta(k)v(k) + E_\delta(k)u(k-1).$$

Conversely, knowing the control input $u(k)$ and the scheduling decision $\delta(k)$, the vector of control commands $v(k)$ that were sent through the bus may be determined.

Let $[M_\delta(k)]_{1 \le i \le b_w, 1 \le j \le m}$ the matrix defined by

$$\begin{cases} \left[M_\delta(k)\right]_{ij} = 1 & \text{if } \delta_j(k) = 1 \text{ and } \sum_{l=1}^{j} \delta_l(k) = i, \\ \left[M_\delta(k)\right]_{ij} = 0 & \text{otherwise,} \end{cases}$$

then

$$v(k) = M_\delta(k)u(k). \tag{3.9}$$

A schematic representation illustrating this mapping is given in Fig. 3.2.

In the same way, the input $\eta(k)$ to the controller is defined by

$$\begin{cases} \eta_i(k) = y_j(k) & \text{if } \sigma_j(k) = 1 \text{ and } \sum_{l=1}^{j} \sigma_l(k) = i, \\ \eta_i(k) = 0 & \text{otherwise.} \end{cases} \tag{3.10}$$

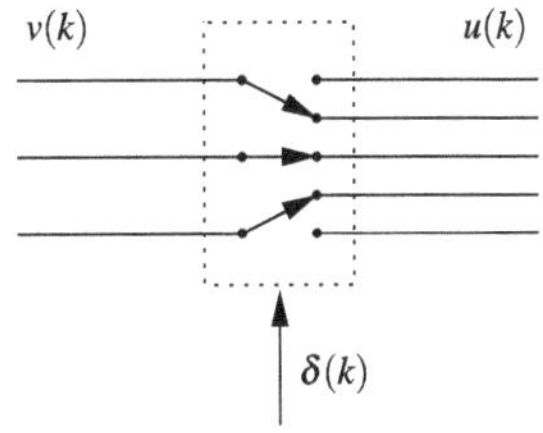


**Fig. 3.2** Schematic representation of the mapping of $v(k) \in \mathbb{R}^3$ to $u(k) \in \mathbb{R}^5$ corresponding to the scheduling vector $\delta(k)$ verifying $\delta_1(k) = 0$, $\delta_2(k) = 1$, $\delta_3(k) = 1$, $\delta_4(k) = 1$ and $\delta_5(k) = 0$

This mapping may be also written in an appropriate matrix form. Let $[M_\sigma(k)]_{1 \le i \le b_r, 1 \le j \le p}$ the matrix defined by

$$\begin{cases} \left[M_\sigma(k)\right]_{ij} = 1 & \text{if } \sigma_j(k) = 1 \text{ and } \sum_{l=1}^{j} \sigma_l(k) = i, \\ \left[M_\sigma(k)\right]_{ij} = 0 & \text{otherwise,} \end{cases}$$

then we have the relation

$$\eta(k) = M_\sigma(k) y(k). \tag{3.11}$$

Equations (3.4a), (3.4b), (3.5), (3.6), (3.8) and (3.10) describe a model where dynamics and plant performance are tightly coupled with the assignment of communication resources. In the particular case where $b_r = p$, $b_w = m$, $\sigma(k) = 1_{p,1}$ and $\delta(k) = 1_{m,1}$, for all $k \in \mathbb{N}$, this model coincides with the classical model of a sampled-data system. The presence of the communication bus moves the classical frontier between "the plant" and "the controller". In fact, for sampled-data systems, this frontier lies at the digital-to-analog and analog-to-digital converters. In the considered model, this frontier moves to the communication bus interface. We will call *resource-constrained system* the entity constituted by the sampled-data model of the plant and the communication bus. The formal definition of a resource-constrained system is given thereafter.

**Definition 3.3** A resource-constrained system is a mixed logical dynamical system having three inputs: the command input $v(k)$, the scheduling vector of the sensors-to-controller link $\sigma(k)$ and the scheduling vector of the controller-to-actuators link $\delta(k)$. It has one output denoted $\eta(k)$. Its mathematical model is defined by:

- recurrent equations (3.4a), (3.4b) describing the sampled dynamics of the plant,
- inequality constraints (3.5) and (3.6) expressing the limitations of the communication medium,
- logic formulas (3.8) describing the mapping of the computed controller outputs $v(k)$ to plant inputs $u(k)$, knowing the scheduling decisions $\delta(k)$,
- logic formulas (3.10) describing the mapping of the sampled plant outputs $y(k)$ to the controller's inputs $\eta(k)$, knowing the scheduling decisions $\sigma(k)$.

The particularity of a resource-constrained system, compared to a sampled-data system, is that at each sampling period, it is important to determine:

- the measures that should be acquired (it is only possible to acquire at most $b_r$ measures, defined by the scheduling function $\sigma(k)$),
- the control commands that should be applied (it is only possible to apply at most $b_w$ control commands, defined by the scheduling function $\delta(k)$),
- the value of the applied control commands.

## 3.3 Notion of Communication Sequence

The notion of communication sequence was introduced by *Brokett* [43] and generalized by *Hristu* [122] in order to quantify the notion of *attention* [44]. It characterizes the "allocation" of the bus resources to the different inputs and outputs of the system, that is, the attention that must be given to each input and output. There are two types of communication sequences: finite communication sequences and periodic infinite communication sequences.

**Definition 3.4** [122] A finite communication sequence $\rho^{N-1}$ of length $N$ and width $l$ is a finite sequence $\rho^{N-1} = (\rho(0), \ldots, \rho(N-1))$ of elements of $\{0, 1\}^l$.

**Definition 3.5** [122] A periodic communication $\rho^{T-1}$ sequence of period $T$ and width $l$ is an infinite sequence $\rho^{T-1} = (\rho(0), \ldots, \rho(T-1))$ of elements of $\{0, 1\}^l$ verifying $\rho(k+iT) = \rho(k)\ \forall i \in \mathbb{N}$.

In resource-constrained system, a communication sequence is a sequence of scheduling vectors. We may distinguish between two types of communication sequences:

- sequences of sensors-to-controller scheduling vectors, which will be called *measurements communication sequence*,
- sequences of controller-to-actuator scheduling vectors, which will be called *commands communication sequence*.

Naturally, each element of a finite or periodic communication sequence must respect the communication and fairness constraints.

**Definition 3.6** [122] Let $\mathscr{S}$ be a resource-constrained system, characterized by its resource limitations $b_r$ and $b_w$. The measurements communication sequence $\sigma^{N-1}$ (resp. the commands communication sequence $\delta^{N-1}$) is called *admissible* if

- $\forall k \in \mathbb{N}$, $0 < \|\sigma(k)\|_2^2 \leq b_r$ (resp. $0 < \|\delta(k)\|_2^2 \leq b_w$) (non-surpassing of the bus capacity),
- $\mathrm{Span}\{\sigma(0), \ldots, \sigma(N-1)\} = \mathbb{R}^p$ (resp. $\mathrm{Span}\{\delta(0), \ldots, \delta(N-1)\} = \mathbb{R}^m$) (during any period $N$, each controller input (resp. plant input) is updated at least one time).

We introduce the notion of maximal communication sequence, in order to characterize the communication sequences that use all the available communication resources. This notion will be helpful to simplify the definition, formulation as well as to find solutions to the problems that will be tackled thereafter.

**Definition 3.7** Let $\mathscr{S}$ be a resource-constrained system characterized by its resource limitations $b_r$ and $b_w$. The measurements communication sequence $\sigma^{N-1}$ (resp. the control communication sequence $\delta^{N-1}$) is called *maximal* if

- $\forall k \in \mathbb{N}$, $0 < \|\sigma(k)\|_2^2 = b_r$ (resp. $0 < \|\delta(k)\|_2^2 = b_w$) (usage of all the available communication resources).

## 3.4 State Representation of Resource-Constrained Systems

For predefined scheduling vectors $\sigma$ and $\delta$, a resource constrained system $\mathscr{S}$ may be viewed, between its input $v(k)$ and its output $\eta(k)$, as a linear time-varying system. Based on the previous definitions, and denoting

$$\chi(k) = u(k-1)$$

and

$$\tilde{x}(k) = \begin{bmatrix} x(k) \\ \chi(k) \end{bmatrix},$$

the linear sampled-data and time-varying model of system $\mathscr{S}$ is given by

$$\tilde{x}(k+1) = \tilde{A}(k)\tilde{x}(k) + \tilde{B}(k)v(k), \tag{3.12a}$$

$$\eta(k) = \tilde{C}(k)\tilde{x}(k), \tag{3.12b}$$

where

$$\tilde{A}(k) = \begin{bmatrix} A & BE_\delta(k) \\ 0_{m,n} & E_\delta(k) \end{bmatrix},$$

$$\tilde{B}(k) = \begin{bmatrix} BD_\delta(k) \\ D_\delta(k) \end{bmatrix},$$

and

$$\tilde{C}(k) = M_\sigma(k) \begin{bmatrix} C & 0_{p,m} \end{bmatrix}.$$

## 3.5 Stabilization with Limited Resources

Given a resource-constrained system and a fixed periodic scheduling defined by two periodic admissible communication sequences $\delta^{T-1} = (\delta(0), \ldots, \delta(T-1))$

(commands communication sequence) and $\sigma^{T-1} = (\sigma(0), \ldots, \sigma(T-1))$ (measurements communication sequence), an interesting question is to determine whether it is possible of stabilize the discrete-time system (3.4a), (3.4b), subject to the communication constraints (3.5), (3.6), (3.8) and (3.10), using a constant output feedback gain $\Sigma \in \mathbb{R}^{b_w \times p}$ such that $v(k) = \Sigma \bar{y}(k)$, where $\bar{y}(k) \in \mathbb{R}^p$ verifies $\bar{y}(k) = D_\sigma(k)\eta(k) + E_\sigma(k)\bar{y}(k-1)$.

**Problem 3.1** [122] Given a resource-constrained system $\mathscr{S}$, two admissible $T$-periodic communication sequences $\sigma^{T-1}$ and $\delta^{T-1}$, find a constant output feedback gain $\Sigma \in \mathbb{R}^{b_w \times p}$ that stabilizes system (3.4a), (3.4b), under the communication constraints (3.5), (3.6), (3.8) and (3.10).

In [122], *Hristu* proves that this problem is equivalent to the following NP-hard problem:

**Problem 3.2** [122] Given a collection of matrices $\mathscr{M}_i \in \mathbb{R}^{\beta \times \beta}$, $0 \leq i \leq i_{\max}$ and scalars $\theta_1, \ldots, \theta_{i_{\max}}$, find a stable element of the affine subspace:

$$\mathscr{M} = \mathscr{M}_0 + \sum_{i=1}^{i_{\max}} \theta_i \mathscr{M}_i, \tag{3.13}$$

where $\beta = (2T^2 - T)n$ and $i_{\max} = mp$.

The proof essentially exploits the periodicity of the communication sequences in order to represent the periodic discrete-time model by an equivalent discrete-time invariant model of higher dimension ($\beta = (2T^2 - T)n$). The procedure allowing to obtain this higher dimension invariant model (and to prove the equivalence) was called "extensification". Using this equivalence, the output-feedback stabilization of the resource-constrained system, based on two fixed $T$-periodic communication sequences $\sigma^{T-1}$ and $\delta^{T-1}$, amounts to find the scalars $\theta_i$ for which the spectral radius of the extensive form lies in the unit circle. A simulated annealing-based heuristic, aiming at minimizing the spectral radius of the extensive form, was proposed.

## 3.6 Trajectory Tracking with Limited Resources

The $N$-steps output tracking problem was also studied in [122]. The considered problem is a feed-forward control problem. All communication resources are allocated to the transmission of the control commands ($b = b_w$ and $b_r = 0$).

**Problem 3.3** [122] Given a resource constrained system $\mathscr{S}$ with $m \leq p$, an integer $N$, an admissible finite commands communication sequence $\delta^{N-1}$ of length $N$ and width $m$, and a desired output $y_d \in \mathscr{L}_2^p([0, NT_s])$, find the optimal sequence of control commands $u^{N-1^*} = (u^*(0), \ldots, u^*(N-1))$ that minimizes $\|y - y_d\|_2$.

An analytic solution was proposed in [122]. Let $\Upsilon$ be the operator defined by

$$\begin{aligned}\Upsilon : l_2^m\big(\{0,\ldots,N-1\}\big) &\longrightarrow \mathscr{L}_2^p[0,T],\\ u &\longmapsto y(t)=\Upsilon u\end{aligned}$$

such that

$$\Upsilon u=\sum_{k=0}^{N-1}\Theta_{T_s}(t-kT_s)u(k),$$

with

$$\Theta_{T_s}(t)=\begin{cases}\int_0^{\min(t,T_s)} C_c e^{A_c(t-\tau)}B_c\,d\tau & t\geq 0,\\ 0 & t<0.\end{cases}$$

Let $\Upsilon^*$ be the adjoint operator of $\Upsilon$ in the sense of the Euclidian dot or inner product. $\Upsilon^*$ is defined by

$$\begin{aligned}\Upsilon^* : \mathscr{L}_2^p[0,T] &\longrightarrow l_2^m\big(\{0,\ldots,N-1\}\big),\\ y &\longmapsto \Upsilon^* y\end{aligned}$$

such that

$$\big(\Upsilon^* y\big)(j)=\int_0^{NT_s}\Theta_{T_s}^T(t-jT_s)y(t)\,dt,\quad j=0,\ldots,N-1.$$

For the finite admissible communication sequence $\delta^{N-1}$, let $\Omega(\delta)_{1\leq i\leq mN,1\leq j\leq b_w N}$ the matrix defined by

$$\begin{cases}\big[\Omega(\delta)\big]_{ij}=1 & \text{if } \left\lfloor\frac{i-1}{m}\right\rfloor\geq\left\lfloor\frac{j-1}{b_w}\right\rfloor, \delta\left(\left\lfloor\frac{j-1}{b_w}\right\rfloor\right)_{i-\lfloor\frac{i}{m+1}\rfloor m}=1 \text{ and}\\ & \displaystyle\sum_{q=1}^{i-\lfloor\frac{i}{m+1}\rfloor m}\delta\left(\left\lfloor\frac{j-1}{b_w}\right\rfloor\right)_q=j-\left\lfloor\frac{j}{b_w+1}\right\rfloor b_w,\\ \big[\Omega(\delta)\big]_{ij}=0 & \text{otherwise,}\end{cases}$$

then we have the following result.

**Theorem 3.1** [122] *The optimal solution $u^*$ of Problem 3.3 is*

$$u^*=\big(\Omega(\delta)^T\Upsilon^*\Upsilon\Omega(\delta)\big)^{-1}\Omega(\delta)^T\Upsilon^*(y_d-y_{ic}),$$

*where*

$$y_{ic}(t)=C_c e^{A_c t}x(0)+\int_0^t C_c e^{A_c(t-\tau)}B_c u_{ic}\,d\tau$$

*and $u_{ic}$ is the initial conditions of the elements of $u(0)$ that are not updated at $t=0$.*

## 3.7 Reachability and Observability with Limited Resources

Reachability and observability are structural properties of systems that are first imposed in control practice. Given a resource-constrained system $\mathscr{S}$, we may ask the following questions:

- If the discrete-time model (3.4a), (3.4b) without communication constraints is reachable (resp. observable), what are the conditions allowing the resource-constrained system to keep these properties?
- Is it possible to construct periodic communication sequences guaranteeing these properties? If the answer is positive, under what conditions?

These questions were raised and answered by *Zhang and Hristu* in [262] for a particular model of resource-constrained systems. In the considered model, if a control command is not received, then the corresponding actuator applies zero value, instead of maintaining constant the last received control command. The generalization of these results to resource-constrained systems, using zero-order holders (corresponding to the model that was considered in this document), was performed by *Ionete and Cela* in [131]. The resource-constrained system $\mathscr{S}$, viewed between its input $v(k)$ and its output $\eta(k)$ (as represented by Eqs. (3.12a), (3.12b)) being linear time-varying, it is necessary to use reachability and observability notions that are suitable for linear time-varying systems.

**Definition 3.8** [204] A linear discrete-time system is called $l$-step reachable (resp. $l$-step observable) if $l$ is a positive integer such that the system is reachable (resp. observable) on $[i, i+l]$, for any $i$.

**Theorem 3.2** [131] *If $A$ is invertible and the pair $(A, B)$ is reachable, then for any integer $b_w$ such that $1 \leq b_w \leq m$, there exist integers $l, T > 0$ and a maximal $T$-periodic communication sequence of width $m$ such that system* (3.12a), (3.12b) *is $l$-step reachable.*

In practice, the matrix $A$ is obtained by digitalization of matrix $A_c$, which is generally invertible. The proof of this theorem is obtained by construction. Consequently, it is possible to employ it in order to construct periodic communication sequences guaranteeing the reachability. A similar result is obtained for the observability.

**Theorem 3.3** [131] *If $A$ is invertible and the pair $(A, C)$ is observable. Then for any $b_r$ such that $1 \leq b_r \leq p$, there exist integers $l, T > 0$ and a maximal $T$-periodic communication sequence of width $p$ such that system* (3.12a), (3.12b) *is $l$-step observable.*

## 3.8 Notes and Comments

In this chapter, we have presented an abstract model of a distributed control and embedded system operating under communication constraints. We have proposed a systematic approach allowing to formally describing it in the *MLD* framework. The originality of the considered model comes from the fact that communication resources allocation is viewed as an input. The interaction between control and scheduling is thus explicitly taken into account. We then reviewed the various theoretical results established in the literature on this model and which include the problems of stabilization, of trajectory tracking, reachability and observability.

Thereafter, we will focus on the possibility that is offered by this model to assign at the same time the control and scheduling. In the next chapter, we will consider the problem of optimal integrated control and scheduling, disregarding practical implementation constraints. The resource-constrained systems being an extension of sampled-data systems, the problem of the joint optimization of control and scheduling may be seen like a natural extension of classical optimal control problems.

# Part II
# Optimal Co-design of Distributed Control and Embedded Systems

# Chapter 4
# Optimal Integrated Control and Scheduling of Resource-Constrained Systems

The general model of resource-constrained systems allows the on-line assignment of the sensors-to-controller and the controller-to-actuators scheduling vectors. This assignment may be based on a pre-computed off-line schedule or on an on-line scheduling algorithm. The problem of the optimal integrated control and *off-line* scheduling of the sensors-to-controller link is the dual problem of the optimal integrated control and off-line scheduling of the controller-to-actuators link, and may be solved in a similar way. However, the problem of the optimal integrated control and *on-line* scheduling of the sensors-to-controller link is different from the problem of the optimal integrated control and on-line scheduling of the controller-to-actuators link, and its formulation and solving are extremely dependant on the used technology (possibility of on-line arbitration based on the comparison of messages identifiers like in *CAN* networks for example). Furthermore, it is not clear whether an on-line scheduling algorithm of the sensors-to-controller link will be better (from a control performance point of view) than a predefined off-line scheduling algorithm, especially, because the sensors are physically distributed whereas the optimal on-line assignment of the sensors-to-controller link requires the knowledge of all sensors values, which may not be possible due to the communication constraints. To the best of the authors knowledge, this last question remains an open problem.

For that reason, in this chapter, as well as in the forthcoming Chaps. 5 and 6, we will assume that the state of the plant is available to the controller at each sampling period. This assumption will allow us to mainly focus on the problem of the optimal integrated control and scheduling of the controller-to-actuators link. Note that assuming that the state of the plant is available to the controller at each sampling period does not necessarily mean that we are restricted to a particular architecture, but rather that "state of the art methods", such as an off-line scheduling, are deployed to obtain a satisfactory estimate of the state at the controller, using an adequate part of the bandwidth. Let $\mathscr{S}$ be a resource-constrained system verifying this assumption. Furthermore, $\mathscr{S}$ verifies $p = n = b_r$, $\sigma(k) = 1_{n,1}$, $\forall k \in \mathbb{N}$ and $0 < b_w \leq m$. To simplify the notation, let $b = b_w$. In the model considered in the sequel, the resource-constrained system $\mathscr{S}$ has two types of inputs: control inputs and scheduling inputs

A. Çela et al., *Optimal Design of Distributed Control and Embedded Systems*, Communications and Control Engineering, DOI 10.1007/978-3-319-02729-6_4,

of the controller-to-actuators link. The fundamental problem that will be addressed is the problem of the joint optimization of control and scheduling. This problem may be viewed as the generalization of optimal control problems for linear sampled-data system. It naturally appears as having a hybrid character, since continuous aspects (system dynamics) as well as logical aspects (scheduling) interact.

Using the basic results of optimal control theory, we first describe the solution of the optimal control problem given a fixed communication sequence, over finite and infinite horizons. Then the problem of the optimal integrated control and scheduling of resource-constrained systems is formulated and solved. The formulation and solving of this problem are based on the existing theoretical tools from hybrid systems theory and especially the *MLD* framework [21], where this problem may be perfectly modeled. Finally, based on a numerical example, the solutions of this problem are studied.

## 4.1 Performance Index Definition

In order to introduce an appropriate measure of the *"quality"* of the control and scheduling, inspired by optimal control metrics [14], a quadratic cost function is associated to the system (3.2) (and implicitly to the resource-constrained system $\mathscr{S}$).

$$J_c(x_c, u_c, 0, T_f) = \int_0^{T_f} \left(x_c^T(t) Q_c x_c(t) + u_c^T(t) R_c u_c(t)\right) dt + x_c^T(T_f) S_c x_c(T_f) \tag{4.1}$$

where $T_f = NT_s$ and $Q_c$, $R_c$ and $S_c$ are positive definite matrices. These matrices define the design specifications of the ideal continuous-time controller. The sampled-data representation of the cost function $J_c(x_c, u_c, 0, T_f)$ at the sampling period $T_s$ is

$$J(x, u, 0, N) = \sum_{k=0}^{N-1} \begin{bmatrix} x(k) \\ u(k) \end{bmatrix}^T \begin{bmatrix} Q_1 & Q_{12} \\ Q_{12}^T & Q_2 \end{bmatrix} \begin{bmatrix} x(k) \\ u(k) \end{bmatrix} + x^T(N) Q_0 x(N). \tag{4.2}$$

The expressions of $Q_1$, $Q_2$, $Q_{12}$ and $Q_0$ may be found in ([14], pp. 411–412). In the following, it is assumed that $Q$, $Q_2$ and $Q_0$ are positive definite matrices, where

$$Q = \begin{bmatrix} Q_1 & Q_{12} \\ Q_{12}^T & Q_2 \end{bmatrix}.$$

Note that this representation does not involve any approximation and it is exact. Using this representation, the inter-sample behavior is taken into account.

## 4.2 Optimal Control over a Finite Horizon for a Fixed Communication Sequence

The problem of the optimal control, over a finite-time $N$, for a fixed admissible and maximal communication sequence $\delta^{N-1}$, may be formulated as follows:

**Problem 4.1** Given an initial state $x(0)$ and a final time $N$, find the optimal control sequence $v^{N-1} = (v(0), \ldots, v(N-1))$ that minimizes the cost function:

$$J(x, u, 0, N) = \sum_{k=0}^{N-1} \begin{bmatrix} x(k) \\ u(k) \end{bmatrix}^T \begin{bmatrix} Q_1 & Q_{12} \\ Q_{12}^T & Q_2 \end{bmatrix} \begin{bmatrix} x(k) \\ u(k) \end{bmatrix} + x^T(N) Q_0 x(N),$$

subject to

$$x(k+1) = Ax(k) + Bu(k),$$

$$u_i(k) = v_j(k), \quad \text{if } \delta_i(k) = 1 \text{ and } \sum_{l=1}^{i} \delta_l(k) = j,$$

$$u_i(k) = u_i(k-1), \quad \text{otherwise.}$$

In order to solve this problem, we reconsider the state representation (3.12) of system $\mathscr{S}$, which was established in Sect. 3.4 of Chap. 3. For a given maximal controller-to-actuators scheduling $\delta$, system $\mathscr{S}$ is described by the state equation (3.12a). The cost function $J(x, u, 0, N)$ may be rewritten in the form

$$J(x, u, 0, N) = J(\tilde{x}, v, 0, N) = \sum_{k=0}^{N-1} \begin{bmatrix} \tilde{x}(k) \\ v(k) \end{bmatrix}^T \tilde{Q}(k) \begin{bmatrix} \tilde{x}(k) \\ v(k) \end{bmatrix} + \tilde{x}^T(N) \tilde{Q}_0 \tilde{x}(N),$$

where

$$\tilde{Q}(k) = \begin{bmatrix} Q_1 & Q_{12} E_\delta(k) & Q_{12} D_\delta(k) \\ E_\delta^T(k) Q_{12}^T & E_\delta^T(k) Q_2 E_\delta(k) & E_\delta^T(k) Q_2 D_\delta(k) \\ D_\delta^T(k) Q_{12}^T & D_\delta^T(k) Q_2 E_\delta(k) & D_\delta^T(k) Q_2 D_\delta(k) \end{bmatrix},$$

and

$$\tilde{Q}_0 = \begin{bmatrix} Q_0 & 0_{n,m} \\ 0_{m,n} & 0_{m,m} \end{bmatrix}.$$

Problem 4.1 is equivalent to the optimal control problem of a discrete linear time-varying system, given as follows:

**Problem 4.2** Given an initial state $x(0)$ and a final time $N$, find the optimal control sequence $v^{N-1} = (v(0), \ldots, v(N-1))$ that minimizes the cost function

$$J(\tilde{x}, v, 0, N) = \sum_{k=0}^{N-1} \begin{bmatrix} \tilde{x}(k) \\ v(k) \end{bmatrix}^T \tilde{Q}(k) \begin{bmatrix} \tilde{x}(k) \\ v(k) \end{bmatrix} + \tilde{x}^T(N)\tilde{Q}_0\tilde{x}(N),$$

subject to

$$\tilde{x}(k+1) = \tilde{A}(k)\tilde{x}(k) + \tilde{B}(k)v(k). \tag{4.3}$$

Problem 4.2 is a classical optimal control problem of a discrete linear time-varying system. Different methods for its resolution were developed in control textbooks (see, for instance, [14]). Let:

$$\tilde{Q}_1(k) = \begin{bmatrix} Q_1 & Q_{12}E_\delta(k) \\ E_\delta^T(k)Q_{12}^T & E_\delta^T(k)Q_2E_\delta(k) \end{bmatrix},$$

$$\tilde{Q}_{12}(k) = \begin{bmatrix} Q_{12}D_\delta(k) \\ E_\delta^T(k)Q_2D_\delta(k) \end{bmatrix},$$

and

$$\tilde{Q}_2(k) = D_\delta^T(k)Q_2D_\delta(k).$$

The solution of this problem closely depends on the solution of an algebraic equation, involving the variable $\tilde{S}(k)$ and described by

$$\begin{aligned} \tilde{S}(k) = {} & \tilde{A}^T(k)\tilde{S}(k+1)\tilde{A}(k) + \tilde{Q}_1(k) - \big(\tilde{A}^T(k)\tilde{S}(k+1)\tilde{B}(k) + \tilde{Q}_{12}(k)\big) \\ & \times \big(\tilde{B}^T(k)\tilde{S}(k+1)\tilde{B}(k) + \tilde{Q}_2(k)\big)^{-1}\big(\tilde{B}^T(k)\tilde{S}(k+1)\tilde{A}(k) + \tilde{Q}_{12}^T(k)\big) \end{aligned} \tag{4.4}$$

under the terminal condition

$$\tilde{S}(N) = \tilde{Q}_0. \tag{4.5}$$

Equation (4.4) is the discrete algebraic Riccati equation associated to the Problem 4.2. Knowing that $\tilde{Q}_0$ is semi-definite positive and $\tilde{Q}_2(k)$ is definite positive (because $D_\delta(k)$ is injective when $\delta$ is a maximal admissible communication sequence), than this equation admits a unique positive semi-definite solution. Consequently, Problem 4.2 admits a unique solution [14] defined by

$$v(k) = -\tilde{K}(k)\tilde{x}(k), \tag{4.6}$$

with

$$\tilde{K}(k) = \big(\tilde{Q}_2(k) + \tilde{B}^T(k)\tilde{S}(k+1)\tilde{B}(k)\big)^{-1}\big(\tilde{B}^T(k)\tilde{S}(k+1)\tilde{A}(k) + \tilde{Q}_{12}^T(k)\big). \tag{4.7}$$

## 4.3 Optimal Control over an Infinite Horizon for a Fixed Communication Sequence

Consider a $T$-periodic maximal communication sequence $\delta^{T-1}$ defined by

$$\delta^{T-1} = \big(\delta(0), \ldots, \delta(T-1)\big)$$

and verifying $\delta(k+T) = \delta(k)$. Assume furthermore that $\delta \in \mathscr{S}^c$, where $\mathscr{S}^c$ is the set of communication sequences that guarantee the reachability of system (3.12). The periodicity of the communication sequence induces the periodicity of the resource-constrained system $\mathscr{S}$. As a result, matrices $\tilde{A}(k)$, $\tilde{B}(k)$ and $\tilde{Q}(k)$ satisfy $\tilde{A}(k+T) = \tilde{A}(k)$, $\tilde{B}(k+T) = \tilde{B}(k)$ and $\tilde{Q}(k) = \tilde{Q}(k+T)$.

Let $\iota$ be a discrete time instant and $H$ a positive integer, assume that $N = \iota + HT$ and consider the optimal control problem:

$$\begin{cases} \min\limits_{v} \quad J(\tilde{x}, v, \iota, N) = \sum\limits_{k=\iota}^{N-1} \begin{bmatrix} \tilde{x}(k) \\ v(k) \end{bmatrix}^T \tilde{Q}(k) \begin{bmatrix} \tilde{x}(k) \\ v(k) \end{bmatrix} + \tilde{x}^T(N)\tilde{Q}_0\tilde{x}(N) \\ \text{subject to} \quad \tilde{x}(k+1) = \tilde{A}(k)\tilde{x}(k) + \tilde{B}(k)v(k). \end{cases} \tag{4.8}$$

As illustrated in [35], a time invariant reformulation of the optimal control problem (4.8) may be obtained by using the lifting technique. The time invariant reformulation may be seen as a down sampled representation of system (3.12) with periodicity $T$, having an augmented input vector. In the following, the formulation of the time invariant representation is described.

Let $\Phi$ be the transition matrix associated with the state matrix $\tilde{A}$ defined by:

$$\begin{cases} \Phi(l,s) = \tilde{A}(l-1)\tilde{A}(l-2)\cdots\tilde{A}(s) & \text{if } l > s, \\ \Phi(l,l) = I_{n+m}. \end{cases}$$

Let $\Gamma$ the matrix defined for $s < l < s+T$ by

$$\Gamma(l,s) = \begin{bmatrix} \Phi(l,s+1)\tilde{B}(s) \; \Phi(l,s+2)\tilde{B}(s+1) \; \cdots \; \Phi(l,l)\tilde{B}(l-1) \; \underbrace{0_{n+m,b} \; \cdots \; 0_{n+m,b}}_{T-l-s} \end{bmatrix}$$

and for $s = l$ by

$$\Gamma(s,s) = [0_{n+m,b} \quad \cdots \quad 0_{n+m,b}].$$

Let

$$\bar{x}_\iota(q) = \tilde{x}(\iota + qT),$$

and

$$\bar{v}_\iota(q) = \begin{bmatrix} v(\iota + qT) \\ \vdots \\ v(\iota + (q+1)T - 1) \end{bmatrix},$$

then for $0 \leq i \leq T$:

$$\tilde{x}(\iota + qT + i) = \Phi(\iota + i, \iota)\bar{x}_\iota(q) + \Gamma(\iota + i, \iota)\bar{v}_\iota(q).$$

In particular, let

$$\bar{A}_\iota = \Phi(\iota + T, \iota),$$

and

$$\bar{B}_\iota = \Gamma(\iota + T, \iota),$$

then the following relation is obtained:

$$\bar{x}_\iota(q+1) = \bar{A}_\iota \bar{x}_\iota(q) + \bar{B}_\iota \bar{v}_\iota(q).$$

Let $\Lambda(i)$ the matrix defined for $0 \leq i < T$ by

$$\Lambda(i) = \left[ \underbrace{0_{b,b} \quad \cdots \quad 0_{b,b}}_{i} \; I_b \quad \underbrace{0_{b,b} \quad \cdots \quad 0_{b,b}}_{T-i-1} \right],$$

then the cost function may be written

$$J(\tilde{x}, v, \iota, N) = J(\bar{x}_\iota, \bar{v}_\iota, 0, H) = \sum_{q=0}^{H-1} \begin{bmatrix} \bar{x}_\iota(q) \\ \bar{v}_\iota(q) \end{bmatrix}^T \bar{Q}_\iota \begin{bmatrix} \bar{x}_\iota(q) \\ \bar{v}_\iota(q) \end{bmatrix} + \bar{x}_\iota^T(H)(\bar{Q}_\iota)_0 \bar{x}_\iota(H)$$

where

$$\bar{Q}_\iota = \sum_{i=0}^{T-1} F^T(i)\tilde{Q}(\iota + i)F(i),$$

$$F(i) = \begin{bmatrix} \Phi(\iota + i, \iota) & \Gamma(\iota + i, \iota) \\ 0_{b,n+m} & \Lambda(i) \end{bmatrix}$$

and $(\bar{Q}\iota)_0 = \tilde{Q}_0$. Finally, the following optimal control problem is obtained

$$\begin{cases} \min\limits_{\bar{v}_\iota} \quad J(\bar{x}_\iota, \bar{v}_\iota, 0, H) = \bar{x}_\iota^T(H)(\bar{Q}\iota)_0 \bar{x}_\iota(H) + \sum\limits_{q=0}^{H-1} \begin{bmatrix} \bar{x}_\iota(q) \\ \bar{v}_\iota(q) \end{bmatrix}^T \bar{Q}_\iota \begin{bmatrix} \bar{x}_\iota(q) \\ \bar{v}_\iota(q) \end{bmatrix} \\ \text{subject to} \quad \bar{x}_\iota(q+1) = \bar{A}_\iota \bar{x}_\iota(q) + \bar{B}_\iota \bar{v}_\iota(q). \end{cases} \tag{4.9}$$

The corresponding discrete algebraic Riccati equation is given by:

$$\begin{aligned} \bar{S}_\iota(q) = {} & \bar{A}_\iota^T \bar{S}_\iota(q+1)\bar{A}_\iota + \bar{Q}_{\iota_1} - \left(\bar{A}_\iota^T \bar{S}_\iota(q+1)\bar{B}_\iota + \bar{Q}_{\iota_{12}}\right) \\ & \times \left(\bar{B}_\iota^T \bar{S}_\iota(q+1)\bar{B}_\iota + \bar{Q}_{\iota_2}(q)\right)^{-1} \left(\bar{B}_\iota^T \bar{S}_\iota(q+1)\bar{A}_\iota + \bar{Q}_{\iota_{12}}^T\right), \end{aligned} \tag{4.10}$$

with the terminal condition

$$\bar{S}_\iota(H) = (\bar{Q}_\iota)_0, \tag{4.11}$$

where

$$\bar{Q}_\iota = \begin{bmatrix} \bar{Q}_{\iota 1} & \bar{Q}_{\iota 12} \\ \bar{Q}_{\iota 12}^T & \bar{Q}_{\iota 2} \end{bmatrix}.$$

The relationship between the solutions of Riccati equations (4.4) and (4.10), which are respectively, associated to the optimal control problems (4.8) and (4.9), is formalized by the following result.

**Lemma 4.1** [35] *If* $\bar{S}_\iota(H) = \tilde{S}(\iota + NT) = \tilde{Q}_0$, *then* $\bar{S}_\iota(q) = \tilde{S}(\iota + qT)$ *for all* $q \leq H$.

This result follows from the fact that optimal control problems (4.8) and (4.9) are similar. By imposing the terminal condition $\bar{S}_\iota(H) = \tilde{S}(\iota + NT) = \tilde{Q}_0$, called *periodic generator*, the optimal costs must be identical, which implies that the solutions of the Riccati equations (4.4) and (4.10) must be the same. As a result, when $H \to +\infty$, $\bar{S}_\iota(q)$ converges to a constant solution $\bar{S}_\iota$. Consequently, $\tilde{S}(k)$ converges to a periodic solution, defined by $\tilde{S}(k) = \bar{S}_{(k \bmod T)}$. This periodic solution may be obtained by solving the algebraic Riccati equation associated with problem (4.9) when $H \to +\infty$ and for $\iota \in \{1, \ldots, T\}$. The optimal control gains may then be deduced from the relation (4.7) and are described by the sequence $(\tilde{K}(0), \ldots, \tilde{K}(T-1))$.

This formulation will be of practical importance at the next chapters when the problems of the optimal integrated control and scheduling of resource-constrained systems will be tackled.

## 4.4 Finite-Time Optimal Integrated Control and Scheduling

In this paragraph, the problem of the finite-time optimal control and scheduling is formulated and translated into the mixed-integer quadratic programming (MIQP) formulation. We assume in the following that $u(k) = 0$ and $v(k) = 0$ for $k < 0$, and that control commands $u(k)$ and $v(k)$ are subject to saturation constraints

$$\begin{cases} L_i \leq u_i(k) \leq U_i, \\ L_i \leq v_i(k) \leq U_i \end{cases} \tag{4.12}$$

where $L_i < 0$ and $U_i > 0$.

### 4.4.1 Problem Formulation

The finite-time optimal control and scheduling problem may be formalized as follows:

**Problem 4.3** Given an initial state $x(0)$ and a final time $N$, find the optimal control sequence $v^{N-1} = (v(0), \ldots, v(N-1))$ as well as the optimal communication sequence $\delta^{N-1} = (\delta(0), \ldots, \delta(N-1))$ which minimizes the performance index:

$$J(x, u, 0, N) = \sum_{k=0}^{N-1} \begin{bmatrix} x(k) \\ u(k) \end{bmatrix}^T \begin{bmatrix} Q_1 & Q_{12} \\ Q_{12}^T & Q_2 \end{bmatrix} \begin{bmatrix} x(k) \\ u(k) \end{bmatrix} + x^T(N) Q_0 x(N)$$

subject to:

$$\begin{aligned} & x(k+1) = Ax(k) + Bu(k), \\ & \sum_{i=1}^{m} \delta_i(k) = b, \\ & u_i(k) = v_j(k), \quad \text{if } \delta_i(k) = 1 \text{ and } \sum_{l=1}^{i} \delta_l(k) = j, \\ & u_i(k) = u_i(k-1), \quad \text{otherwise}, \\ & L_i \leq v_i(k) \leq U_i. \end{aligned}$$

The problem of finding the optimal control sequence $v^{N-1}$ for a given fixed communication sequence $\delta^{N-1}$ is a quadratic programming (*QP*) problem. The number of possible communication sequences is finite. The resolution of Problem 4.3 may be reduced to the exploration of all the feasible maximal communication sequences and the solving of a *QP* problem for each fixed communication sequence. However, in practice, the number of feasible communication sequences grows exponentially with $N$, which means that exhaustive search may not be applied to problems with large values of $N$.

The solution of the Problem 4.3 may be obtained by solving a simpler optimization problem, which may be seen as a constrained control problem, where the variables $v^{N-1}$ are eliminated and the constraint (3.8) is replaced by (3.7). Let $u^{N-1} = (u(0), \ldots, u(N-1))$, this problem may be stated as follows:

**Problem 4.4** Given an initial state $x(0)$ and a final time $N$, find the optimal control sequence $u^{N-1}$ as well as the optimal scheduling sequence $\delta^{N-1}$ that minimizes the performance index:

$$J(x, u, 0, N) = \sum_{k=0}^{N-1} \begin{bmatrix} x(k) \\ u(k) \end{bmatrix}^T \begin{bmatrix} Q_1 & Q_{12} \\ Q_{12}^T & Q_2 \end{bmatrix} \begin{bmatrix} x(k) \\ u(k) \end{bmatrix} + x^T(N) Q_0 x(N),$$

subject to

$$
\begin{aligned}
&x(k+1) = Ax(k) + Bu(k),\\
&\sum_{i=1}^{m} \delta_i(k) = b,\\
&\delta_i(k) = 0 \quad \Longrightarrow \quad u_i(k) = u_i(k-1),\\
&L_i \le v_i(k) \le U_i.
\end{aligned}
$$

We observe that the constraints of optimal scheduling problem are composed of a set of linear equalities and inequalities as well as of the following logical formula:

$$\delta_i(k) = 0 \quad \Longrightarrow \quad u_i(k) = u_i(k-1). \tag{4.13}$$

In order to solve this problem, it is necessary to translate the logical formula (4.13) into linear inequalities. The connective "$\Longrightarrow$" may be eliminated if (4.13) is rewritten in the following equivalent form

$$u_i(k) - u_i(k-1) = \delta_i(k)u_i(k) - \delta_i(k)u_i(k-1). \tag{4.14}$$

However, Eq. (4.14) contains terms which are the product of logical variables and continuous variables. Using the procedure described in [21], this product is translated into an equivalent conjunction of linear inequalities. For example, let

$$\xi_i(k) = \delta_i(k)u_i(k). \tag{4.15}$$

Then (4.14) may be rewritten in the equivalent form:

$$
\begin{cases}
\xi_i(k) \le U_i\delta_i(k),\\
\xi_i(k) \ge L_i\delta_i(k),\\
\xi_i(k) \le u_i(k) - L_i\big(1-\delta_i(k)\big),\\
\xi_i(k) \ge u_i(k) - U_i\big(1-\delta_i(k)\big).
\end{cases} \tag{4.16}
$$

Note that the same procedure may be applied to

$$o_i(k) = \delta_i(k)u_i(k-1). \tag{4.17}$$

Let

$$
\check{\Delta} = \begin{bmatrix} \delta(0) \\ \vdots \\ \delta(N-1) \end{bmatrix}; \qquad \check{U} = \begin{bmatrix} u(0) \\ \vdots \\ u(N-1) \end{bmatrix},
$$

$$\check{X} = \begin{bmatrix} x(0) \\ \vdots \\ x(N) \end{bmatrix}; \qquad \check{\Xi} = \begin{bmatrix} \xi(0) \\ \vdots \\ \xi(N-1) \end{bmatrix},$$

and

$$\check{O} = \begin{bmatrix} o(0) \\ \vdots \\ o(N-1) \end{bmatrix}; \qquad \mathscr{V} = \begin{bmatrix} \check{\Delta} \\ \check{U} \\ \check{X} \\ \check{\Xi} \\ \check{O} \end{bmatrix},$$

then Problem 4.4 may be written in the form

$$\begin{cases} \min\limits_{\mathscr{V}} \ \frac{1}{2}\mathscr{V}^T \mathscr{F} \mathscr{V} + \mathscr{G}^T \mathscr{V} \\ \mathscr{A}\mathscr{V} \leq \mathscr{B}, \\ \mathscr{V}_i \in \{0, 1\}, \quad \forall i \in \{1, \ldots, mN\}, \end{cases} \tag{4.18}$$

where the expressions $\mathscr{F}$, $\mathscr{G}$, $\mathscr{A}$ and $\mathscr{B}$ are defined in the sequel.

Matrices $\mathscr{A}$ and $\mathscr{B}$ include both the equality and the inequality constraints of the problem, and may be written in the form:

$$\mathscr{A} = \begin{bmatrix} \mathscr{A}_{\text{eq}} \\ -\mathscr{A}_{\text{eq}} \\ \mathscr{A}_{\text{in}} \end{bmatrix},$$

$$\mathscr{B} = \begin{bmatrix} \mathscr{B}_{\text{eq}} \\ -\mathscr{B}_{\text{eq}} \\ \mathscr{B}_{\text{in}} \end{bmatrix}.$$

It easy to see that the relation

$$\mathscr{A}_{\text{eq}}\mathscr{V} = \mathscr{B}_{\text{eq}}$$

is equivalent to

$$(\mathscr{A}_{\text{eq}}\mathscr{V} \leq \mathscr{B}_{\text{eq}}) \wedge (-\mathscr{A}_{\text{eq}}\mathscr{V} \leq -\mathscr{B}_{\text{eq}}).$$

Matrices $\mathscr{A}_{\text{eq}}$ and $\mathscr{B}_{\text{eq}}$ describe equalities (3.4a), (4.14) and impose to the scheduling sequence of the controller-to-actuators link to be maximal. Matrix $\mathscr{A}_{\text{eq}}$ is defined by

$$\mathscr{A}_{\text{eq}} = \begin{bmatrix} SC & 0_{N,n(N+1)} \ 0_{N,mN} & 0_{N,mN} & 0_{N,mN} \\ 0_{n(N+1),mN} & ST & 0_{n(N+1),mN} & 0_{n(N+1),mN} \\ 0_{mN,mN} & 0_{mN,n(N+1)} \ UU & -I_{mN} & I_{mN} \end{bmatrix}$$

where

- $SC$ is a $N \times mN$ matrix described by

$$SC = \begin{bmatrix} 1_{1,m} & 0_{1,m} & \cdots & 0_{1,m} \\ 0_{1,m} & 1_{1,m} & \cdots & 0_{1,m} \\ & & \ddots & \\ 0_{1,m} & 0_{1,m} & \cdots & 1_{1,m} \end{bmatrix}.$$

- $ST$ is a $n(N+1) \times (n(N+1)+mN)$ matrix that is used to represent the constraint (3.4a):

$$ST = \begin{bmatrix} I_n & 0_{n,n} & 0_{n,n} & \cdots & 0_{n,n} & 0_{n,m} & 0_{n,m} & \cdots & 0_{n,m} \\ -A & I_n & 0_{n,n} & \cdots & 0_{n,n} & -B & 0_{n,m} & \cdots & 0_{n,m} \\ 0_{n,n} & -A & I_n & \cdots & 0_{n,n} & 0_{n,m} & -B & \cdots & 0_{n,m} \\ & & \ddots & \ddots & & & & \ddots & \\ 0_{n,n} & 0_{n,n} & & -A & I_n & 0_{n,m} & 0_{n,m} & \cdots & -B \end{bmatrix}.$$

- $UU$ is a $mN \times mN$ matrix described by

$$UU = \begin{bmatrix} I_m & 0_{m,m} & 0_{m,m} & \cdots & 0_{m,m} \\ -I_m & I_m & 0_{m,m} & \cdots & 0_{m,m} \\ 0_{m,m} & -I_m & I_m & \cdots & 0_{m,m} \\ & & \ddots & \ddots & \\ 0_{m,m} & 0_{m,m} & & -I_m & I_m \end{bmatrix}.$$

Matrix $\mathscr{B}_{\text{eq}}$ is defined by

$$\mathscr{B}_{\text{eq}} = \begin{bmatrix} b \times 1_{N,1} \\ x(0) \\ 0_{nN,1} \\ 0_{mN,1} \end{bmatrix}.$$

Let $U$ and $L$ the vectors defined by

$$U = \begin{bmatrix} U_1 \\ \vdots \\ U_m \end{bmatrix}; \qquad L = \begin{bmatrix} L_1 \\ \vdots \\ L_m \end{bmatrix}.$$

Matrices $\mathscr{A}_{\text{in}}$ and $\mathscr{B}_{\text{in}}$ describe the constraints defining the variables $\xi(k)$ and $o(k)$ ((4.15) and (4.17)). Matrix $\mathscr{A}_{\text{in}}$ is defined by

$$\mathscr{A}_{\text{in}} = \begin{bmatrix} -\operatorname{Diag}(U) & 0_{mN,n(N+1)} & 0_{mN,mN} & I_{mN} & 0_{mN,mN} \\ \operatorname{Diag}(L) & 0_{mN,n(N+1)} & 0_{mN,mN} & -I_{mN} & 0_{mN,mN} \\ -\operatorname{Diag}(L) & 0_{mN,n(N+1)} & -I_{mN} & I_{mN} & 0_{mN,mN} \\ \operatorname{Diag}(U) & 0_{mN,n(N+1)} & I_{mN} & -I_{mN} & 0_{mN,mN} \\ -\operatorname{Diag}(U) & 0_{mN,n(N+1)} & 0_{mN,mN} & 0_{mN,mN} & I_{mN} \\ \operatorname{Diag}(L) & 0_{mN,n(N+1)} & 0_{mN,mN} & 0_{mN,mN} & -I_{mN} \\ -\operatorname{Diag}(L) & 0_{mN,n(N+1)} & -UM & 0_{mN,mN} & I_{mN} \\ \operatorname{Diag}(U) & 0_{mN,n(N+1)} & UM & 0_{mN,mN} & -I_{mN} \end{bmatrix}$$

where

$$UM = -(UU - I_{mN}).$$

Next, the matrix $\mathscr{B}_{\text{in}}$ is defined by

$$\mathscr{B}_{\text{in}} = \begin{bmatrix} 0_{m,1} \\ 0_{m,1} \\ -L \\ U \\ 0_{m,1} \\ 0_{m,1} \\ -L \\ U \end{bmatrix}.$$

Finally, matrices $\mathscr{F}$ and $\mathscr{G}$ are defined by

$$\mathscr{F} = \begin{bmatrix} 0_{mN,mN} & 0_{mN,n(N+1)+mN} & 0_{mN,2mN} \\ 0_{n(N+1)+mN,mN} & \check{\mathscr{Q}} & 0_{n(N+1)+mN,2mN} \\ 0_{2mN,mN} & 0_{2mN,n(N+1)+mN} & 0_{2mN,2mN} \end{bmatrix}$$

and respectively,

$$\mathscr{G} = 0_{(4m+(n+1))N,1}$$

where

$$\check{\mathscr{Q}} = \begin{bmatrix} Q_1 & 0_{n,n} & \cdots & 0_{n,n} & 0_{n,n} & Q_{12} & 0_{n,m} & \cdots & 0_{n,m} \\ 0_{n,n} & Q_1 & \cdots & 0_{n,n} & 0_{n,n} & 0_{n,m} & Q_{12} & \cdots & 0_{n,m} \\ & & \ddots & & & & & \ddots & \\ 0_{n,n} & 0_{n,n} & \cdots & Q_1 & 0_{n,n} & 0_{n,m} & 0_{n,m} & \cdots & Q_{12} \\ 0_{n,n} & 0_{n,n} & \cdots & 0_{n,n} & Q_0 & 0_{n,m} & 0_{n,m} & \cdots & 0_{n,m} \\ Q_{12}^T & 0_{m,n} & \cdots & 0_{m,n} & 0_{m,n} & Q_2 & 0_{m,m} & \cdots & 0_{m,m} \\ 0_{m,n} & Q_{12}^T & \cdots & 0_{m,n} & 0_{m,n} & 0_{m,m} & Q_2 & \cdots & 0_{m,m} \\ & & \ddots & & & & & \ddots & \\ 0_{m,n} & 0_{m,n} & \cdots & Q_{12}^T & 0_{m,n} & 0_{m,m} & 0_{m,m} & \cdots & Q_2 \end{bmatrix}.$$

The Problem 4.3 is identical to Problem 4.4 augmented with the additional constraint

$$v(k) = u^f(k)$$

where the vector $u^f(k) \in \mathbb{R}^b$ containing the $b$ "free" elements of $u(k)$ (i.e., the elements of $u(k)$ whose indices $i$ satisfy $\delta_i(k) = 1$) is arranged according to the increasing order of their indices. As a consequence, the optimal solutions of Problem 4.3 may be deduced from the optimal solutions of Problem 4.4 using the mapping (3.9).

The problem (4.18) is a mixed-integer quadratic program. It may be solved using many efficient academic and commercial solvers. In this monograph, this problem was solved using the solver CPLEX, whose MIP solver is based on the branch and bound method.

### 4.4.2 The Branch and Bound Method

The branch and bound method is a general search method for finding optimal solutions of discrete and combinatorial optimization problems. It was introduced in 1960 by *Land and Doig* [145] for the solving of the traveling salesman problem. This method aims at exploring, in an intelligent way, the space of feasible solutions of the problem. That's why it may be classified among the *implicit enumeration* methods. In the following, the principles of this algorithm will be described in the case of a minimization. In order to simplify our explanation, we will suppose that considered problem admits at least one optimal solution. Of course, the other cases may be easily taken into account.

#### 4.4.2.1 General Concepts

As its name indicates it, this method is based on two complementary mechanisms: *branching* and *bounding*.

- Branching makes it possible to decompose a given problem into subproblems (by adding additional constraints), such that the union of the feasible solutions of these subproblems forms a partition (in the worst case a covering) of the feasible solutions of the original problem. In this manner, the resolution of the original problem is reduced to the resolution of the subproblems obtained by its branching. Branching induces a hierarchical relation between the different subproblems. This relation may be described and represented using concepts from the graph theory. In fact, in the branch and bound method, branching is applied in a recursive way to the subproblems where it may be possible (at a given stage of the execution of the algorithm), to find an optimal solution of the initial problem. As a result, the subproblems obtained by branching may be seen as the

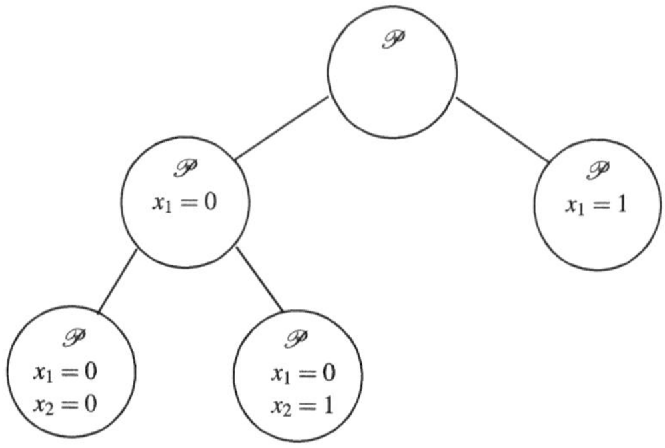


**Fig. 4.1** Search tree of problem $\mathscr{P}$

*child nodes* of the problem to which branching was applied, which is called *parent node*. Thus, all these nodes form a rooted tree, whose *root node* represents the initial problem to solve. This tree is usually called *search tree* or *decision tree*.

- Bounding consists in computing an upper and a lower bound of the optimal solution of a given node. The bounding stage allows the branch and bound to avoid exploring the nodes where it is possible to certify that they do not contain any optimal solution. In fact, if the upper bound of a subproblem **A** is larger than the lower bound of another subproblem **B**, then the optimal solution cannot lie in the feasible set of solutions of subproblem **A**. For that reason, it becomes useless to branch node **A**. Node **A** is then *pruned.*

A node is called *solved* if an optimal solution of the associated subproblem was obtained. This may occur, for example, when the constraints defining it (and which were added progressively along the different branching steps), reduce its set of feasible solutions to a singleton. In other situations, branching may make the subproblem sufficiently simple to be solved by polynomial algorithms. The solved nodes are the *final nodes* or *leave nodes* of the decision tree. The algorithm finishes when all the nodes were either solved or pruned.

*Example 4.1* Figure 4.1 describes the search tree of problem $\mathscr{P}$ defined by

$$\mathscr{P}\begin{cases}\min\limits_{x} \ 4x_1^2 - 3x_2^2\\ x_1 = 0,\\ x_1, x_2 \in \{0, 1\}.\end{cases}$$

The branch and bound algorithm may be parameterized using the following four rules:

- *branching rules*, describing how to divide a given problem into subproblems,

- *bounding rules*, defining how to compute the upper and lower bounds of a given subproblem,
- *selection rules*, stating how to select the next problem to consider,
- *elimination rules*, describing how to recognize the subproblems that do not contain optimal solutions and that should be eliminated.

#### 4.4.2.2 Application to Mixed-Integer Programming

To the best of the authors' knowledge, the application of the branch and bound method to solve mixed-integer nonlinear programs was proposed for the first time by *Dakin* [68]. The resolution of mixed-integer quadratic programs was studied by many authors, for example, [85, 86, 146]. In the paper by *Fletcher and Leyffer* [85], the branch and bound method was applied to solve mixed-integer quadratic programs, allowing to obtain better experimental results than the other methods.

In this paragraph, we consider programs in the form:

$$\mathscr{P} \begin{cases} \min\limits_{\mathscr{V}} \quad f(\mathscr{V}) = \dfrac{1}{2}\mathscr{V}^T \mathscr{F} \mathscr{V} + \mathscr{G}^T \mathscr{V} \\ \mathscr{A}\mathscr{V} \leq \mathscr{B}, \\ \mathscr{V}_i \in \{0, 1\}, \quad \forall i \in I \end{cases}$$

where $f$ is a positive semi-definite function and $I$ is a set of indices indexing the Boolean components of $\mathscr{V}$. A basic branch and bound algorithm for solving program $\mathscr{P}$ is given in the following listing (Algorithm 4.1).

In Algorithm 4.1, $U$ represents the cost of the best feasible solution of problem $\mathscr{P}$ at a given stage of the execution of the algorithm and $L$ a lower bound of the optimal cost of $\mathscr{P}$. The set $\mathscr{M}$, used for the computation of $L$, represents the set of nodes that have been already evaluated (i.e., bounded) and that may contain the optimal solution. Problem $\mathscr{P}'$ denotes the problem obtained from $\mathscr{P}$ by relaxing the integrality constraints that are imposed on the variables indexed by $I$, that is, replacing the constraints

$$\mathscr{V}_i \in \{0, 1\}, \quad \forall i \in I$$

by the constraints

$$\mathscr{V}_i \in [0, 1], \quad \forall i \in I.$$

These variables are thus considered as continuous variables in $[0, 1]$. Thus, $\mathscr{P}'$ is a quadratic program, and may be solved efficiently. Its solving is the most important part of the bounding or evaluation phase. In fact, the optimal cost of $\mathscr{P}'$ represents a lower bound of the optimal cost of $\mathscr{P}$.

**Algorithm 4.1:** Basic branch and bound algorithm

$\mathscr{L} := \{\mathscr{P}\}$;
$\mathscr{M} := \{\mathscr{P}\}$;
$U := +\infty$;
$L := -\infty$;
**while** $\mathscr{L} \neq \emptyset$ *or* $U - L > \varepsilon$ **do**
  select a node $\mathscr{Q}$ from list $\mathscr{L}$;
  remove the selected node $\mathscr{Q}$ from $\mathscr{L}$;
  solve the relaxed program $\mathscr{Q}'$ (bounding or evaluation of $\mathscr{Q}$);
  **if** $\mathscr{Q}'$ *admits an optimal solution* $x^*(\mathscr{Q}')$ *of cost* $J^*(\mathscr{Q}')$ *and* $J^*(\mathscr{Q}') < U$ **then**
    lowerbound($\mathscr{Q}$) := $J^*(\mathscr{Q}')$;
    **if** $x^*(\mathscr{Q}')$ *is integer-feasible* **then**
      $U := J^*(\mathscr{Q}')$;
      $x^*(\mathscr{P}) := x^*(\mathscr{Q}')$;
      $J^*(\mathscr{P}) := J^*(\mathscr{Q}')$;
    **else**
      branch node $\mathscr{Q}$ and update list $\mathscr{L}$ with its child nodes;
    **end if**
    **if** *all the brother nodes of* $\mathscr{Q}$ *were evaluated* **then**
      remove the father node of $\mathscr{Q}$ from $\mathscr{M}$;
      put $\mathscr{Q}$ and its brother nodes in $\mathscr{M}$;
      $L = \min\{\text{lowerbound}(\mathscr{Q}), \mathscr{Q} \in \mathscr{M}\}$;
    **end if**
  **else**
    lowerbound($\mathscr{Q}$) := $+\infty$ (pruning of $\mathscr{Q}$);
  **end if**
**end while**

The algorithm begins by solving $\mathscr{P}'$ and giving its solution $x^*(\mathscr{P}')$. If this solution respects the integrality constraints, then it is also an optimal solution for $\mathscr{P}$ and the algorithm ends. Otherwise, there exists at least one non Boolean variable $\mathscr{V}_j$, $j \in I$ in $x^*(\mathscr{P}')$. The algorithm proceeds by branching problem $\mathscr{P}$ into two subproblems

$$\mathscr{Q}_1 \begin{cases} \min\limits_{\mathscr{V}} \ \frac{1}{2}\mathscr{V}^T \mathscr{F} \mathscr{V} + \mathscr{G}^T \mathscr{V} \\ \mathscr{A}\mathscr{V} \leq \mathscr{B}, \\ \mathscr{V}_j = 0, \\ \mathscr{V}_i \in \{0, 1\}, \quad \forall i \in I_1 = I - \{j\} \end{cases}$$

and

$$\mathcal{Q}_2 \begin{cases} \min_{\mathcal{V}} & \frac{1}{2}\mathcal{V}^T \mathcal{F} \mathcal{V} + \mathcal{G}^T \mathcal{V} \\ & \mathcal{A}\mathcal{V} \leq \mathcal{B}, \\ & \mathcal{V}_j = 1, \\ & \mathcal{V}_i \in \{0, 1\}, \quad \forall i \in I_2 = I - \{j\} \end{cases}$$

that are added to the list $\mathcal{L}$. The procedure is reiterated by the selection of a subproblem $\mathcal{Q}$ from $\mathcal{L}$, based on the predefined selection rules. It may be possible that the relaxed subproblem $\mathcal{Q}'$ does not have any feasible solution. In this situation, it is pruned from the list $\mathcal{L}$ without being branched. The algorithm finishes when the list $\mathcal{L}$ becomes empty or when a suboptimal solution with a predefined absolute tolerance (defined by $\varepsilon$) is found.

### 4.4.3 An Illustrative Numerical Example

Consider the collection of three continuous-time linear time invariant (*LTI*) subsystems defined by

$$A_c^{(1)} = \begin{bmatrix} 0 & 130 \\ -800 & 10 \end{bmatrix}, \qquad B_c^{(1)} = \begin{bmatrix} 0 \\ 224 \end{bmatrix}, \qquad S^{(1)}$$

$$A_c^{(2)} = \begin{bmatrix} 0 & 14 \\ -250 & -200 \end{bmatrix}, \qquad B_c^{(2)} = \begin{bmatrix} 0 \\ 620 \end{bmatrix}, \qquad S^{(2)},$$

$$A_c^{(3)} = \begin{bmatrix} 0 & 0 & 0 & 100 \\ 0 & 0 & 100 & 0 \\ 0 & 0 & -10 & 0 \\ 11.6 & 0 & 1.184 & 0 \end{bmatrix}, \qquad B_c^{(3)} = \begin{bmatrix} 0 \\ 0 \\ 10 \\ 10.18 \end{bmatrix}, \qquad S^{(3)}$$

where the subsystems $S^{(1)}$ and $S^{(3)}$ are open-loop unstable in opposition to the subsystem $S^{(2)}$ which is open-loop stable. The global system $S$ composed by $S^{(1)}$, $S^{(2)}$ and $S^{(3)}$ may be described by the state matrix $A_c$ and the input matrix $B_c$ defined by

$$A_c = \mathrm{Diag}\big(A_c^{(1)}, A_c^{(2)}, A_c^{(3)}\big)$$

and

$$B_c = \mathrm{Diag}\big(B_c^{(1)}, B_c^{(2)}, B_c^{(3)}\big).$$

Assume that the global system $S$ is controlled by a discrete-time controller executed at the sampling period $T_s = 2$ ms. The control commands are sent to the actuators through a bus that can carry at most one control command every 2 ms ($b = b_w = 1$). Assume now that the design criteria of the optimal continuous time controller for each subsystem are defined by matrices $Q_c^{(1)} = \mathrm{Diag}(100, 10)$, $R_c^{(1)} = 1$, $Q_c^{(2)} = \mathrm{Diag}(1000, 10)$, $R_c^{(2)} = 1$, $Q_c^{(3)} = \mathrm{Diag}(1000, 1000, 1, 1)$ and $R_c^{(3)} = 1$ and the desired closed-loop specifications for the global system are described by the closed-loop weighting matrices:

$$Q_c = \mathrm{Diag}\big(\mu_1 Q_c^{(1)}, \mu_2 Q_c^{(2)}, \mu_3 Q_c^{(3)}\big),$$
$$R_c = \mathrm{Diag}\big(\mu_1 R_c^{(1)}, \mu_2 R_c^{(2)}, \mu_3 R_c^{(3)}\big)$$

and

$$S_c = Q_c,$$

where $\mu_1$, $\mu_2$ and $\mu_3$ are the weighting coefficients. In this example, $\mu_1$, $\mu_2$ and $\mu_3$ were chosen equal to the inverse of steady state performance index of each separate subsystem controlled through a bus having an infinite bandwidth ($\mu_1 = 2.9$, $\mu_2 = 0.13$ and $\mu_3 = 0.016$).

*Remark 4.1* The use of a resource-limited shared communication medium for the transmission of the control commands to the distributed actuators introduces a coupling between the three subsystems, which requires the weighting of relative importance of each subsystem using coefficients $\mu_1$, $\mu_2$ and $\mu_3$. This contrasts with the optimal control problem without communication constraints, where these constants have no impact on the optimal control of the three independent subsystems.

The length of the optimal control and communication sequences is $N = 100$. An optimal solution with an error bound of $1 \times 10^{-5}$ was required. The global system is started from the initial state $x(0) = [1\ 0\ 1\ 0\ 1\ 0\ 0\ 0]^T$. Its responses are depicted in Figs. 4.2 and 4.3. The optimal schedule is depicted in Fig. 4.4. In this schedule, $\delta_i = 1$ means that the bus is dedicated to the transmission of control signal $u_i$.

The first network slots are mainly dedicated to subsystem $S^{(1)}$ until it is stabilized. It may be observed that subsystem $S^{(2)}$, which is open-loop stable and whose response time is larger than $S^{(1)}$, needs only three time slots to be stabilized. After the stabilization of subsystems $S^{(1)}$ and $S^{(2)}$, the network resources are entirely dedicated to the stabilization of subsystem $S^{(3)}$. When the subsystem $S^{(3)}$ is close to the equilibrium state (from $t = 0.13$ s), then its control signals changes are minor. Consequently, the scheduling has no significant impact on control since the control signals of the three subsystems are relatively constant explaining thus the shape of the scheduling diagram, after $t = 0.13$ s. However, this optimal schedule is *dependent* on the *initial conditions*. If the initial condition $x(0)$ is modified, then the optimal control and schedule would be different. It is clear that such an open-loop schedule, which is generated off-line, cannot be applied at runtime as static schedule. In fact, assume that subsystem $S^{(1)}$ is disturbed at $t = 0.024$ s. Observing

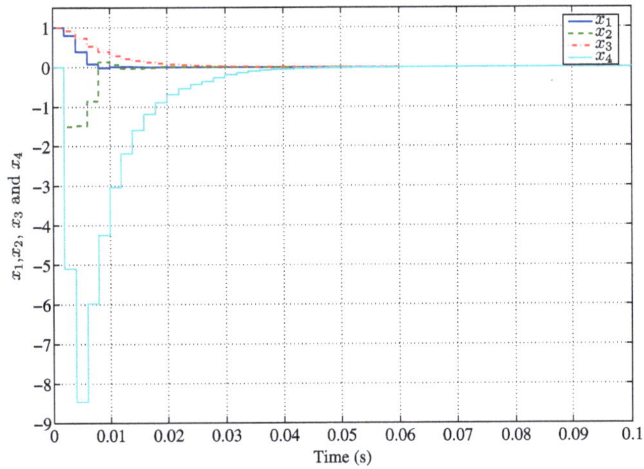


**Fig. 4.2** Global system response—states $x_1, x_2$ (subsystem $S^{(1)}$) and $x_3, x_4$ (subsystem $S^{(2)}$) from $t = 0$ s to $t = 0.1$ s

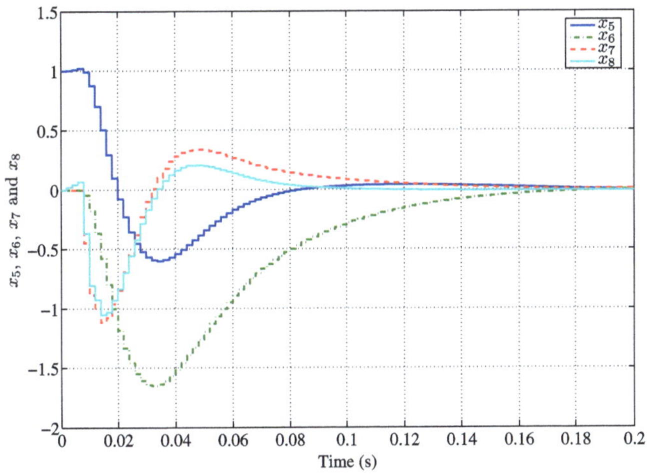


**Fig. 4.3** Global system response—states $x_5$, $x_6$, $x_7$ and $x_8$ (subsystem $S^{(3)}$)

the schedule, no slots are allocated to the transmission of the messages of subsystem $S^{(1)}$ between $t = 0.024$ s and $t = 0.128$ s, which would induce performance degradations.

These observations show that in the same way as the optimal control, the optimal scheduling depends on the current state of the system. Consequently, fixed schedules may not be optimal if the system is started from another initial state or if it is disturbed at runtime.

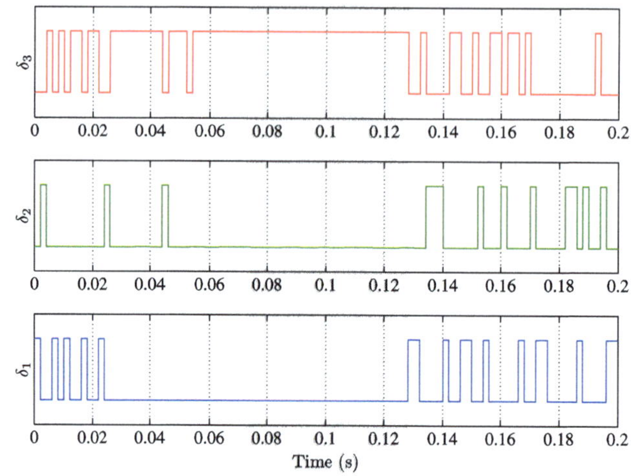


**Fig. 4.4** Optimal scheduling of the controller-to-actuators link

## 4.5 Notes and Comments

In this chapter, we have started by studying the problem of optimal control for a given fixed finite communication sequence. Then, we have formulated and solved the problem of the joint optimization of control and scheduling, for a given initial state. The numerical examples illustrating this method have shown that the obtained optimal schedule depends on the chosen initial state $x(0)$. In other words, similarly to the standard optimal control case, the optimal scheduling depends on the current state of the system. However, from a computer science point of view, off-line scheduling has many advantages, essentially because it consumes a few computing resources and does not induce execution overheads. In order to obtain off-line schedules that are optimal from a certain point of view, it is necessary to use performance criteria that depend on the intrinsic characteristics of the system, not on a particular initial state. This will be the objective of the next chapter.

Nevertheless, this dependency between the optimal schedule and the plant state may be seen as a promising way for improving the quality of control by means of plant state-based scheduling algorithms. The design of such algorithms will be studied in Chap. 6.

# Chapter 5
# Optimal Integrated Control and Off-line Scheduling of Resource-Constrained Systems

## 5.1 Preliminaries

In order to formulate the joint problem of the optimal control and off-line scheduling, in addition to the modeling of the resource limitations and the representation of the system's dynamics, it is necessary to choose an adequate criterion of performance. The previous studies (illustrated in Chap. 4), which were carried out on the joint problem of the optimal control and scheduling, starting from a given initial condition, have shown that the optimal schedule depends on the chosen initial state of the controlled dynamical system. This dependence may be exploited by the on-line scheduling algorithms in order to improve the control performance. But when only a fixed schedule is desired, it is necessary to use performance criteria that depend on the intrinsic characteristics of the system, not on a particular evolution or initial state. The use of the well-known $\mathscr{H}_2$ norm provides a solution to meet these objectives. In fact, using this performance index, the obtained off-line schedules will be independent from any initial condition. Moreover, the results may be easily transposed to an *LQG* context [78]. Intuitively, a $\mathscr{H}_2$ optimal off-line schedule may be seen as a "mean square schedule", obtained when all the system's components are uniformly disturbed by a zero mean unit intensity Gaussian white noise. Off-line schedules are generally periodic. For that reason, we will focus on *$T$-periodic resource constrained systems*. A $T$-periodic resource constrained system represents a resource-constrained system whose scheduling is performed according to a $T$-periodic communication sequence and whose control is ensured by a $T$-periodic linear discrete-time controller.

In this chapter, we start by defining the $\mathscr{H}_2$ norm of $T$-periodic resource-constrained systems. Based on this definition, we propose a method for the joint control and off-line scheduling in the sense of the $\mathscr{H}_2$ criterion. We show that this problem may be decomposed into two sub-problems, that can be solved separately. The first sub-problem aims in determining the optimal off-line scheduling in the sense of the $\mathscr{H}_2$ criterion and can be solved by using the branch and bound method. The second sub-problem aims in determining the optimal control gains and can be solved by using existing results for deriving an appropriate optimal periodic control.

A. Çela et al., *Optimal Design of Distributed Control and Embedded Systems*, Communications and Control Engineering, DOI 10.1007/978-3-319-02729-6_5,

The proposed approach is illustrated by a numerical example and methods for the improvement of its computational complexity are proposed and discussed.

## 5.2 On Characterizing $\mathscr{H}_2$ Norm of Resource-Constrained Systems

In this section, we describe the different steps allowing the definition and the computation of the $\mathscr{H}_2$ norm of a $T$-periodic resource-constrained system. Taking into account the communication constraints, a $T$-periodic resource-constrained system may be viewed as a sampled-data model of a continuous-time LTI system controlled by a $T$-periodic linear discrete-time controller. In order to compute its $\mathscr{H}_2$ norm, a sampled-data model of system (3.2) is first derived. The particularity of this sampled-data model is that its $\mathscr{H}_2$ norm (computed in the discrete-time domain) is identical to the $\mathscr{H}_2$ norm of the continuous-time LTI system (3.2) (computed in the continuous-time domain), when they are both controlled by a discrete-time LTI controller. Based on this sampled-data model, and on the periodicity of the scheduling and control, the $\mathscr{H}_2$ norm of a $T$-periodic resource-constrained system is appropriately defined. Before presenting the definition of the $\mathscr{H}_2$ norm of a $T$-periodic resource-constrained system, the basic definitions of the $\mathscr{H}_2$ norm of a continuous-time LTI, discrete-time LTI and sampled-data systems are presented.

### *5.2.1 Standard Extended Model Definition*

Consider the continuous-time LTI plant defined by Eqs. (3.2) (Chap. 3) and the performance criterion (4.1) that was assigned to it in Sect. 4.1 of Chap. 4. The performances of the controlled plant may be expressed as the power of continuous-time signal $z_c(t)$. In fact, let $\breve{Q}_c$ and $\breve{R}_c$ be the matrices obtained by the Cholesky decomposition of $Q_c$ and $R_c$ (defined in Eq. (4.1) in Chap. 4). Then the following relation is obtained

$$\breve{Q}_c^T \breve{Q}_c = Q_c$$

and

$$\breve{R}_c^T \breve{R}_c = R_c.$$

Let $C_{1_c}$ and $D_{12_c}$ be the matrices defined by

$$C_{1_c} = \begin{bmatrix} \breve{Q}_c \\ 0_{m,n} \end{bmatrix}$$

and

$$D_{12_c} = \begin{bmatrix} 0_{n,m} \\ \breve{R}_c \end{bmatrix},$$

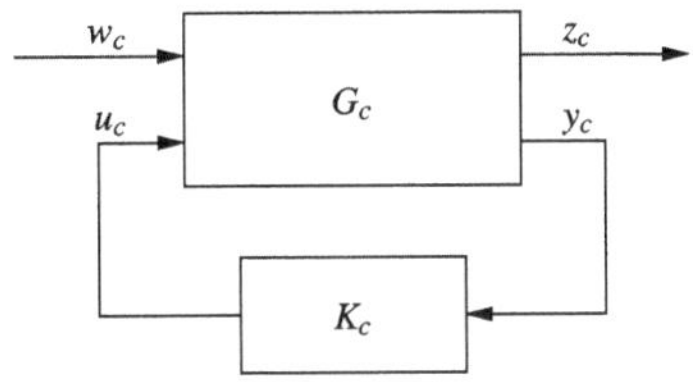


**Fig. 5.1** Standard representation

then the signal $z_c$ may be written as

$$z_c(t) = C_{1_c} x_c(t) + D_{12_c} u_c(t).$$

Consequently

$$\int_0^{T_f} \left(x_c^T(t) Q_c x_c(t) + u_c^T(t) R_c u_c(t)\right) dt = \int_0^{T_f} z_c^T(t) z_c(t)\, dt.$$

An extended model that takes into account both the state and output equations (3.2) as well as the quality output $z_c$ is given by:

$$\dot{x}_c(t) = A_c x_c(t) + B_c u_c(t) + W_c w_c(t), \tag{5.1a}$$

$$y_c(t) = C_c x_c(t) \tag{5.1b}$$

$$z_c(t) = C_{1_c} x_c(t) + D_{12_c} u_c(t). \tag{5.1c}$$

The representation (5.1a)–(5.1c) may be seen as a linear operator $G_c$:

$$G_c : \mathscr{L}_2^r(\mathbb{R}) \times \mathscr{L}_2^m(\mathbb{R}) \longrightarrow \mathscr{L}_2^{n+m}(\mathbb{R}) \times \mathscr{L}_2^p(\mathbb{R}),$$
$$(w_c, u_c) \longmapsto (z_c, y_c).$$

First, assume that $G_c$ is controlled by a continuous-time LTI controller $K_c$. The controller may also be seen as a linear operator $K_c$ defined by

$$K_c : \mathscr{L}_2^p(\mathbb{R}) \longrightarrow \mathscr{L}_2^m(\mathbb{R}),$$
$$y_c \longmapsto u_c.$$

In $\mathscr{H}_2$ optimal control problems, the control system is described following the standard representation from Fig. 5.1. In this representation,

- $w_c$ represents the exogenous inputs,
- $u_c$ represents the control inputs,
- $z_c$ represents the controlled outputs,
- $y_c$ represents the measured outputs.

Naturally, as $G_c$ is linear and possesses two types of inputs $(w_c, u_c)$ and two types of outputs $(z_c, y_c)$, it may be partitioned as follows:

$$z_c = G_{11_c} w_c + G_{12_c} u_c, \tag{5.2a}$$

$$y_c = G_{21_c} w_c + G_{22_c} u_c. \tag{5.2b}$$

If we denote by $\hat{H}(s)$ the transfer function (in the Laplace domain) corresponding to the linear operator $H$, then the expressions of the transfer matrices of $G_{11_c}$, $G_{12_c}$, $G_{21_c}$ and $G_{22_c}$ are given by

$$\begin{aligned}
\hat{G}_{11_c}(s) &= C_c(sI - A_c)^{-1}B_c,\\
\hat{G}_{12_c}(s) &= C_c(sI - A_c)^{-1}W_c,\\
\hat{G}_{21_c}(s) &= C_{1_c}(sI - A_c)^{-1}B_c + D_{12_c},\\
\hat{G}_{22_c}(s) &= C_{1_c}(sI - A_c)^{-1}W_c.
\end{aligned}$$

### 5.2.2 *Standard $\mathscr{H}_2$ Norm of a Continuous/Discrete-Time Linear Time Invariant (LTI) System*

Let $T_{z_c w_c}$ be the transfer operator between the exogenous inputs $w_c$ and the controlled outputs $z_c$ of the closed-loop system. Then, the signals $w_c$ and $z_c$ are linked by the relation

$$z_c = T_{z_c w_c} w_c.$$

The transfer matrix $\hat{T}_{z_c w_c}(s)$ of $T_{z_c w_c}$ is given by:

$$\hat{T}_{z_c w_c}(s) = \hat{G}_{11_c}(s) + \hat{G}_{12_c}(s)\big(Is - \hat{K}_c(s)\hat{G}_{22_c}(s)\big)^{-1}\hat{K}_c(s)\hat{G}_{21_c}(s).$$

Let $(e_i)_{1 \le i \le r}$ be the canonical basis vectors in $\mathbb{R}^r$ and $\delta^c$ the continuous-time Dirac impulse. A Dirac impulse applied to the $i$th exogenous input of system $G_c$ is obtained by applying an input $w_c(t) = \delta^c(t)e_i$. The produced output, assuming zero initial conditions, is $z_c(t) = T_{z_c w_c}\delta^c e_i(t)$.

**Definition 5.1** The $\mathscr{H}_2$ norm of a continuous-time *LTI* system $G_c$ is defined over all stabilizing continuous-time *LTI* controllers $K_c$ by

$$\|G_c\|_{\mathscr{H}_2} = \left(\sum_{i=1}^{r} \|T_{z_c w_c}\delta^c e_i\|_{\mathscr{L}_2}^2\right)^{1/2}. \tag{5.3}$$

The $\mathscr{H}_2$ norm of system $G_c$ is the quadratic mean of the $\mathscr{L}_2$ norms of the responses to $r$ Dirac impulses, each impulse affecting only one exogenous input $i$, $1 \le i \le r$. In a similar way to the continuous-time case, consider now the discrete-time *LTI* plant $G$ controlled by a discrete-time *LTI* controller $K$. Let $\delta^d$ be the discrete-time Dirac impulse. A Dirac impulse applied to the $i$th exogenous input of system $G$ is obtained by applying the input $w(k) = \delta^d e_i(k)$. The produced output, under the assumption of zero initial conditions, is $z(k) = T_{zw}\delta^d e_i(k)$.

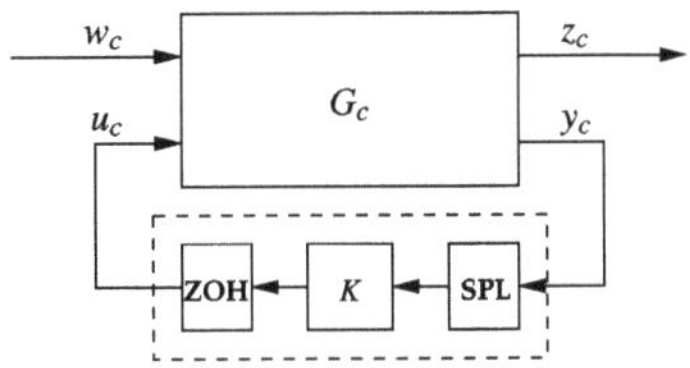


**Fig. 5.2** Standard representation of a sampled-data system

**Definition 5.2** The $\mathscr{H}_2$ norm of a discrete-time *LTI* system $G$ is defined over all stabilizing discrete-time *LTI* controllers $K$ by

$$\|G\|_{\mathscr{H}_2} = \left( \sum_{i=1}^{r} \| T_{z_c w_c} \delta^d e_i \|_{\mathscr{L}_2}^2 \right)^{1/2}. \tag{5.4}$$

### *5.2.3 Computing $\mathscr{H}_2$ Norm of a Sampled-Data System*

Figure 5.2 represents a continuous-time *LTI* plant $G_c$ controlled by a discrete-time *LTI* controller $K$, the discrete controller being preceded by a sampler *SPL* and followed by a zero-order holder *ZOH*. In this setting, the previous definition of the $\mathscr{H}_2$ norm of a continuous-time *LTI* system does not make sense since the transfer operator $T_{z_c w_c}$ is no longer invariant but periodically time-varying with the period $T_s$ (imposed by the sampler *SPL*). In fact, the response of this sampled-data system to a Dirac impulse $\delta^c$ applied at instant $t = 0$ will not be necessarily identical to the response to a Dirac impulse $\delta^c_\tau$ applied at instant $0 < \tau < T_s$ (i.e., defined by $\delta^c_\tau(t) = \delta^c(t - \tau)$). This problem was pointed out and tackled in [19, 62, 137].

**Definition 5.3** The generalized $\mathscr{H}_2$ norm of a sampled-data system is defined over all the stabilizing discrete-time *LTI* controllers $K$ by:

$$\|G_c\|_{\mathscr{H}_2} = \left( \frac{1}{T_s} \int_0^{T_s} \left( \sum_{i=1}^{r} \| T_{z_c w_c} \delta^c_\tau e_i \|_{\mathscr{L}_2}^2 \right) d\tau \right)^{1/2}. \tag{5.5}$$

This generalization is based on the observation that the periodicity implies that $T_{z_c w_c}$ is completely determined by its responses to Dirac impulses $\delta^c_\tau(t) = \delta^c(t - \tau)$, applied at instants $\tau$ verifying $0 \leq \tau \leq T_s$. The $\mathscr{H}_2$ norm of sampled-data systems may be seen as the quadratic mean (in the continuous-time domain) of the previously defined $\mathscr{H}_2$ of continuous-time *LTI* systems corresponding to Dirac impulses applied from instant $\tau = 0$ to instant $\tau = T_s$. The computation of the $\mathscr{H}_2$ norm of a sampled-data system may be reduced to the computation of the $\mathscr{H}_2$ norm of a discrete-time *LTI* system. In the sequel, we present the method proposed in [137], allowing the computation of the $\mathscr{H}_2$ norm of a sampled-data system from the $\mathscr{H}_2$ norm of an equivalent discrete-time *LTI* system. We will describe thereafter the different steps allowing to build the corresponding state-space model of this equivalent discrete-time system.

Let $S_1$ and $S_2$ the $n \times s_1$ and $r \times s_2$ matrices such that

$$S_1 S_1^T = \frac{1}{T_s} \int_0^{T_s} e^{A_c \tau} W_c W_c^T e^{A_c^T \tau} \, d\tau$$

and

$$S_2 S_2^T = \frac{1}{T_s} \int_0^{T_s} \int_0^t C_{1_c} e^{A_c \tau} W_c W_c^T e^{A_c^T \tau} C_{1_c}^T \, d\tau \, dt.$$

Let $\mathscr{W}$ the $s_3 \times (n+m)$ matrix verifying

$$\begin{aligned} \mathscr{W}^T \mathscr{W} = \int_0^{T_s} & \left[ C_{1_c} e^{A_c t} \quad C_{1_c} (\textstyle\int_0^t e^{A_c \tau} d\tau) B_c + D_{12_c} \right]^T \\ & \times \left[ C_{1_c} e^{A_c t} \quad C_{1_c} (\textstyle\int_0^t e^{A_c \tau} d\tau) B_c + D_{12_c} \right] dt. \end{aligned}$$

Consider the partition of $\mathscr{W}$ such that

$$\mathscr{W} = \begin{bmatrix} S_3 & S_4 \end{bmatrix}$$

where $S_3$ and $S_4$ are the matrices of size $s_3 \times n$ and $s_3 \times m$. The equivalent discrete-time system may be described by the following state-space model

$$x(k+1) = Ax(k) + B_1 w(k) + B_2 u(k), \tag{5.6a}$$

$$y(k) = Cx(k), \tag{5.6b}$$

$$z(k) = C_1 x(k) + D_{11} w(k) + D_{12} u(k) \tag{5.6c}$$

where

$$\begin{aligned} A &= e^{A_c T_s}, \\ B_1 &= \begin{bmatrix} S_1 & 0_{n,s_2} \end{bmatrix}, \\ B_2 &= \int_0^{T_s} e^{A_c \tau} \, d\tau \, B_c, \\ C_1 &= \begin{bmatrix} S_3 \\ 0_{r,n} \end{bmatrix}, \\ D_{11} &= \begin{bmatrix} 0_{s_3,s_1} & 0_{s_3,s_2} \\ 0_{r,s_1} & S_2 \end{bmatrix}, \\ D_{12} &= \begin{bmatrix} S_4 \\ 0_{r,m} \end{bmatrix}. \end{aligned}$$

### 5.2.4 Introducing the $\mathscr{H}_2$ Norm of Periodically Scheduled Resource-Constrained Systems

Taking into account the communication constraints, and assuming that the full state vector is available to the controller at each sampling period and that the scheduling of the controller-to-actuators link is performed by a maximal $T$-periodic communication sequence, the open-loop model of a $T$-periodic resource-constrained system can be described by the following extended model:

$$x(k+1) = Ax(k) + B_1 w(k) + B_2 u(k), \tag{5.7a}$$

$$z(k) = C_1 x(k) + D_{12} u(k) + D_{22} w(k), \tag{5.7b}$$

$$\delta_i(k) = 0 \quad \Rightarrow \quad u_i(k) = u_i(k-1). \tag{5.7c}$$

Here, the matrices $A$, $B_1$, $B_2$, $C_1$, $D_{12}$ and $D_{22}$ are computed by using the discretization approach mentioned in the previous paragraph. We will assume, in this chapter, that there exists a maximal $T$-periodic communication sequence ensuring the reachability of system (5.7a)–(5.7c). Since the system (5.7a)–(5.7c) being $T$-periodic, it becomes then natural to control it by using a $T$-periodic controller. Consequently, the global system obtained by the interconnection of the plant and the controller will be also $T$-periodic. In order to evaluate its $\mathscr{H}_2$ norm, the definition $\mathscr{H}_2$ norm of discrete-time periodic system proposed by [256, 266] was adopted. This definition generalizes the well-known definition of the $\mathscr{H}_2$ norm for the discrete *LTI* systems, previously introduced.

Let $\delta_k^d$ be the discrete-time Dirac impulse applied at instant $k$, $0 \leq k \leq T-1$. Based on these notations, $\delta_k^d e_i$ represents the Dirac impulse applied to the $i$th exogenous input at instant $k$, $1 \leq i \leq s_1 + s_2$. To simplify the notation, introduce $\bar{r} = s_1 + s_2$. Let $z^{ik}$ the resulting controlled output under the assumption of zero initial conditions.

**Definition 5.4** The $\mathscr{H}_2$ norm of discrete-time linear $T$-periodic system $G$ is defined over all the stabilizing discrete-time linear $T$-periodic controllers by:

$$\|G\|_2 = \sqrt{\frac{1}{T}\sum_{k=0}^{T-1}\sum_{i=1}^{\bar{r}} \|z^{ik}\|_{l_2}^2}. \tag{5.8}$$

This definition will be used in the sequel in order to compute the $\mathscr{H}_2$ norm of the $T$-periodic resource-constrained system (5.7a)–(5.7c). This involves the computation of $\|z^{ik}\|_{l_2}$, which requires the observation of the system's response over an infinite time horizon. In this work, a very close approximation of $\|z^{ik}\|_{l_2}$ is obtained through a finite horizon $H$ from the instant when the impulse is applied. For that reason, it is necessary to choose $\mathscr{H}$ greater than the response time of the system. In the practical implementation of the algorithm, it is possible to visualize the responses to the different Dirac impulses, and to evaluate how much these responses, which correspond to exponentially stable trajectories, are close to zero. The $\mathscr{H}_2$ norm of the system is the square root of the sum of the squares of the $l_2$ norms corresponding to these responses. Consequently, the "finite-horizon computed $\mathscr{H}_2$ norm" $\|G\|(H)$ converges asymptotically to the true $\mathscr{H}_2$ norm $\|G\|(\infty)$ computed over an infinite horizon, when $H \longrightarrow +\infty$. Consequently, for a desired precision, specified by the maximal absolute error $\varepsilon_{\mathscr{H}_2}$ between the true $\mathscr{H}_2$ norm $\|G\|(\infty)$ computed over an infinite horizon and the "finite-horizon computed $\mathscr{H}_2$ norm" $\|G\|(H)$, there exists an appropriate horizon, $H_{\min}$, such that, for all $H \geq H_{\min}$, $|\|G\|(\infty) - \|G\|(H)| \leq \varepsilon_{\mathscr{H}_2}$. Based on this remark, and on the visualization tools which were implemented, the horizon $H_{\min}$ may be determined iteratively.

## 5.3 Introducing the $\mathscr{H}_2$ Optimal Integrated Control and Off-line Scheduling Problem

### *5.3.1 Solving the Optimal Scheduling Subproblem*

In this paragraph, the optimal scheduling subproblem in the sense of the $\mathscr{H}_2$ performance index is considered. The formulation of this problem requires first the definition of its constraints, which may be classified into two groups:

- the first group includes the constraints that appear in the computation of all the impulsive responses $z^{ik}$, $0 \leq k \leq T-1$ and $1 \leq i \leq \bar{r}$,
- the second group contains the constraints that are specific to a given impulsive response $z^{ik}$.

The first group includes the resource constraints (3.6)

$$\sum_{i=1}^{m} \delta_i(k) = b$$

as well as the communication sequence periodicity constraints

$$\delta(k) = \delta(k+T), \quad \text{for } 0 \leq k \leq \mathscr{H} - T - 1. \tag{5.9}$$

As a consequence, the constraints belonging to the first category may be written as:

$$\mathscr{A}_s \check{\Delta} \leq \mathscr{B}_s \tag{5.10}$$

where $\check{\Delta}$ is the vector

$$\check{\Delta} = \begin{bmatrix} \delta(0) \\ \vdots \\ \delta(H-1) \end{bmatrix}.$$

The second group is related to the computation of the impulsive responses $z^{ik}$, for $0 \leq k \leq T-1$ and $1 \leq i \leq \bar{r}$, over the horizon $H$, knowing the communication constraints previously defined. Let $u^{ik}$, $x^{ik}$, $z^{ik}$, $\xi^{ik}$ and $o^{ik}$ be respectively the values of the control input, the state, the controlled output and the auxiliary variables (introduced in the Sect. 4.4.1 of Chap. 4) corresponding to a *Dirac impulse* applied at instant $k$ to the $i$th exogenous input of system (5.7a)–(5.7c). Let $S^{ik}$ be the set of the involved constraints, for a given response $z^{ik}$ (excepting the constraints belonging to the first group). Then $S^{ik}$ includes:

- the constraints defined by the model (5.7a)–(5.7c) and whose translation into mixed integer quadratic programs was addressed in the previous chapter,
- the constraints defining the *Dirac impulses*, which have to satisfy

$$w_i^{ik}(k) = 1 \tag{5.11}$$

  and

$$w_j^{ik}(l) = 0 \quad \text{for } j \neq i \text{ or } l \neq k, \tag{5.12}$$

- the constraints that have to be added to the problem to ensure the causality of the response

$$u^{ik}(l) = 0 \quad \text{for } l < k. \tag{5.13}$$

Let

$$\check{U}^{ik} = \begin{bmatrix} u^{ik}(0) \\ \vdots \\ u^{ik}(H-1) \end{bmatrix},$$

$$\check{X}^{ik} = \begin{bmatrix} x^{ik}(0) \\ \vdots \\ x^{ik}(H-1) \end{bmatrix},$$

$$\check{Z}^{ik} = \begin{bmatrix} z^{ik}(0) \\ \vdots \\ z^{ik}(H-1) \end{bmatrix},$$

$$\check{\Xi}^{ik} = \begin{bmatrix} \xi^{ik}(0) \\ \vdots \\ \xi^{ik}(H-1) \end{bmatrix},$$

$$\check{O}^{ik} = \begin{bmatrix} o^{ik}(0) \\ \vdots \\ o^{ik}(H-1) \end{bmatrix}$$

and

$$\mathscr{V}^{ik} = \begin{bmatrix} \check{U}^{ik} \\ \check{X}^{ik} \\ \check{Z}^{ik} \\ \check{\Xi}^{ik} \\ \check{O}^{ik} \end{bmatrix}.$$

Then the set of constraints $S^{ik}$ may be described by

$$\mathscr{A}^{ik} \begin{bmatrix} \check{\Delta} \\ \mathscr{V}^{ik} \end{bmatrix} \leq \mathscr{B}^{ik}. \tag{5.14}$$

Consequently, the subproblem of the optimal scheduling in the sense of the $\mathscr{H}_2$ performance index may be written as:

$$\begin{cases} \min\limits_{\check{\Delta},(\mathscr{V}^{ik})_{1\leq i\leq \bar{r},0\leq k\leq T-1}} \displaystyle\sum_{k=0}^{T-1}\sum_{i=1}^{\bar{r}} z^{ik^T} z^{ik} \\ \mathscr{A}_s \check{\Delta} \leq \mathscr{B}_s, \\ \mathscr{A}^{ik} \begin{bmatrix} \check{\Delta} \\ \mathscr{V}^{ik} \end{bmatrix} \leq \mathscr{B}^{ik}, \quad \text{for } 1 \leq i \leq \bar{r},\ 0 \leq k \leq T-1. \end{cases} \tag{5.15}$$

The problem (5.15) is a mixed integer quadratic program and may be solved using the branch and bound method, which was described in the previous chapter. Its resolution provides the optimal $T$-periodic off-line communication sequence $\delta^{T-1^*}$.

### 5.3.2 *Solving the Optimal Control Subproblem*

When the controller has full access to the state, and when the disturbances cannot be measured, the optimal $\mathscr{H}_2$ controller becomes identical to the optimal *LQR* controller [63]. Thus, by knowing the optimal off-line communication sequence $\delta^{T-1^*}$, finding the optimal control boils down to solve the optimal control problem over an infinite horizon for a fixed communication sequence. The general solution to this problem was described in the Sect. 4.3 of Chap. 4. Solving this problem leads to the optimal sequence of control gains $\tilde{K}^{T-1} = (\tilde{K}(0), \ldots, \tilde{K}(T-1))$.

### 5.3.3 *An Illustrative Numerical Example*

Consider the collection of 2 open-loop unstable continuous-time *LTI* subsystems defined by

$$A_c^{(1)} = \begin{bmatrix} 0 & 26.6 \\ -250 & 2 \end{bmatrix}, \quad W_c^{(1)} = \begin{bmatrix} 1.5811 \\ 37.9473 \end{bmatrix}, \quad B_c^{(1)} = \begin{bmatrix} 0 \\ 1000 \end{bmatrix}, \quad S^{(1)},$$

$$A^{(2)} = \begin{bmatrix} 0 & 0 & 0 & 532.3 \\ 0 & 0 & 243.3 & 0 \\ 0 & 0 & -272.67 & 0 \\ 434.86 & 0 & 10.184 & 0 \end{bmatrix}, \quad W_c^{(2)} = \begin{bmatrix} 0 \\ 0 \\ 0.0632 \\ 0.4743 \end{bmatrix},$$

$$B_c^{(2)} = \begin{bmatrix} 0 \\ 0 \\ 10 \\ 10.184 \end{bmatrix}, \quad S^{(2)}.$$

The global system $S$ composed by $S^{(1)}$ and $S^{(2)}$ may be described by the state matrix $A_c$, the exogenous input matrix $W_c$ and the control input matrix $B_c$ defined by

$$A_c = \text{Diag}\big(A_c^{(1)}, A_c^{(2)}\big),$$
$$W_c = \text{Diag}\big(W_c^{(1)}, W_c^{(2)}\big)$$

and

$$B_c = \text{Diag}\big(B_c^{(1)}, B_c^{(2)}\big).$$

Assume that the global system $S$ is controlled by a discrete-time controller executed at the sampling period $T_s = 1$ ms. The control commands are sent to the actuators through a bus with that can carry at most one control command every 1 ms ($b = b_w = 1$). Assume further that the design criteria of the optimal continuous-time controller for each subsystem are defined by the matrices $Q_c^{(1)} = \text{Diag}(8000, 10)$, $R_c^{(1)} = 1$, $Q_c^{(2)} = \text{Diag}(100, 1000, 1, 1)$ and $R_c^{(2)} = 1$, and the desired closed-loop specifications for the global system are described by the following closed-loop weighting matrices:

$$Q_c = \text{Diag}\big(Q_c^{(1)}, Q_c^{(2)}\big)$$

and

$$R_c = \text{Diag}\big(R_c^{(1)}, R_c^{(2)}\big).$$

By using the approach proposed in Sect. 5.2.1, the equivalent sampled-data model at the sampling period $T_s = 1$ ms can be derived. This equivalent model is explicitly used in the $H_2$ optimization.

The optimal solutions, corresponding to different choices of the period $T$ are illustrated in Table 5.1. The relative optimality gap of the used branch and bound algorithm is equal to $10^{-5}$, which means that the best-obtained solution will be considered as an *optimal* solution if the difference between its cost and the lower

**Table 5.1** Optimal $\mathscr{H}_2$ norm as a function of the period $T$—exact discretization

| Period $T$ | $\mathscr{H}$ | $\mathscr{H}_2$ norm | Optimal schedule | CPU time (s) |
|---|---|---|---|---|
| 2 | 28 | 10.3271 | 12... | 20 |
| 3 | 30 | 9.0312 | 112... | 144 |
| 4 | 28 | 9.7701 | 1112... | 206 |
| 5 | 30 | 9.4861 | 11212... | 343 |
| 6 | 30 | 9.0312 | 112112... | 541 |
| 7 | 28 | 9.3481 | 1121121... | 696 |
| 8 | 32 | 9.3176 | 11211212... | 1746 |
| 9 | 27 | 9.0312 | 112112112... | 1102 |
| 10 | 30 | 9.2538 | 1121121112... | 3191 |

bound of the *true optimal* cost is less than 0.01 %. The computations were performed on a PC equipped with a 3.6 GHz Intel Pentium processor and 1 GB of RAM. The optimization problem was solved by using the solver CPLEX (Release 9.1.0) from ILOG. In this particular implementation of the optimization algorithm, $\mathscr{H}$ must be a multiple of $T$. It is sufficient to choose $\mathscr{H}$ greater than 27 in order to guarantee a maximal absolute error $\varepsilon_{\mathscr{H}_2} = 10^{-4}$.

In Table 5.1, the column *CPU Time* indicates the time needed by the optimization algorithm to end. The algorithm stops when it proves that the best obtained solution is *close enough* to the lower bound of the true optimal solution (i.e., the relative difference between the best obtained solution and the true optimal one is less than the specified relative optimality gap). Theses results indicate that the minimal optimal schedule is of length $T = 3$. The length of the optimal schedules which gives the best $\mathscr{H}_2$ norm ($\mathscr{H}_2 = 9.0312$) is a multiple of 3. The resource allocation depends on the dynamics of the subsystems and on their sensitivity to the Dirac impulse disturbance of the $\mathscr{H}_2$ performance evaluation. It is clear from the used definition of the $\mathscr{H}_2$ norm that a circular permutation of an optimal schedule remains optimal.

When the required the CPU time in the last column of Table 5.1 is analyzed, it is easy to see that the contribution of the time needed to solve the relaxed root problem is very important. For example, the continuous relaxation of the root problem in the case $T = 2$ takes 13 s, whereas the total CPU time is 20 s. For $T = 3$, it takes 139 s whereas the total CPU time is 144 s. Since our objective is to find primarily an optimal off-line schedule, it is worth questioning if such an extreme precision in the discretization is necessary for our objective. In fact, the exact discretization causes the loss of the sparsity of the original problem. This loss of sparsity increases the time needed to perform the continuous relaxation of the obtained subproblems, which constitutes the major part of the bounding phase of the branch and bound algorithm. For example, the exact discretization of the continuous-time cost function (4.1) leads to the matrices $C_1$ and $D_{12}$ such that

**Table 5.2** Optimal $\mathscr{H}_2$ norm as a function of the period $T$—approximate discretization

| Period $T$ | $\mathscr{H}$ | $\mathscr{H}_2$ norm | Optimal schedule | CPU time (s) |
|---|---|---|---|---|
| 2 | 28 | 11.2703 | 12... | 6 |
| 3 | 30 | 9.7650 | 112... | 11 |
| 4 | 28 | 10.2593 | 1112... | 18 |
| 5 | 30 | 10.3372 | 11212... | 38 |
| 6 | 30 | 9.7650 | 112112... | 65 |
| 7 | 28 | 9.9447 | 1121121... | 86 |
| 8 | 32 | 10.1259 | 11211212... | 182 |
| 9 | 27 | 9.7650 | 112112112... | 299 |
| 10 | 30 | 9.8908 | 1121121112... | 400 |

$$Q = \begin{bmatrix} C_1^T \\ D_{12}^T \end{bmatrix} \begin{bmatrix} C_1 & D_{12} \end{bmatrix}$$

$$= \begin{bmatrix} 7.9825 & 0.1050 & 0 & 0 & 0 & 0 & 0.0346 & 0 \\ 0.1050 & 0.0119 & 0 & 0 & 0 & 0 & 0.0057 & 0 \\ 0 & 0 & 0.1081 & 0 & 0.0001 & 0.0290 & 0 & 0.0001 \\ 0 & 0 & 0 & 1.0000 & 0.1113 & 0 & 0 & 0.0004 \\ 0 & 0 & 0.0001 & 0.1113 & 0.0169 & 0.0000 & 0 & 0.0001 \\ 0 & 0 & 0.0290 & 0 & 0.0000 & 0.0110 & 0 & 0.0000 \\ 0.0346 & 0.0057 & 0 & 0 & 0 & 0 & 0.0046 & 0 \\ 0 & 0 & 0.0001 & 0.0004 & 0.0001 & 0.0000 & 0 & 0.001 \end{bmatrix},$$

whereas the original continuous-time cost matrix is the diagonal matrix

$$\text{Diag}(Q_c, R_c) = \text{Diag}(8000, 10, 100, 1000, 1, 1, 1, 1).$$

We may remark that these matrices contain new elements with very low magnitude. The original matrix $W_c$ is $6 \times 2$ whereas the discretized matrix $B_1$ is $6 \times 6$. In the sequel, the optimization procedure is performed by considering the approximation $\hat{Q}$ of $Q$ such that:

$$\hat{Q} = T_s \ \text{Diag}(Q_c, R_c).$$

The optimization results are illustrated in Table 5.2. The obtained optimal schedules are identical to the results of the optimization by using the exact discretization. The column *CPU Time* indicates the required CPU time using the default parameters of the solver. The computation time is essentially spent in order to prove that the obtained solution is optimal. In comparison to the exact discretization case, the relaxation of the root problem takes 0.25 s for $T = 2$ and 0.47 s for $T = 3$ when the approximate discretization is employed. The improvements in computation time that are due to the use of the approximate discretization are very significant and vary from 70 % to 92 %. In general, further improvements may be obtained if a feasible initial solution is used to prune the tree. This will be illustrated in Chap. 6. In this particular example, since the optimal solution is found quickly by the algorithm, the use of these initial solutions does not contribute to a significant improvement of the computation time.

## 5.4 Notes and Comments

In this chapter, we have motivated the use of the $\mathscr{H}_2$ performance index as an appropriate criterion allowing the optimal integrated control and off-line scheduling of resource-constrained systems. The $\mathscr{H}_2$ norm of a periodically off-line scheduled resource-constrained system was defined. Based on this definition, a new method for solving this problem was proposed. This method relies on the decomposition of the optimal control and off-line scheduling problem into two independent subproblems. The first subproblem aims in finding the optimal cyclic schedule and is solved by using the branch and bound method. The second sub-problem makes use of the result of the first sub-problem to determine the optimal control gains by applying the lifting technique. This method was evaluated through a numerical example and various techniques allowing to improve its efficiency were proposed.

The problem of optimal control and off-line scheduling of the controller-to-actuators link was studied in *Rehbinder and Sanfridson* [200]. In the proposed model, the control commands are sent to the actuators through a shared *TDMA* bus. At each slot, only one control command can be sent, the remaining commands for the other actuators are held constant. The choice of which actuator to update at each slot was handled by using the notion of communication sequence introduced by *Brockett* [43]. Only periodic communication sequences were considered. A quadratic cost function is associated to each communication sequence, corresponding to the worst-case initial condition and worst-case sequence permutation. Control commands and periodic communication sequences are obtained by solving a complex combinatorial optimization problem. The problem of the optimal control and scheduling in the sense of *LQG* was introduced and developed in *Lincoln and Bernhardsson* [155]. The relaxed dynamic programming method was applied for its resolution leading to a more efficient search heuristics. It is worth mentioning that an heuristic approach for the problem of the optimal control and off-line scheduling of the sensors-to-controller link in the sense of the $\mathscr{H}_\infty$ performance index was proposed in *Lu* [161]. The application of branch and bound algorithm to this problem was performed in *Ben Gaid et al.* [22]. Finally, the application of genetic algorithms and particle swarm optimization to this problem was undertaken by *Longo et al.* in [159]. A generalization of this approach to tackle uncertainties of the plant model was undertaken by *Al-Areqi et al.* in [2], and to cope with non linearities in *Su et al.* [221].

# Chapter 6
# Optimal Integrated Control and On-line Scheduling of Resource-Constrained Systems

Numerical results of the optimal integrated control and scheduling problem show that the optimal scheduling is closely dependent on the dynamical state of the controlled systems. This dependence confirms the intuition that the most disturbed plants have always more important "needs" in terms of communication resources than the plants that are near to the equilibrium. By using on-line scheduling algorithms, it is possible to exploit this dependency to achieve a better control performance, thanks to a more efficient use of the available resources.

In this chapter, we consider a resource-constrained system $\mathscr{S}$ where the full state vector $x(k)$ is available to the controller at each sampling period. We first propose the use of the model predictive control approach as an algorithmic solution allowing to compute on-line, at the same time, the optimal values of the control signals and the communication scheduling of resource-constrained systems. In opposition to [191], control signals that could not be updated are held constant (and not set to zero). Furthermore, a quadratic cost function (and not linear function) is used in order to evaluate the control performance as well as the ability of the adaptive scheduling to improve the performance of sampled-data systems (instead of discrete-time systems) is demonstrated. However, the on-line solving of the optimization algorithm, which is required by the *MPC* approach, is very costly. For that reason, an on-line scheduling algorithm, called *optimal pointer placement* (*OPP*) is proposed. While being based on a pre-computed optimal off-line schedule, *OPP* makes possible to allocate on-line the communication resources, based on the state of the controlled dynamical systems. It is shown that, under mild conditions, *OPP* ensures the asymptotic stability of the controlled systems and enables in all the situations to improve the control performance compared to the basic static scheduling. Furthermore, under these conditions, the determination of *OPP* control and scheduling amounts by comparing a limited number of quadratic functions of the state. Finally, *OPP* is applied to two typical examples of distributed control and embedded systems: the active suspension of car and the attitude control of a quadrotor.

A. Çela et al., *Optimal Design of Distributed Control and Embedded Systems*,
Communications and Control Engineering, DOI 10.1007/978-3-319-02729-6_6,

## 6.1 Model Predictive Control (*MPC*) of Resource-Constrained Systems

### 6.1.1 Problem Formulation

Open-loop optimization problems, similar to the ones described in Chap. 4, constitute the cornerstone of a successful control method: the model predictive control (*MPC*). *MPC* has strong theoretical foundations, and many interesting properties that make it suitable for addressing constrained control problems. However, its main drawback is that, unfortunately, it requires very expensive computational resources, which make it only applicable to slow systems, as, for example, chemical processes. To the best of the authors' knowledge, model predictive control is the standard approach to control *MLD* systems [21]. Among others, its application to this particular problem was motivated by:

- The need to optimize simultaneously control actions and network scheduling, in order to achieve a better quality of control with respect to the one obtained in the case of static network allocation schemes.
- The need for a control law that changes on-line the "actuation period" in order to improve the quality of control. This requires that these variations are taken into account by the control law [168].

When using *MPC*, an optimal control problem is solved on-line at each sampling period $T_s$. It aims in finding the optimal control values sequence $\hat{u}^{N-1^*} = (\hat{u}^*(0), \ldots, \hat{u}^*(N-1))$ and the optimal communication sequence $\hat{\delta}^{N-1^*} = (\hat{\delta}^*(0), \ldots, \hat{\delta}^*(N-1))$, which are solutions of the following optimization problem:

$$\left\{\begin{array}{l}\displaystyle\min_{\hat{u}^{N-1},\hat{\delta}^{N-1}} \sum_{h=0}^{N-1} \begin{bmatrix}\hat{x}(h)\\ \hat{u}(h)\end{bmatrix}^T Q \begin{bmatrix}\hat{x}(h)\\ \hat{u}(h)\end{bmatrix} + \hat{x}^T(N) Q_0 \hat{x}(N)\\ \text{subject to}\\ \quad \hat{x}(0) = x(k),\\ \quad \hat{x}(h+1) = A\hat{x}(h) + B\hat{u}(h), \quad h \in \{0, \ldots, N-1\},\\ \quad \displaystyle\sum_{i=1}^{m} \hat{\delta}_i(h) = b, \quad h \in \{0, \ldots, N-1\},\\ \quad \hat{\delta}_i(0) = 0 \implies \hat{u}_i(0) = u_i(k-1),\\ \quad \hat{\delta}_i(h) = 0 \implies \hat{u}_i(h) = \hat{u}_i(h-1), \quad h \in \{1, \ldots, N-1\}.\end{array}\right. \tag{6.1}$$

The solution of this problem is based on the prediction of the future evolution of the system over an appropriate horizon of $N$ sampling periods. This predicted evolution is calculated according to the model of the plant, knowing the current state $x(k)$ of the system. The variables $\hat{x}(h)$, $h \in \{0, \ldots, N\}$ represent the predicted values of system states $x(k+h)$. The sequences $(\hat{u}(0), \ldots, \hat{u}(N-1))$ (virtual control sequence) and $(\hat{\delta}(0), \ldots, \hat{\delta}(N-1))$ (virtual communication sequence) are called *virtual sequences*, since they are based on the predicted evolution of the system. The resolution of this problem aims in finding the optimal virtual control sequence $(\hat{u}^*(0), \ldots, \hat{u}^*(N-1))$ as well as the optimal virtual communication sequence $(\hat{\delta}^*(0), \ldots, \hat{\delta}^*(N-1))$ that minimize a quadratic cost function over a finite horizon of $N$ sampling periods. Assuming that the optimal virtual sequences exist, the actual control commands are obtained by setting:

$$v(k) = M_{\hat{\delta}^*}(0)\hat{u}^*(0) \tag{6.2}$$

and

$$\delta(k) = \hat{\delta}^*(0) \tag{6.3}$$

and disregarding the remaining elements $(\hat{u}^*(1), \ldots, \hat{u}^*(N-1))$ and $(\hat{\delta}^*(1), \ldots, \hat{\delta}^*(N-1))$. At the next sampling period (step $k+1$), the whole optimization procedure is repeated, based on $x(k+1)$.

An important issue concerns the stability of the proposed model predictive controller. If the following constraint is added to problem (6.1):

$$\hat{x}(N) = 0, \tag{6.4}$$

the following result is obtained.

**Theorem 6.1** [27] *If at* $k = 0$, *a feasible solution exists for the problem* (6.1) *augmented with the additional constraint* (6.4), *then* $\forall Q = Q^T > 0$, *the* MPC *law* (6.1), (6.4) *stabilizes the system* $\mathscr{S}$ *such that*

$$\lim_{k \to +\infty} x(k) = x_e = 0,$$

$$\lim_{k \to +\infty} u(k) = u_e = 0.$$

*Proof* The proof may be easily performed by following the same ideas as in the proof of the sufficient stability conditions for the model predictive control of MLD systems stated and discussed in [21]. □

### 6.1.2 Optimality

The optimality of the model predictive controller may be proved if an infinite horizon cost function $J(\tilde{x}, v, \delta, 0, +\infty) = J(x, u, 0, +\infty)$ is used and if the prediction horizon $N$ is chosen infinite. More precisely, at each time step $k$, and for any extended state $\tilde{x}(k)$, the model predictive controller over an infinite horizon computes the optimal solutions $v^*(k)$ and $\delta^*(k)$ that minimizes the cost function $J(\tilde{x}, v, \delta, k, +\infty)$, subject to the communication constraints. Its optimality directly results from the Bellman optimality principle, which states: "*An optimal policy has* *the property that whatever the initial state and the initial decision are, the remaining decisions must constitute an optimal policy with respect to the state resulting from the first decision.*"

In many practical situations, it is sufficient to choose the prediction horizon $N$ large enough with respect to the response time of the system in order to get a performance that is close to the optimality. This is possible when the virtual sequences of the optimal control commands, which are computed at each sampling period, converge exponentially to zero as the horizon increases. The obtained finite horizon solution will then approximate the optimal infinite horizon solution.

### 6.1.3 An Illustrative Numerical Example

In order to illustrate the proposed approach and to study the interdependency of control and scheduling, especially the relationship between the state space vector of the plant and the optimal network bandwidth allocation, consider the continuous-time LTI system described by the state matrix:

$$A_c = \begin{bmatrix} A_c^{(1)} & 0 \\ 0 & A_c^{(2)} \end{bmatrix}$$

and the input matrix:

$$B_c = \begin{bmatrix} B_c^{(1)} & 0 \\ 0 & B_c^{(2)} \end{bmatrix}$$

with

$$A_c^{(1)} = A_c^{(2)} = \begin{bmatrix} 0 & 130 \\ -800 & 10 \end{bmatrix}$$

and

$$B_c^{(1)} = B_c^{(2)} = \begin{bmatrix} 0 \\ 230 \end{bmatrix}.$$

The global system consists in two identical and independent subsystems, $S^{(1)}$ and $S^{(2)}$, which are open-loop unstable. The two control inputs $u_1$ (corresponding to subsystem $S^{(1)}$) and $u_2$ (corresponding to subsystem $S^{(2)}$) are sent to the corresponding actuators through a shared communication bus. Assume now that the bus bandwidth is such that only a control command may be transmitted at each sampling period $T_s = 2$ ms, thus $b = 1$. Furthermore, assume that the specifications of the optimal continuous-time controller are described by the matrices $Q_c = S_c = \text{Diag}(100, 10, 100, 10)$ and $R_c = \text{Diag}(1, 1)$.

In order to take into account the bandwidth limitations, a simple strategy by sending "alternately" control commands $u_1$ and $u_2$ over the bus may be considered. This

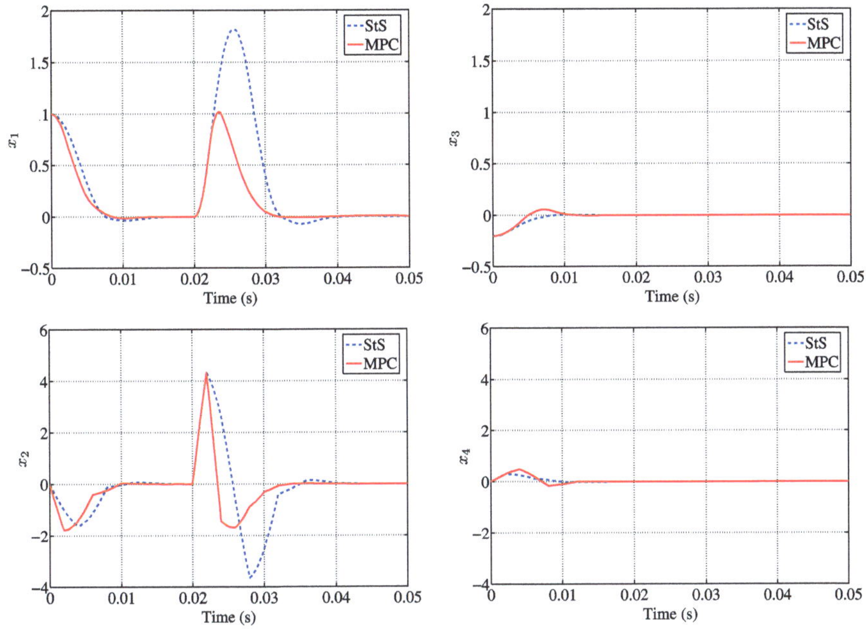


**Fig. 6.1** Subsystems $S^{(1)}$ and $S^{(2)}$ responses

strategy consists in applying an off-line schedule, defined by the periodic communication sequence

$$\left(\begin{bmatrix}1\\0\end{bmatrix},\begin{bmatrix}0\\1\end{bmatrix}\right).$$

Since the subsystems $S^{(1)}$ and $S^{(2)}$ are identical, an optimal off-line schedule should fairly share the available resources between them. In order to apply this strategy, optimal sampled-data controllers (at the sampling period of 4 ms) are considered.

This static scheduling strategy is compared to an adaptive scheduling strategy based on the model predictive control algorithm. The *MPC* relies on a plant discrete-time model at the period of 2 ms. The prediction horizon $N$ is equal to 14 (and is greater than the global system response time). A sub-optimal solution with a relative error bound of $1 \times 10^{-5}$ was required for the branch and bound solver, which is used by the *MPC*.

Figure 6.1 compares the evolution of the state variables of subsystem $S^{(1)}$ ($x_1$ and $x_2$) and those of subsystem $S^{(2)}$ ($x_3$ and $x_4$), corresponding respectively, to the application of the static strategy (StS), and to the use of the model predictive controller (*MPC*). The accumulated cost functions corresponding to these responses are depicted in Fig. 6.2. The considered initial state is $[1\ 0\ {-0.2}\ 0]^T$. At instant $t = 20$ ms, subsystem $S^{(1)}$ is severely disturbed.

These results show that significant improvements in control performance are achieved by the adaptive scheduling scheme, compared to the static fair network

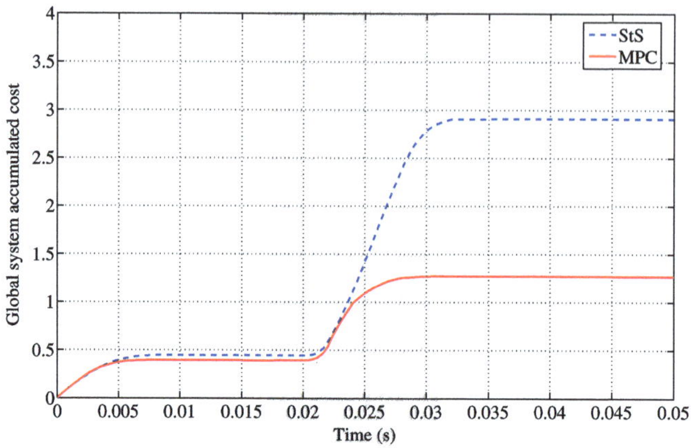


**Fig. 6.2** Accumulated continuous-time cost functions

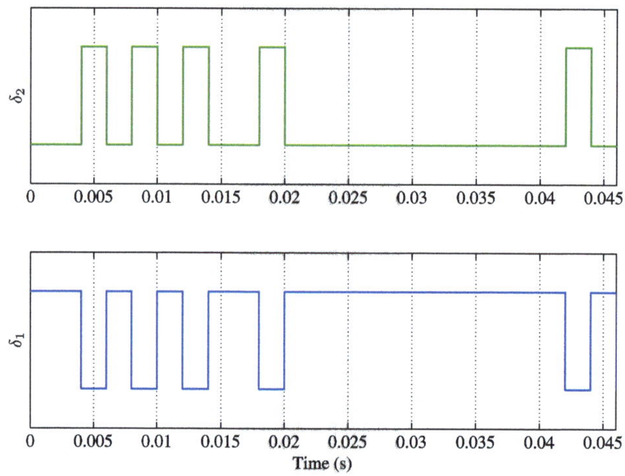


**Fig. 6.3** Adaptive schedule (*MPC*)

allocation. In order to understand the reasons behind these improvements, it is necessary to analyze the operation of the model predictive controller. For that, consider the controller-to-actuators schedule corresponding the use of the *MPC* algorithm (in Fig. 6.3). Figure 6.4 describes the static off-line schedule. At $t = 0$ s, the subsystem $S^{(1)}$ has the greatest deviation from the equilibrium position. In order to optimize the cost function, the *MPC* allocates the two first slots to the transmission of the control signal $u_1$, which contrasts with the static strategy, where resources are pre-allocated independently on the dynamical state of the controlled plants. Next, when the two subsystems reach approximately the same "distance" from the equilibrium, the network bandwidth is allocated fairly. At $t = 20$ ms, when the subsystem $S^{(1)}$ is disturbed (subsystem $S^{(2)}$ being at the equilibrium), the model predictive controller

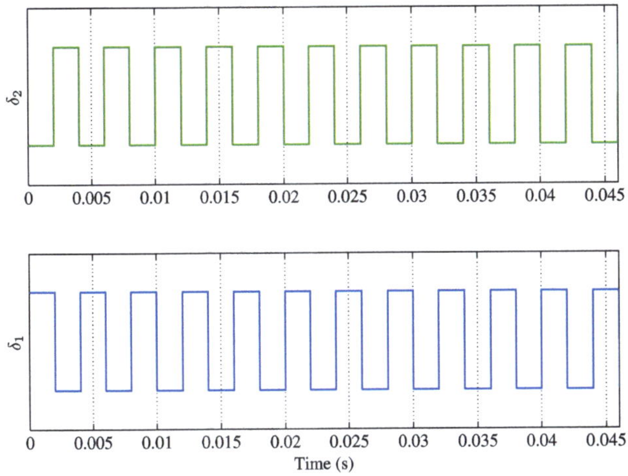


**Fig. 6.4** Static schedule (StS)

reacts "extremely" fast by allocating 11 consecutive slots to the transmission of the control command $u_1$. This contrasts with the operation of the static strategy, where half of the bandwidth is allocated to subsystem $S^{(2)}$, which is at the equilibrium. This aptitude of the model predictive controller to change the "actuation period" comes from its capacity to compensate the control commands for these changes. This dynamic allocation of the resources (as well as the automatic compensation of the control commands) makes possible to reject disturbances in a much more powerful way, contributing thus to the significant improvement of the quality of control (as illustrated to the Fig. 6.2).

These results illustrate the ability of the model predictive controller to play the role of an adaptive controller-scheduler, as well as its aptitudes to improve the quality of control compared to any off-line scheduling strategy. These improvements manifest themselves by better disturbance rejection capabilities, which are due to the ability of the adaptive scheduling scheme to react earlier and faster to the disturbances, compared to any off-line scheduling strategy. These results are in accordance with those of [167], where the empirical study of the relationship between resource allocation and control performance was undertaken.

## 6.2 Optimal Pointer Placement (*OPP*) Scheduling

### *6.2.1 Problem Formulation*

The motivation behind the optimal pointer placement (*OPP*) scheduling algorithm presented in this section is to be a "trade-off" between the advantages of the on-line scheduling (control performance) and those of the off-line scheduling (a very limited usage of computing resources).

**Fig. 6.5** Illustration of the notion of pointer

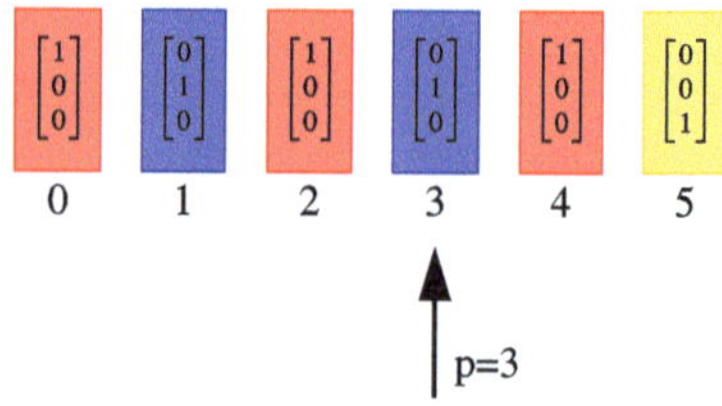


Consider now a static off-line control and scheduling strategy, which is based on:

- a $T$-periodic controller defined by a sequence of state-feedback control gains $\tilde{K}^{T-1} = (\tilde{K}(0), \ldots, \tilde{K}(T-1))$,
- a static off-line schedule defined by a $T$-periodic communication sequence $\gamma^{T-1} = (\gamma(0), \ldots, \gamma(T-1))$.

Note that the optimal off-line selection of $\tilde{K}^{T-1}$ and $\gamma^{T-1}$ was described and discussed in the previous chapter.

At runtime, the execution of the periodic off-line controller and scheduler may be described by using the notion of pointer. The pointer may be seen as a variable that contains the index of the control gain to use and the scheduling to apply. The pointer is incremented at each sampling period. If it reaches the end of the sequence, its position is reset. More formally, if the pointer is started at position $p$ ($0 \leq p < T$), its expression $\mathscr{I}_p(k)$ at the $k$th sampling period is given by

$$\mathscr{I}_p(k) = (k+p) \mod T. \tag{6.5}$$

According to the off-line strategy, the control commands that are transmitted as well as the scheduling decisions are chosen such that:

$$v(k) = \tilde{K}\big(\mathscr{I}_p(k)\big)\tilde{x}(k) \tag{6.6}$$

and

$$\delta(k) = \gamma\big(\mathscr{I}_p(k)\big). \tag{6.7}$$

*Example 6.1* Consider the periodic communication sequence $\gamma^5$ defined by

$$\gamma^5 = \left(\begin{bmatrix}1\\0\\0\end{bmatrix}, \begin{bmatrix}0\\1\\0\end{bmatrix}, \begin{bmatrix}1\\0\\0\end{bmatrix}, \begin{bmatrix}0\\1\\0\end{bmatrix}, \begin{bmatrix}1\\0\\0\end{bmatrix}, \begin{bmatrix}0\\0\\1\end{bmatrix}\right)$$

and the associated control gains sequence

$$\tilde{K}^5 = \big(\tilde{K}(0), \tilde{K}(1), \tilde{K}(2), \tilde{K}(3), \tilde{K}(4), \tilde{K}(5)\big).$$

Figure 6.5 graphically illustrates the notion of pointer. The communication sequence $\gamma^5$ as well as the pointer (located at position $p = 3$) are depicted.

The idea behind the *OPP* scheduling heuristic is that instead of finding an optimal solution to problem (6.1), the search is restricted to find a sub-optimal solution, based on an optimal off-line schedule, over a horizon $N$ (which is assumed to be a multiple of $T$), according to the following problem:

$$\begin{cases} \min\limits_{p} \sum\limits_{h=0}^{N-1} \begin{bmatrix} \hat{x}(h) \\ \hat{u}(h) \end{bmatrix}^T Q \begin{bmatrix} \hat{x}(h) \\ \hat{u}(h) \end{bmatrix} + \hat{x}^T(N) Q_0 \hat{x}(N) \\ \text{subject to} \\ \quad \hat{x}(0) = x(k), \\ \quad \hat{u}(-1) = u(k-1), \quad \text{and for all } h \in \{0, \ldots, N-1\}, \\ \quad \hat{v}(h) = \tilde{K}\big(\mathscr{I}_p(h)\big) \begin{bmatrix} \hat{x}(h) \\ \hat{u}(h-1) \end{bmatrix}, \\ \quad \hat{u}(h) = D_\gamma\big(\mathscr{I}_p(h)\big)\hat{v}(h) + E_\gamma\big(\mathscr{I}_p(h)\big)\hat{u}(h-1), \\ \quad \hat{x}(h+1) = A\hat{x}(h) + B\hat{u}(h). \end{cases} \tag{6.8}$$

The cost function is computed according to a prediction of the future evolution of the system (described by $\hat{x}(h)$). This evolution is calculated by assuming that the sequences $\tilde{K}^{T-1}$ and $\gamma^{T-1}$ are started from position $p$. For example, if the pointer is located at position $p = 3$, as in Fig. 6.5, the evolution of the system is predicted based on the communication sequence

$$\big(\gamma(3), \gamma(4), \gamma(5), \gamma(0), \gamma(1), \gamma(2), \gamma(3), \ldots\big)$$

and the control gains sequence

$$\big(\tilde{K}(3), \tilde{K}(4), \tilde{K}(5), \tilde{K}(0), \tilde{K}(1), \tilde{K}(2), \tilde{K}(3), \ldots\big).$$

The solution of problem (6.8) is the pointer's position $p^*$ that minimizes the cost function $\hat{J}(\tilde{x}(k), p)$, subject to the constraints expressed above, where

$$\hat{J}\big(\tilde{x}(k), p\big) = \sum_{h=0}^{N-1} \begin{bmatrix} \hat{x}(h) \\ \hat{u}(h) \end{bmatrix}^T Q \begin{bmatrix} \hat{x}(h) \\ \hat{u}(h) \end{bmatrix} + \hat{x}^T(N) Q_0 \hat{x}(N).$$

The control command $v(k) = \tilde{K}(p^*)\tilde{x}(k)$ is sent according to the scheduling vector $\delta(k) = \gamma(p^*)$. At the next sampling period, this procedure is reiterated in a receding horizon manner.

### 6.2.2 An Illustrative Numerical Example

In order to illustrate the effectiveness of the *OPP* scheduling algorithm, the example proposed in Sect. 4.4.3 of Chap. 4 is reconsidered. The control performances

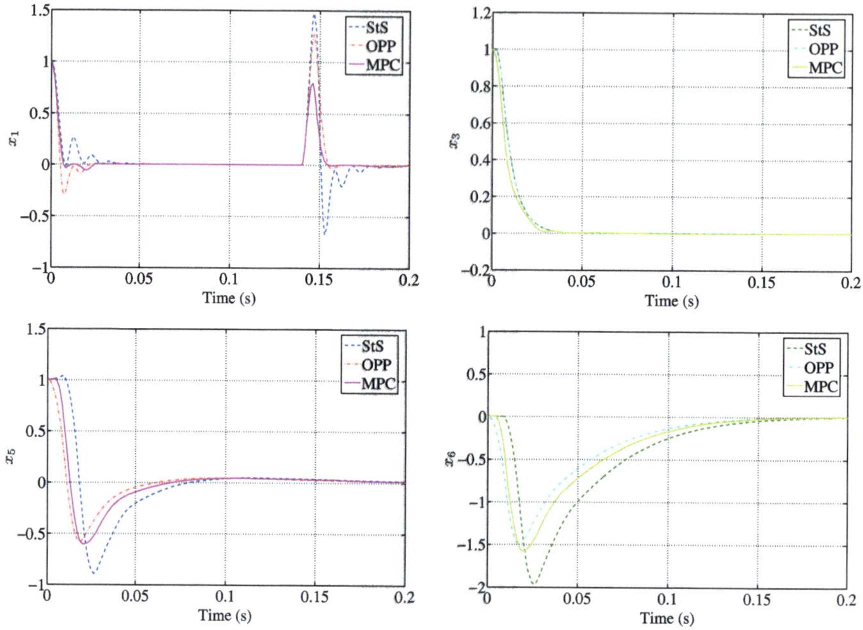


**Fig. 6.6** Global system responses—states $x_1$ and $x_3$

corresponding to the use of a static scheduling (StS) algorithm, *OPP* and *MPC* are compared. Static scheduling and *OPP* algorithms use the communication sequence

$$\gamma^5 = \left( \begin{bmatrix} 1 \\ 0 \\ 0 \end{bmatrix}, \begin{bmatrix} 0 \\ 1 \\ 0 \end{bmatrix}, \begin{bmatrix} 1 \\ 0 \\ 0 \end{bmatrix}, \begin{bmatrix} 0 \\ 1 \\ 0 \end{bmatrix}, \begin{bmatrix} 1 \\ 0 \\ 0 \end{bmatrix}, \begin{bmatrix} 0 \\ 0 \\ 1 \end{bmatrix} \right)$$

that together with the control gains sequence $\tilde{K}^5 = (\tilde{K}_1, \tilde{K}_2, \tilde{K}_1, \tilde{K}_2, \tilde{K}_1, \tilde{K}_3)$ guarantee the asymptotic stability of the global system, where

$$\begin{aligned}
\tilde{K}_1 &= [-1.14 \quad -1.04 \quad 0 \quad 0 \quad 0 \quad 0 \quad 0 \quad 0 \quad 0 \quad 0 \quad 0], \\
\tilde{K}_2 &= [0 \quad 0 \quad 4.12 \quad 0.36 \quad 0 \quad 0 \quad 0 \quad 0 \quad 0 \quad 0 \quad 0], \\
\tilde{K}_3 &= [0 \quad 0 \quad 0 \quad 0 \quad 17.80 \quad -6.28 \quad -33.28 \quad 48.91 \quad 0 \quad 0 \quad 0].
\end{aligned}$$

The period $T$ of the schedule is equal to 6 and the horizon $N$ of the *OPP* and *MPC* algorithms is equal to 60. A sub-optimal solution with a relative error of $1 \times 10^{-5}$ was required for the *MPC* algorithm. Global system responses corresponding to the state variables $x_1, x_3, x_5$ and $x_6$ (positions) are depicted in Fig. 6.6. The accumulated continuous-time cost functions corresponding to these responses are illustrated in Fig. 6.7.

It is assumed that the global system starts from the initial state $[1\ 0\ 1\ 0\ 1\ 0\ 0\ 0]^T$. The three subsystems $S^{(1)}$, $S^{(2)}$ and $S^{(3)}$ converge progressively to the steady state.

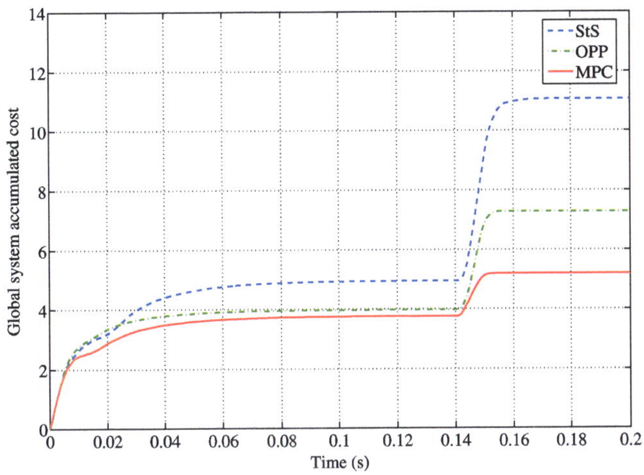


**Fig. 6.7** Accumulated continuous-time cost functions

Next, we also assume that, at $t = 0.14$ s, the subsystem $S^{(1)}$ is disturbed. This deviation is fast corrected. The best response is achieved by the *MPC*. However, this algorithm cannot be implemented in practice, because the required computation time is too long (in this case, a few seconds at each step on a PC equipped with an Intel Celeron processor cadenced at 2.2 GHz). For $N = 60$, the number of admissible maximal communication sequences is $3^{60} = 4.23 \times 10^{28}$. The fact that the used branch and bound algorithm stops in a few seconds shows its efficiency to address this particular problem. The *OPP* algorithm significantly improves the control performance with respect to the static scheduling algorithm, requiring fewer computing resources than the *MPC*. In fact, when using the *OPP* scheduling algorithm, the maximum number of possible communication sequences is equal to $T = 6$.

The schedule obtained by the *OPP* and *MPC* algorithms are respectively depicted in Figs. 6.8 and 6.9. At $t = 0$ s, the *OPP* scheduling algorithm chooses to reserve the first slot to subsystem $S^{(3)}$. This choice contributes to the improvement of its performance as well as the performance of the global system (shown in Fig. 6.6). It is important to point out that both *MPC* and *OPP* have the ability to change on-line the "actuation period" of each subsystem. Consequently, the scheduling pattern is irregular. Network slots are allocated in order to improve the control performance. When a subsystem is closer to the equilibrium, then its control commands remain constant and they may be sent less frequently over the network. This explains the reason why the *MPC* algorithm does not allocate network slots to subsystems in the equilibrium, when other subsystems are "far" form the steady state. At $t = 0.14$ s, when the subsystem $S^{(1)}$ is disturbed, *OPP* and *MPC* algorithms allocate the network slots mainly for the transmission of the control command of subsystem $S^{(1)}$, allowing to react earlier and faster to the disturbance. This contrasts with the static scheduling algorithm, where the allocation

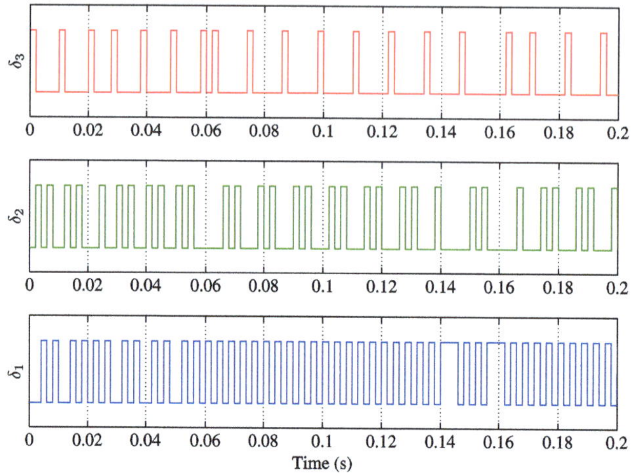


**Fig. 6.8** *OPP* schedule

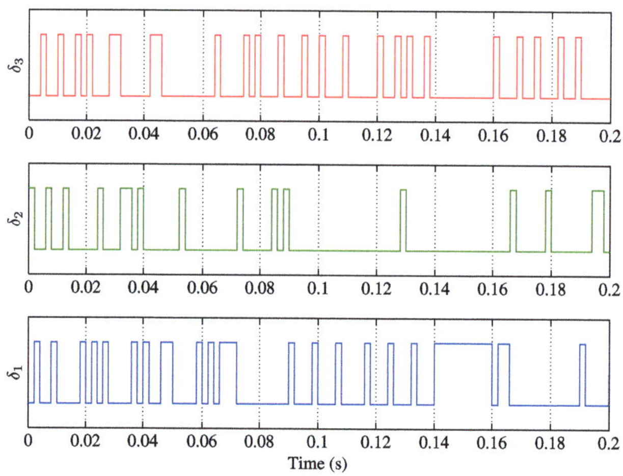


**Fig. 6.9** *MPC* schedule

of the network resources is independent from the dynamical state of the subsystems.

Finally, assume now that the three subsystems are disturbed with a band limited white noise characterized by a noise power of 0.1 and a correlation time of $1 \times 10^{-5}$. Simulation results are depicted in Fig. 6.10. The performance improvements using the *OPP* and the *MPC* algorithms are similar to those observed in the previous simulations.

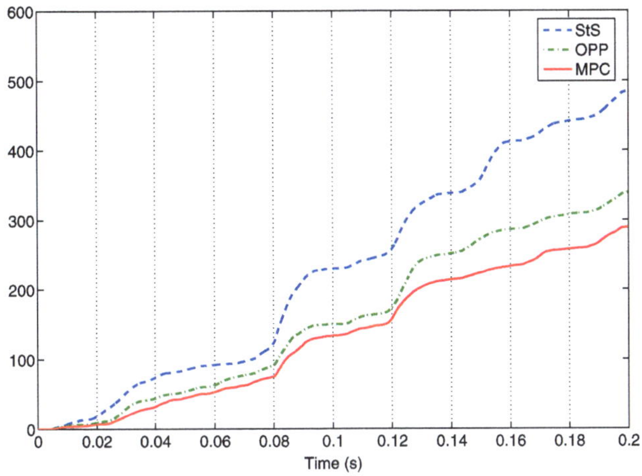


**Fig. 6.10** Accumulated continuous-time cost functions resulting from a band limited white noise disturbance

### 6.2.3 Optimal Pointer Placement over Infinite Horizon

If the horizon $N$ is infinite, the *OPP* scheduling algorithm presents some interesting properties: its implementation becomes simpler and a formal proof of its stability may be given. Moreover, it may also be proven that the quality of control obtained when using *OPP* is always better or similar (in the worst-case) compared to the quality of control obtained by using its basic static scheduling algorithm.

The simplification of *OPP* comes from the relation between the expression of the cost function over an infinite horizon (corresponding to the use of a $T$-periodic communication sequence) and the solution of the discrete algebraic periodic Riccati equation over an infinite horizon described in Sect. 4.3 of Chap. 4. This relation is formalized by

$$\hat{J}\left(\tilde{x}(k), p\right) = \tilde{x}^T(k)\tilde{S}(p)\tilde{x}(k).$$

Based on this relation, the implementation of *OPP* becomes simpler and may be described by the following algorithm:

$$\text{Find } p^* = \operatorname*{argmin}_p \tilde{x}^T(k)\tilde{S}(p)\tilde{x}(k),$$

$$v(k) = \tilde{K}\left(p^*\right)\tilde{x}(k) \quad \text{and} \quad \delta(k) = \gamma\left(p^*\right).$$

For $N = +\infty$, the stability of the *OPP* scheduling algorithm is stated in the following:

**Theorem 6.2** [27] *If the asymptotic stability of the resource-constrained system $\mathscr{S}$ is guarantied by the off-line control and scheduling using the control gains sequence $\tilde{K}^{T-1}$ and the network scheduling sequence $\gamma^{T-1}$, then it is also ensured by the* OPP *scheduling algorithm.*

*Proof* The proof is based on the comparison of the trajectory of the system scheduled by the static scheduling algorithm (denoted by $\tilde{x}^{\text{sts}}$) and that of the system scheduled by the *OPP* algorithm (denoted by $\tilde{x}^{\text{opp}}$), starting from the same arbitrary initial condition $\tilde{x}^{\text{sts}}(0) = \tilde{x}^{\text{opp}}(0) = \tilde{x}(0)$.

Let $J^{\text{sts}}(\tilde{x}(i), i, f, p)$ be the cost function corresponding to an evolution from instant $k = i$ to instant $k = f$ starting from the state $\tilde{x}(i)$ where the static scheduling algorithm is applied and where the pointer at instant $i$ is placed at position $p$:

$$J^{\text{sts}}\big(\tilde{x}(i), i, f, p\big) = \sum_{k=i}^{f} \begin{bmatrix} x^{\text{sts}}(k) \\ u^{\text{sts}}(k) \end{bmatrix}^T Q \begin{bmatrix} x^{\text{sts}}(k) \\ u^{\text{sts}}(k) \end{bmatrix}.$$

Let $J^{\text{opp}}(\tilde{x}(i), i, f)$ be the cost function corresponding to an evolution from instant $k = i$ to instant $k = f$ starting from the state $\tilde{x}(i)$ where the *OPP* algorithm is applied:

$$J^{\text{opp}}\big(\tilde{x}(i), i, f\big) = \sum_{k=i}^{f} \begin{bmatrix} x^{\text{opp}}(k) \\ u^{\text{opp}}(k) \end{bmatrix}^T Q \begin{bmatrix} x^{\text{opp}}(k) \\ u^{\text{opp}}(k) \end{bmatrix}.$$

In order to prove the stability of the system scheduled by using the *OPP* algorithm, we must prove that for all $p_0 \in \{0, \ldots, T-1\}$

$$J^{\text{opp}}\big(\tilde{x}(0), 0, +\infty\big) \leq J^{\text{sts}}\big(\tilde{x}(0), 0, +\infty, p_0\big). \tag{6.9}$$

In fact, $Q$ is definite positive. As a consequence, $J^{\text{sts}}(\tilde{x}(0), 0, +\infty, p_0)$ (resp. $J^{\text{opp}}(\tilde{x}(0), 0, +\infty)$) is finite if and only if the system scheduled using the static scheduling algorithm (resp. the *OPP* algorithm) is asymptotically stable.

Let $J^{\text{opp}-\text{sts}}(\tilde{x}(0), l)$ be the cost function corresponding to an evolution starting from the initial state $\tilde{x}(0)$ where *OPP* is applied form $k = 0$ to $k = l$ and then followed by the static scheduling algorithm (which is applied from $k = l + 1$ to $k = +\infty$). Then it is easy to see that

$$J^{\text{opp}}\big(\tilde{x}(0), 0, +\infty\big) = \lim_{l \to +\infty} J^{\text{opp}-\text{sts}}\big(\tilde{x}(0), l\big). \tag{6.10}$$

Consequently, in order to prove (6.9), it is sufficient to verify that, for all $p_0 \in \{0, \ldots, T-1\}$ and for all $l \in [0, +\infty)$

$$J^{\text{opp}-\text{sts}}\big(\tilde{x}(0), l\big) \leq J^{\text{sts}}\big(\tilde{x}(0), 0, +\infty, p_0\big). \tag{6.11}$$

This proof may be done by recurrence on $l$. Let $p_0 \in \{0, \ldots, T-1\}$ an arbitrary start position of the static scheduling algorithm.

At the stage $l = 0$, the *OPP* scheduling algorithm will choose the pointer position such that

$$p^*(0) = \underset{p}{\operatorname{argmin}}\, J^{\text{sts}}\big(\tilde{x}(0), 0, +\infty, p\big). \tag{6.12}$$

By knowing that

$$J^{\text{sts}}\big(\tilde{x}(0), 0, +\infty, p^*(0)\big) = J^{\text{opp}-\text{sts}}\big(\tilde{x}(0), 0\big)$$

it follows that

$$J^{\text{opp}-\text{sts}}\big(\tilde{x}(0),0\big) \le J^{\text{sts}}\big(\tilde{x}(0),0,+\infty,p_0\big).$$

Thus, the property is then valid at stage 0.

Assume now that (6.11) is valid at stage $l-1$ with $l>0$. We have to prove that (6.11) is also valid at stage $l$.

At stage $l$, according to the *OPP* strategy, the pointer position $p^*(l)$ will be chosen such that

$$p^*(l) = \operatorname*{argmin}_p J^{\text{sts}}\big(\tilde{x}^{\text{opp}}(l),l,+\infty,p\big). \tag{6.13}$$

Consequently

$$J^{\text{sts}}\big(\tilde{x}^{\text{opp}}(l),l,+\infty,p^*(l)\big) \le J^{\text{sts}}\big(\tilde{x}^{\text{opp}}(l),l,+\infty,I_{p^*(l-1)}(1)\big), \tag{6.14}$$

where $p^*(l-1)$ is the optimal pointer position at instant $l-1$. Adding $J^{\text{opp}}(\tilde{x}(0), 0, l-1)$ to both left and right terms of the previous inequality, we get

$$J^{\text{opp}-\text{sts}}\big(\tilde{x}(0),l\big) \le J^{\text{opp}-\text{sts}}\big(\tilde{x}(0),l-1\big). \tag{6.15}$$

Using the recurrence assumption

$$J^{\text{opp}-\text{sts}}\big(\tilde{x}(0),l-1\big) \le J^{\text{sts}}\big(\tilde{x}(0),0,+\infty,p_0\big), \tag{6.16}$$

and the inequality (6.15), the theorem is proved. □

## 6.3 Real-Time Implementation Aspects of the *OPP* over Infinite Horizon Algorithm

The *OPP* over infinite horizon algorithm requires the on-line evaluation of $T$ quadratic functions of the extended state vector $\tilde{x}(k)$. The computational complexity is linear with respect to the period of the sequence, quadratic with respect to the extended state (like a classical state-feedback control law) and independent of the size of the horizon. In the case of the suspension example, the additional computational requirements of *OPP* are approximately 4.3 times those required by the state feedback operation $v(k) = \tilde{K}(p)\tilde{x}(k)$. Using *OPP* follows the idea of trading additional computations for a more efficient use of network resources [259]. Further reducing the computational requirements is a both a difficult and interesting research issue. The application of multi-parametric programming techniques to the optimal control of hybrid systems have been recently considered [38]. In [134], an approximate solution to the model predictive control of linear systems with input and state constraints was proposed. This method is based on the partitioning the state space onto hypercubes which may be further partitioned in order to meet on the cost function approximation error bounds and constraints violations bounds. By imposing an orthogonal search tree on the partition, the on-line computational requirements are significantly reduced with respect to the true optimal explicit *MPC* law. The search method is logarithmic with respect to the number of regions, but this number may augment exponentially with respect to the state vector size. The memory requirements, needed to the storage of this partitioning information and of the state-feedback control law parameters, are also tightly dependent on this number of regions. This problem is tractable for low dimensional systems but it may become larger more complicated for problems similar to the ones treated in this chapter. The difference between this approach and that of [134] resides in the fact that we explore the *set of pointer positions*, which is in general, and particularly in this application, "less" complex than the *state space*. This considerably reduces the computational complexity.

## 6.4 Optimal Pointer Placement Scheduling: Application to a Car Suspension System

In this section, the *OPP* scheduling algorithm is applied to a distributed active suspension controller. The considered controller is based on a full-vehicle model and is implemented on a central processor. The controller sends the control commands to four hydraulic actuators located on the vehicle's corners through a bus subject to bandwidth limitations. In the sequel, the considered active suspension model is described, the control design methodology is illustrated and finally the *OPP* scheduling strategy is evaluated and compared to a fair static scheduling strategy.

### *6.4.1 The Suspension Control System*

The simulated model (Fig. 6.11) was adopted from [60]. It consists in a seven degree-of-freedom system. In this model, the car body, or sprung mass, is free to heave, roll and pitch. In order to obtain a linear model, roll and pitch angles are assumed to be small. The suspension system connects the sprung mass to the four unsprung masses (front–left, front–right, rear–left and rear–right wheels), which are free to bounce vertically with respect to the sprung mass. The suspension system consists in a spring, a shock absorber and a hydraulic actuator at each corner. The shock absorbers are modeled as linear viscous dampers, and the tires are modeled as linear springs in parallel to linear dampers.

In order to describe this system, fifteen variables need to be considered:

$x_{c_1}$: heave velocity of the center of gravity of the sprung mass
$x_{c_2}$: pitch angular velocity of the sprung mass
$x_{c_3}$: roll angular velocity of the sprung mass
$x_{c_4}$: front–left suspension deflection
$x_{c_5}$: rear–left suspension deflection
$x_{c_6}$: rear–right suspension deflection
$x_{c_7}$: front–right suspension deflection

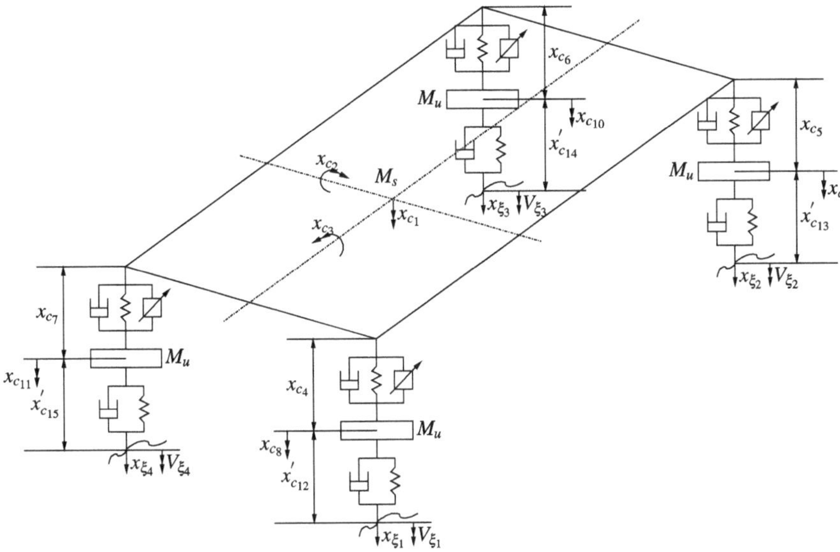


**Fig. 6.11** Full vehicle model

$x_{c_8}$: front–left unsprung mass velocity
$x_{c_9}$: rear–left unsprung mass velocity
$x_{c_{10}}$: rear–right unsprung mass velocity
$x_{c_{11}}$: front–right unsprung mass velocity
$x'_{c_{12}}$: front–left tire deflection
$x'_{c_{13}}$: rear–left tire deflection
$x'_{c_{14}}$: rear–right tire deflection
$x'_{c_{15}}$: front–right tire deflection

Road disturbances acting on the four vehicle wheels consist of height displacement inputs ($x_{\xi_1}, x_{\xi_2}, x_{\xi_3}, x_{\xi_4}$) and height velocity inputs ($V_{\xi_1}, V_{\xi_2}, V_{\xi_3}, V_{\xi_4}$) defined with respect to an inertial reference frame.

As mentioned earlier, the suspension model has seven degrees of freedom. Consequently, only fourteen state variables are needed to describe it. The extra variable may be eliminated if the wheel deflections are expressed as a function of three state variables $x_{c_{12}}$, $x_{c_{13}}$ and $x_{c_{14}}$ and of the road disturbances $x_{\xi_1}$, $x_{\xi_4}$, $x_{\xi_3}$, $x_{\xi_4}$ as illustrated in [60].

Applying a force-balance analysis to the model in Fig. 6.11, the state equation may be derived from the equations of motion and is given by

$$\dot{x}_c(t) = A_c x_c(t) + B_c u_c(t) + W_c w_c(t) \tag{6.17}$$

where

$x_c(t)$ is the state vector (14 variables),

$u_c(t)$ is the control vector: $u_c(t) = [u_{c_1}(t)\ u_{c_2}(t)\ u_{c_3}(t)\ u_{c_4}(t)]^T$, $u_{c_1}$, $u_{c_2}$, $u_{c_3}$ and $u_{c_4}$ represent the control forces applied respectively by the front-left, rear-left, rear-right and front-right hydraulic actuators,

$w_c(t)$ is the vector of road disturbances (displacements and velocities): $w_c(t) = [x_{\xi_1}(t)\ x_{\xi_2}(t)\ x_{\xi_3}(t)\ x_{\xi_4}(t)\ V_{\xi_1}(t)\ V_{\xi_2}(t)\ V_{\xi_3}(t)\ V_{\xi_4}(t)]^T$.

### *6.4.2 Active Suspension Control Law*

The control design for a vehicle's active suspension aims to maximize the driving comfort (as measured by sprung mass accelerations) and the safety (as measured by tire load variations) under packaging constraints (as measured by suspension deflections). However, comfort and safety are two conflicting criteria [201]. We adopt the control design methodology of [60], approach that divides the control design problem for a vehicle's active suspension into two sub-problems:

- The design of the ride controller, whose role is to improve ride comfort by isolating the sprung mass from road disturbances. The ride controller has also to maintain a sufficient contact force between the tires and the road to insure convenient road holding.
- The design of the attitude controller, responsible of maintaining load-leveling, performing convenient load distribution and controlling roll and pitch angles during vehicle maneuvers (roll during cornering and pitch during breaking and acceleration).

In this section, the ride controller part of the suspension controller is considered. The used ride controller is a linear quadratic regulator that aims at minimizing the following cost function:

$$J_c = \int_0^{+\infty} \left\{ \sum_{i=0}^{12} \mu_i y_i^2(t) + \sum_{j=0}^{4} a_j u_{c_j}^2(t) \right\} dt, \tag{6.18}$$

where $\mu_i$ and $a_j$ are weighting factors and

$y_1$: vertical acceleration of the sprung mass
$y_2$: pitch angular acceleration of the sprung mass
$y_3$: roll angular acceleration of the sprung mass
$y_4$: sum of the suspension deflections at the four corners
$y_5$: difference between the suspension deflections at the right- and left-hand sides
$y_6$: difference between the suspension deflections at the front and rear of the vehicle
$y_7$: difference between the suspension deflections at the diagonally opposite corners
$y_8$: sum of the velocities of the unsprung masses
$y_9$: difference between the velocities of the unsprung masses on the left- and right-hand sides

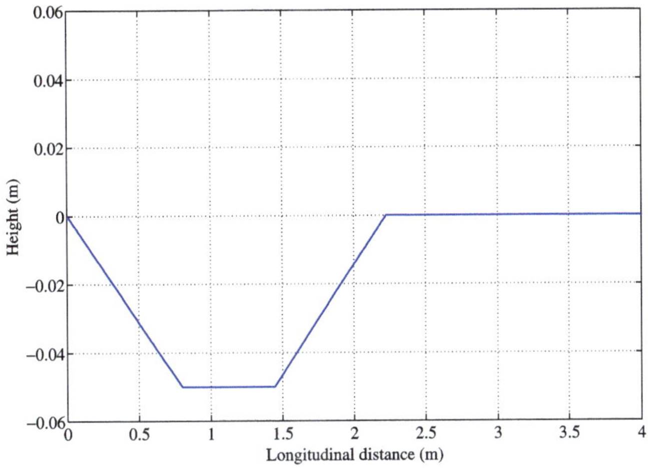


**Fig. 6.12** "Chuck hole" road disturbance

$y_{10}$: difference between the velocities of the unsprung masses at the front and rear of the vehicle

$y_{11}$: difference between the velocities of the unsprung masses at the diagonally opposite corners

$y_{12}$: wrap torque acting on the sprung mass

### 6.4.3 Simulation Setup and Results

Assume that the communication network connecting the controller to the actuators is subject to communication constraints: only a control command can be sent to an actuator every 10 ms. A simple approach to handle this problem is to send the control commands alternately according to the periodic communication sequence

$$\gamma^3 = \left( \begin{bmatrix} 1 \\ 0 \\ 0 \\ 0 \end{bmatrix}, \begin{bmatrix} 0 \\ 1 \\ 0 \\ 0 \end{bmatrix}, \begin{bmatrix} 0 \\ 0 \\ 1 \\ 0 \end{bmatrix}, \begin{bmatrix} 0 \\ 0 \\ 0 \\ 1 \end{bmatrix} \right).$$

By using this communication sequence, the discretized model of the controlled car suspension system becomes periodic. Applying the methodology described in Sect. 4.3 of Chap. 4, the optimal periodic control gains $(\tilde{K}(0), \tilde{K}(1), \tilde{K}(2), \tilde{K}(3))$ may be derived.

The suspension system is evaluated by subjecting the left-hand side of the vehicle to a "chuck hole" discrete road disturbance [60] (Fig. 6.12). The vehicle speed is equal to 40 km/h. First, the performance of the designed active sus-

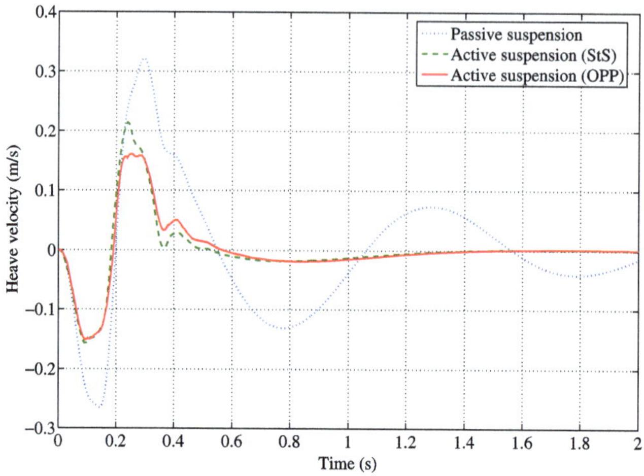


**Fig. 6.13** Heave velocity of the passive and active suspension (controlled with the static scheduling and the *OPP* algorithms)

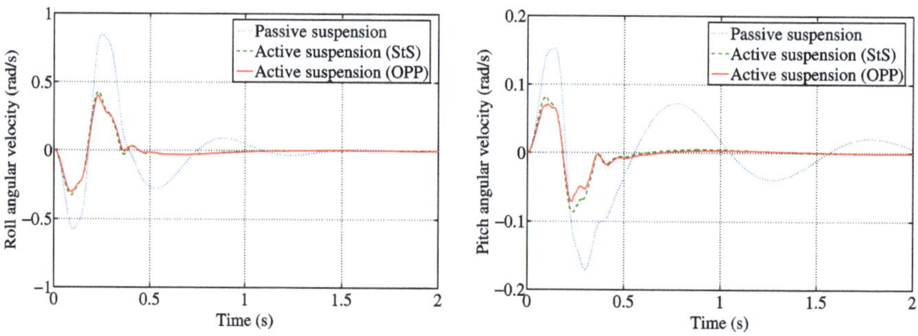


**Fig. 6.14** Roll and pitch velocities of the passive and active suspension (controlled with the static scheduling and the *OPP* algorithms)

pension controller is evaluated and compared to the passive suspension. Then, the performance of the *OPP* algorithm is compared to that obtained when the static scheduling (StS) algorithm is applied. *OPP* and the static scheduling algorithm are both based on the communication sequence and control gains described above. Heave, roll and pitch velocity responses are illustrated in Figs. 6.13 and 6.14.

From these simulation results, it may be seen that the active suspension induces an important improvement of the ride performance compared to the passive suspension (smaller and better damped velocities and thus accelerations). Furthermore, the responses when using the *OPP* algorithm show a slight improvement with respect to the static scheduling (StS) algorithm. The improvements in ride comfort shown by the active suspension are obtained with suspension and tire deflection levels that are close to those obtained with the pas-

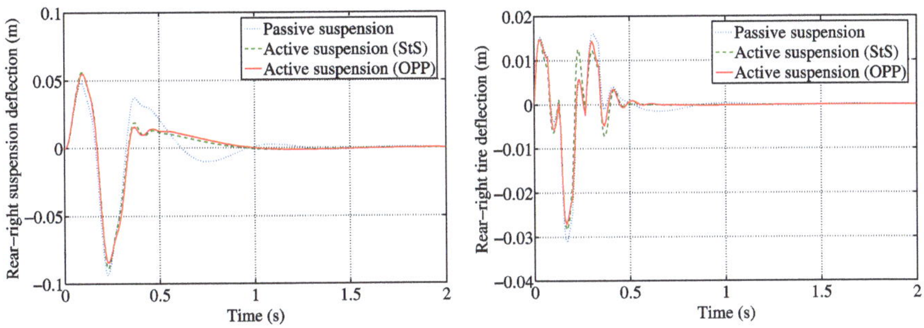


**Fig. 6.15** Suspension and tire deflections of the passive and active suspension (controlled with the static scheduling and the *OPP* algorithms)

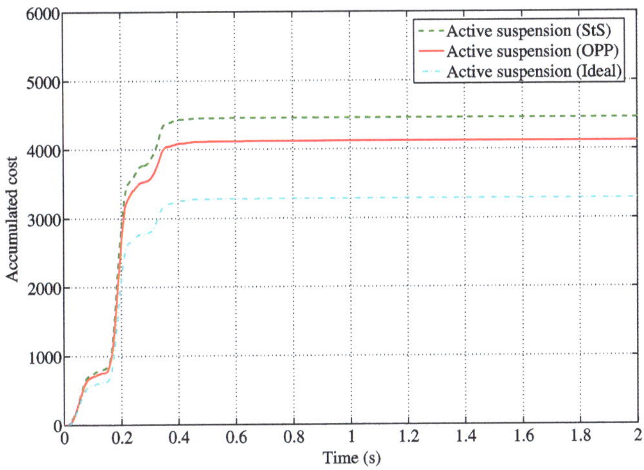


**Fig. 6.16** Quadratic cost functions corresponding to the active suspension (controlled with the static scheduling, with the *OPP* algorithm and using an ideal implementation)

sive suspension (the rear–right suspension and tire deflections are depicted in Fig. 6.15).

Finally, the quadratic cost functions corresponding to the ideal continuous time *LQR* controller, the static scheduling algorithm and the *OPP* scheduling algorithm are compared in Fig. 6.16. The steady state cost function values corresponding to the static scheduling, *OPP* and to the ideal implementation are respectively equal to 4459, 4122 and 3290. Consequently, the improvements in terms of quality of control achieved by the *OPP* algorithm are equal to 28.8 %. These improvements are significant, but not as "spectacular" as those observed in the previous example since the different components of the suspension system are tightly coupled: a disturbance affecting a single wheel influences all the state variables of the system.

### 6.4.4 Embedded Computing Implementation Aspects of the Distributed Suspension Model

The study of the practical implementation of the considered distributed active suspension model, based on state of the art methods, was undertaken in [25, 30, 138]. The Controller Area Network bus was deployed to ensure the information exchange between the distributed components. In [25, 30], the tool TRUETIME [5, 56] was used to simulate these implementations. The impact of messages priorities on control performance was studied in [30]. It was shown that the assignment of some messages priorities, which may be chosen arbitrarily from a real-time scheduling point of view, may have a considerable impact on the robustness of system. In [138], the control performance resulting from different implementation choices of the suspension system were studied. It was shown that synchronous implementations achieve the best control performance. In [25], the impact of the traffic that is generated by the other unrelated network nodes on the control performance of the suspension system was studied, as a function of the available bandwidth resources.

## 6.5 Optimal Pointer Placement Scheduling: Application to a Quadrotor

Consider now the problem of the attitude control of a quadrotor. A quadrotor is a rotor-craft that is lifted and propelled by four rotors. Appendix B provides a detailed description of the used quadrotor model, adopted from [17], as well as a review of the mathematical concepts used to derive it. The attitude (i.e., orientation) of the quadrotor is completely described using a unitary quaternion $q = [q_0 \ \mathbf{q}^T]^T$, with $\|q\|_2 = 1$, where $q_0$ and $\mathbf{q} = [q_1 \ q_2 \ q_3]^T$ are respectively, the scalar and vector parts of the quaternion. The unit quaternion represents the rotation from an inertial frame to the body frame attached to the quadrotor. The dynamic evolution of the attitude quaternion is given by:

$$\begin{pmatrix} \dot{q}_0 \\ \dot{\mathbf{q}} \end{pmatrix} = \frac{1}{2} \begin{pmatrix} -\mathbf{q}^T \\ I_3 q_0 + [\mathbf{q}^\times] \end{pmatrix} \omega, \tag{6.19}$$

where $\omega \in \mathbb{R}^3$ is the angular velocity of the quadrotor in the body frame and $[\mathbf{q}^\times]$ is the skew symmetric tensor associated to $\mathbf{q}$

$$[\mathbf{q}^\times] = \begin{pmatrix} 0 & -q_3 & q_2 \\ q_3 & 0 & -q_1 \\ -q_2 & q_1 & 0 \end{pmatrix}.$$

The rotational motion of the quadrotor (under the assumption of neglected gyroscopic torques) is given by:

$$I_f \dot{\omega} = -[\omega^\times] I_f \omega + \tau_a + \tau_{\text{dist}}, \tag{6.20}$$

where $I_f \in \mathbb{R}^{3\times 3}$ is the inertia matrix of the quadrotor with respect to the body frame. $I_f$ is constant. Here, $\tau_a \in \mathbb{R}^3$ represents the torques resulting from the differences of the relative speeds of the four rotors, and can be written as:

$$\tau_a = \begin{bmatrix} \tau_{\text{roll}} \\ \tau_{\text{pitch}} \\ \tau_{\text{yaw}} \end{bmatrix},$$

and $\tau_{\text{dist}} \in \mathbb{R}^3$ describes the aerodynamic disturbances acting on the quadrotor.

The objective of the attitude controller is to drive the quadrotor attitude to a desired value, which is specified by a unitary quaternion. In practice, this attitude may also be specified by Euler angles $(\phi, \theta, \psi)$, which are more intuitive than quaternions.

To design the attitude controller, consider the linearized model of Eqs. (6.19) and (6.20), and where Euler angles are used instead of quaternions for the description of the attitude. This linearized model is described by

$$\dot{x}_c(t) = A_c x_c(t) + B_c u_c(t) + B_c \tau_{\text{dist}}(t), \tag{6.21}$$

with

$$A_c = \begin{bmatrix} 0_3 & I_3 \\ 0_3 & 0_3 \end{bmatrix}, \quad B_c = \begin{bmatrix} 0_3 \\ I_f^{-1} \end{bmatrix}, \quad x_c = \begin{bmatrix} e \\ w \end{bmatrix},$$

$$e = \begin{bmatrix} \phi \\ \theta \\ \psi \end{bmatrix} \quad \text{and} \quad u_c = \tau_a.$$

The attitude controller acts on $\tau_a$, based on $e$ and $\omega$. It is assumed that $e$ and $\omega$ are measured by an Inertial Measurement Unit (IMU).

The *OPP* algorithm is evaluated by applying it to this non-linear model. Simulation results are depicted in Fig. 6.17.

In this particular case study, *OPP* impact on disturbance rejection abilities is studied. Figure 6.17 illustrates this point, and compares the controlled attitude of quadrotor, described by Euler angles $(\phi, \theta, \psi)$, in two different situations:

- using the static scheduling algorithm,
- using the *OPP* algorithm.

The aerodynamic disturbances $\tau_{\text{dist}}(t)$ are modeled by band-limited white noises, with noise power $10^{-3}$ and period $10^{-1}$, and which are produced by using different seeds. Figure 6.17 shows that significant improvements in control performance result from *OPP*, with respect to the static schedule. The band-limited white noise disturbances are better rejected. Finally, Fig. 6.18 depicts the cumulative cost functions that are associated to the previous simulations. It is worth mentioning that, although the design was performed on the linearized model, these significant improvements are observed on the nonlinear system. This suggests some degree of robustness of the proposed *OPP* design methodology.

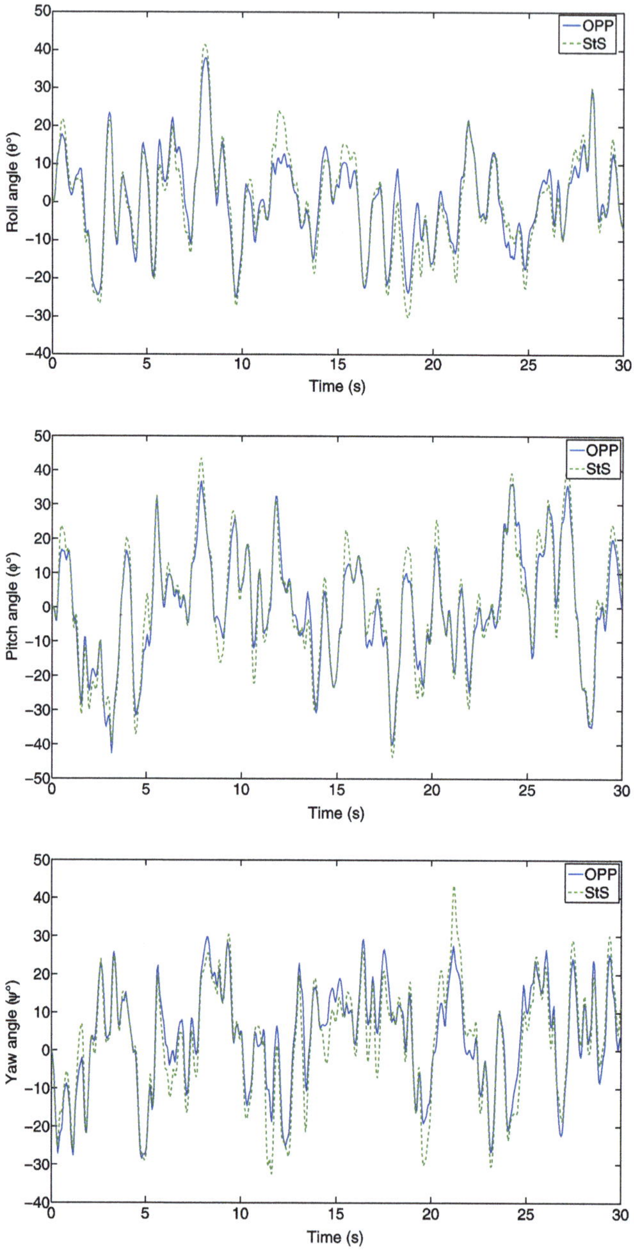


**Fig. 6.17** Quadrotor roll $\phi$, pitch $\theta$ and yaw $\psi$ with *OPP* and StS

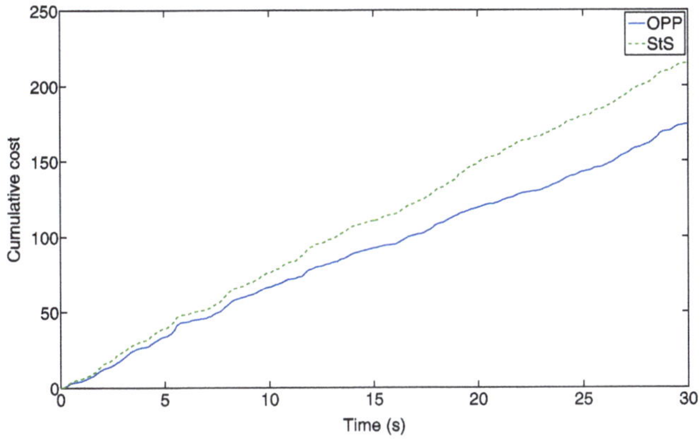


**Fig. 6.18** Cumulative costs with *OPP* and StS

## 6.6 Notes and Comments

In this chapter, we have presented an algorithm, based on the model predictive control approach, allowing to assign on-line the optimal values of the control inputs and the scheduling inputs of resource-constrained systems. Focusing on its practical implementation aspects, such as its computational requirements, we have proposed a more efficient heuristic, called *OPP*. By using a pre-computed off-line schedule, *OPP* is able to assign on-line the values of the control inputs and the scheduling inputs based on the plant state information. The computational requirements of this heuristic are considerably reduced in comparison to those of the model predictive control approach. We have also proved that, if some mild conditions are satisfied, then *OPP* guaranties the stability of the system and performance improvements compared to its basic off-line schedule. We have shown that the design of the control gains based on periodic optimal control theory and used by *OPP* leads to a significant simplification of its implementations, which boils down to the comparison of $T$ quadratic cost functions, $T$ being the period of the basic off-line schedule. The *OPP* scheduling algorithm was finally applied to two distributed control system: the active suspension controller of car and the quadrotor attitude control.

These results show the practical importance of the study of the medium access control as well as scheduling algorithms. The problems of the concurrent access to shared communication resources were studied during the last period within various theoretical frameworks and with various modeling assumptions. We present thereafter a brief summary of the approaches taking into account explicitly the concurrent on-line access to the communication network. These contributions were classified into two categories, according to the class of scheduling algorithms to be considered: the on-line scheduling of the sensors-to-controller link and the on-line scheduling of the controller-to-actuators link.

- *On-line scheduling of the sensors-to-controller link*: The scheduling of sensor measures was studied in *Walsh and Ye* [244]. The addressed configuration consists in a continuous-time plant where the controller is directly connected to the actuators. The network only connects the sensors to the controller. The notion of *MATI* (maximum allowable transfer interval) was introduced, and represents the upper bound on the time between two consecutive sensor messages transmissions that guarantees the stability of the plant. The *MATI* is defined for a given scheduling algorithm. A new on-line scheduling algorithm, called *MEF-TOD* (maximum error first—try once discard), was introduced. In this dynamic priority on-line scheduling algorithm, the priority of a sensor message depends on the error of the measure that it carries; smaller the error is, lower is the assigned priority. The error is defined as the weighted absolute value of the difference between the value of the current measure and the value of the last transmitted measure. If a node fails to send a message, then this message is discarded (dropped from the queue). The authors stated sufficient stability conditions, involving the value of the *MATI*, which ensure the stability of the system, when the *MEF-TOD* algorithm and a round robin like static scheduling algorithm are used. These results are based on the perturbation theory and are very conservative. This approach was generalized to nonlinear systems in *Walsh et al.* [245]. The practical implementation of the *MEF-TOD* algorithm was considered in [243]. This implementation, which was performed on *CAN* networks, is mainly based on the nondestructive bitwise arbitration of *CAN* technology to (dynamically) encode the dynamic priorities. The effects of the quantization of priorities were experimentally studied. Sufficient input/output $\mathscr{L}_p$-stability results for a class of network scheduling protocols, including *MEF-TOD* and static scheduling algorithms were stated and illustrated in *Nesic and Teel* [185]. These results considerably reduces the conservativeness of the results that were initially stated in *Walsh et al.* [244]. The application of the *Rate Monotonic* scheduling algorithm to the networked control systems was investigated in *Branicky et al.* [41].

- *On-line scheduling of the controller-to-actuators link*: On-line scheduling of control commands to the actuators was studied in *Palopoli et al.* [191]. In the proposed model, it is assumed that at every slot, only one command vector can be sent to an actuator group, the other control vectors are set to zero. The stabilization is achieved by using a model predictive controller, which calculates on-line the appropriate control law and the allocation of the shared bus. The cost function used by the *MPC* calculates a weighted sum of the infinity norms of the states and the control commands over a specified horizon. The optimization problem solved at each step by the *MPC* algorithm was proven to be equivalent to the generalized linear complementarity problem (*GLCP*) [258]. The same architecture is considered in *Goodwin et al.* [99]. The considered model assumes that it is possible to send only one message during one sampling period. The actuators, which do not receive their control inputs, maintain constant the last received ones. The control commands are quantized with a fixed precision. The expression of the optimal model predictive controller, in the sense of a quadratic cost function, was established. The optimal solution as well as computationally efficient approach (*OPP*) to the problem of joint control and network scheduling was proposed in *Ben Gaid et al.* [27].

The optimal control and scheduling of networked control systems (*NCS*) where controllers and actuators are connected via a shared communication medium is proposed in *Görges et al.* [101]. The *NCS* is modeled as a discrete-time switched linear system and a receding-horizon control and scheduling (*RHCS*) problem with a quadratic performance criterion is formulated and solved by (relaxed) dynamic programming. A solution expressed as piecewise linear feedback law is proposed and an a posteriori stability criterion based on piecewise quadratic Lyapunov functions is given. A further refinement of the *NCS* model and robustness issues are considered in *Al-Areqi et al.* [2]. A periodic control and on-line scheduling strategy with uncertain but bounded time-varying computation and transmission induced delays is presented. The *NCS* is modeled as a discrete-time switched linear system and a quadratic cost function with infinite time horizon is considered as a performance criterion. The solution of the control and scheduling problem is based on the periodicity assumption and an infinite horizon as given in *Ben Gaid et al.* [27]. The online scheduling subproblem is solved. Its solution is based on some exhaustive search. Furthermore, a method for reducing this inherent complexity is also given. The continuous *Linear Quadratic Regulator* (*LQR*)-based design problem of a sampled-data periodic controller taking into account the limited communication medium and inter-sampling behavior in given in *Longo et al.* [159]. Two stochastic algorithms are proposed to find some optimal or sub-optimal communication sequence. The associated optimal controller is obtained from a discrete algebraic Riccati equation for the given optimal sequence.

# Chapter 7
# Optimal Relation Between Quantization Precision and Sampling Rates

In this chapter, we refine the model of resource-constrained systems to take into account quantization aspects and mainly focus on performance considerations in presence of information limitations. The communication constrains are modeled at the bit level, in bits per second. By using this modeling, we have to determine the control inputs that have to be updated as well as their quantization precision. In order to limit the inherent complexity of the proposed protocol, we suppose that the quantization precision choices of the input control signals belong to a reduced finite set. At each sampling period, quantization possibilities may be chosen from this set. In the case of allocation of communication resources based on the "per symbol" paradigm, the information exchange is modeled at the symbol level. The quantization of measurements and control commands is thus implicitly taken into account. The general model is given in Fig. 7.1.

In this model, the communication channel can transmit at most $R$ bits per time unit. Because of these resource limitations, measurements and control commands must be encoded (as a flow of symbols) before their transmission and decoded at their reception. Various coding techniques may be employed. A fundamental question is to determine the necessary and/or sufficient data-rate allowing the existence of a coder, a decoder and a controller that achieve the stabilization of the system.

In general, increasing the sampling frequency improves the disturbance rejection abilities whereas increasing the quantization precision improves the steady state precision. However, when the bandwidth is limited, increasing the sampling frequency necessitates the reduction of the quantization precision. In the opposite, augmenting the quantization precision requires the lowering of the sampling frequency.

Motivated by these observations, an approach for the dynamical on-line assignment of sampling frequencies and control inputs quantization is proposed. This approach, which is based on the model predictive control (*MPC*) philosophy, enables to choose the sampling frequency and the quantization levels of control signals from a predefined set, in order to optimize the control performance. Naturally, handling dynamically the quantization precision requires some communication resources and some extra computational resources. Consequently, we have to jointly handle the computational complexity, the protocol bandwidth consumption and performance

A. Çela et al., *Optimal Design of Distributed Control and Embedded Systems*, Communications and Control Engineering, DOI 10.1007/978-3-319-02729-6_7,

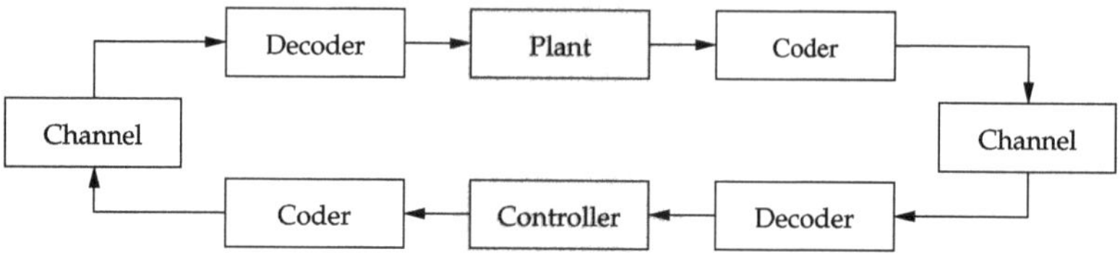


**Fig. 7.1** General model of the information flow in a control system whose control loop is closed through finite bandwidth communication channels

improvements. In order to limit the inherent complexity of the proposed protocol, we suppose that the quantization choices of the input control signals belong to a reduced finite set, which may be chosen by the designer in order to ensure the stability and to comply with the computational requirements. At each sampling period, quantization possibilities may be chosen from this set. This contrasts with the approach of [99], where the quantization precision of control signals is fixed. The proposed approach aims to capture the intuitive notion that high sampling rates improve the disturbance rejection and the transient behavior whereas the fine quantization improves the static precision near the origin [80]. The proposed method allows to dynamically choosing the pertinent control information to send, knowing the plant state and subject to the communication constraints.

## 7.1 Modeling and Computation Issues

Consider the discrete-time LTI system described by the state equation

$$x(k+1) = Ax(k) + Bu(k), \tag{7.1}$$

where $x(k) \in \mathbb{R}^n$ and $u(k) \in \mathbb{R}^m$. We assume that the pair $(A, B)$ is reachable and that the full state vector $x(k)$ is available to the controller at each sampling period.

The controller is connected to the actuators of the plant through a limitedbandwidth communication channel. At each sampling period, at most $R$ bits can be sent to the actuators through the communication channel. The considered communication scheme is described in Fig. 7.2. The control inputs, which are computed by the controller, need to be properly encoded before their transmission over the network. This function is performed by the encoder, which converts these control inputs into a sequence of binary symbols. The transmitted information is then decoded by the decoder and applied to the inputs of the plant. We assume that the inputs of the plant are subject to saturation constraints at the actuators, which are defined by:

$$-U_i \leq u_i(k) \leq U_i, \quad \text{where } U_i > 0 \text{ and } i = 1, \ldots, m. \tag{7.2}$$

Here, $\|.\|$ denotes a given matrix norm, $0_{n,m}$ represents the $n \times m$ matrix whose elements are equal to zero and $I_n$ the $n \times n$ identity matrix.

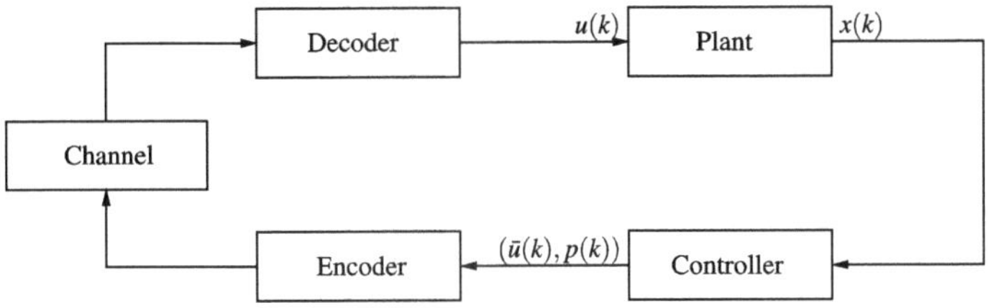


**Fig. 7.2** Information pattern

*Remark 7.1* (Down-link resources limitations) The literature addressing the problem of control under communication constraints mainly consider the down-link resources limitations, rather than the up-link. For that reason, our contribution may be seen as a complementary approach. For its practical implementation on systems with both down-link and up-link limitations, the previously proposed methods in the literature may be used in order to obtain a state estimation at the controller, which could be subsequently used in our approach.

### 7.1.1 Quantization Aspects

Quantization is the process of approximating a continuous range of values into a finite set of discrete values, called *reconstruction levels* (or *quantization levels*). The quantization is performed using *mid-tread uniform quantizers*, which are characterized by an odd number of reconstruction levels (which include the value of zero). Let $u$ be a bounded continuous scalar signal verifying

$$|u| \leq U, \quad U \in \mathbb{R}_+^*. \tag{7.3}$$

The set of reconstruction levels of the mid-tread uniform quantizer which may be encoded using $M$ bits ($M > 1$), given the lower and upper bounds $-U$ and $+U$, is defined by the set-valued function:

$$\mathbb{U}(M, U) = \{-U + lL(M, U),\ l = 0, \ldots, 2^M - 2\}, \tag{7.4}$$

where $l$ is the *quantization index* and $L(M, U)$ defined by

$$L(M, U) = \frac{U}{2^{M-1} - 1} \tag{7.5}$$

is the *quantization step size*.

*Remark 7.2* (Odd number of quantization levels) In the definition of the set valued function $\mathbb{U}(M, U)$, we have intentionally chosen to obtain an odd number of reconstruction levels $0, \ldots, 2^M - 2$ instead of having an even number of reconstruction levels $0, \ldots, 2^M - 1$. This choice was motivated by the need of integrating the zero reconstruction level, which has a particular signification, from the control point of view.

Given $M \in \mathbb{N}^*$ such that $M > 1$ and $U \in \mathbb{R}_+^*$, the quantizer $\mathcal{Q}_{(M,U)}$ is defined by

$$\begin{aligned}\mathcal{Q}_{(M,U)} : \mathbb{R} &\longrightarrow \mathbb{U}(M, U),\\ u &\longmapsto c,\end{aligned}$$

such that

$$c = \mathcal{Q}_{(M,U)}(u) = \begin{cases} +U & \text{if } u > U, \\ -U & \text{if } u \leq -U, \\ -U + \lfloor \frac{u+U}{L(M,U)} + \frac{1}{2} \rfloor L(M, U) & \text{if } -U < u \leq U. \end{cases}$$

The quantizer $\mathcal{Q}_{(M,U)}$ uniquely associates to a real scalar $u \in \mathbb{R}$ a nearest neighbor $c \in \mathbb{U}(M, U)$.

*Remark 7.3* (Quantization when $M = 1$) When $M = 1$, the expression (7.5) is not defined. For the sake of simplicity in the following discussions, we will assume that when $M = 1$, for all $u \in \mathbb{R}$, $\mathcal{Q}_{(1,U)}(u) = 0$.

### *7.1.2 Information Pattern*

The number of reconstruction levels of the components of a given control input may vary over the time according to an optimization policy (which will be developed in Sect. 7.3). Consequently, it is necessary for the decoder to identify how to reconstruct the different components of the control input of the plant from the received binary symbols. To this end, the *precision vector* $p(k) \in \mathbb{N}^m$ is introduced. The $i$th component $p_i(k)$ of the precision vector $p(k)$ describes the number of bits that are required to encode all the reconstruction levels of the quantized control signal $u_i(k)$ (using the quantizer $\mathcal{Q}_{(p_i(k),U_i)}$). A compact representation of the precision vector has to be sent to the decoder, together with the control information. To minimize the necessary decoding information (which is described by $p(k)$), and to be able to produce compact representations of $p(k)$, the set $\mathcal{P}$ of possible values of $p(k)$ should contain a limited number of elements.

Let $R_I$ and $R_D$ be the number of bits which are used to encode respectively the control information and the decoding information. The communication constraints impose:

$$R_I + R_D = R. \tag{7.6}$$

Then $\mathscr{P}$ must verify

$$\mathscr{P} \subseteq \left\{ p \in \mathbb{N}^m \text{ such that } \sum_{i=1}^{m} p_i = R_I \right\}, \tag{7.7}$$

and

$$|\mathscr{P}| \leq 2^{R_D}, \tag{7.8}$$

where $|\mathscr{P}|$ denotes the cardinality of $\mathscr{P}$.

### *7.1.3 Notion of Quantization Sequence*

The notion of communication sequence was introduced by Brockett [43] and generalized by Hristu [122] in order to quantify the notion of *attention* [44]. A communication sequence describes, at each sampling period, which control inputs of the system are updated, assuming an infinite quantization precision, which represents an idealized situation.

**Definition 7.1** (Periodic communication sequence [122]) A periodic communication $\delta^{T-1}$ sequence of period $T$ and width $m$ is an infinite sequence $(\delta(0), \ldots, \delta(T-1), \ldots)$ of elements of $\{0, 1\}^m$ verifying $\forall i \in \mathbb{N}$, $\delta(k + iT) = \delta(k)$. A periodic communication sequence is fully characterized by the sequence $\delta^{T-1} = (\delta(0), \ldots, \delta(T-1))$ corresponding to the first period.

The notion of communication sequence is well suited to model the problems of medium access arbitration, when quantization aspects are disregarded. Since quantization is a main concern, we propose a generalization of the notion of communication sequence, to take into account quantization aspects. This leads to the notion of quantization sequence.

**Definition 7.2** (Periodic quantization sequence) A periodic quantization sequence $s^{T-1}$ of period $T$, width $m$ and maximal precision $M$ is an infinite sequence $(s(0), \ldots, s(T-1), \ldots)$ of elements of $\{0, \ldots, M\}^m$ verifying $\forall i \in \mathbb{N}$, $s(k+iT) = s(k)$. A periodic quantization sequence is fully characterized by the sequence $s^{T-1} = (s(0), \ldots, s(T-1))$ corresponding to the first period.

*Example 7.1* The sequence

$$s^2 = \left( \begin{bmatrix} 2 \\ 1 \\ 4 \end{bmatrix}, \begin{bmatrix} 1 \\ 4 \\ 3 \end{bmatrix}, \begin{bmatrix} 0 \\ 4 \\ 2 \end{bmatrix}, \begin{bmatrix} 2 \\ 1 \\ 4 \end{bmatrix}, \begin{bmatrix} 1 \\ 4 \\ 3 \end{bmatrix}, \begin{bmatrix} 0 \\ 4 \\ 2 \end{bmatrix}, \ldots \right)$$

is a periodic quantization sequence of period $T = 3$, width $m = 3$ and maximal precision $M = 4$.

The precision vectors are restricted to belong to the set $\mathscr{P}$. For that reason, we will only consider quantization sequences whose components are precision vectors in $\mathscr{P}$. These particular periodic quantization sequences are called *admissible periodic quantization sequences*. Since $\mathscr{P}$ contains only precision vectors that respect the communication constraints, an admissible periodic quantization sequence implicitly respects the required communication constraints.

*Remark 7.4* (Packet-dropouts issues) The proposed approach could be easily extended to take into account packet-dropouts which are known in advance (and which may be easily modeled using the notion of quantization sequence).

### 7.1.4 Performance Index Definition

The performance of the controlled system (7.1) is evaluated using a quadratic cost function, which may be seen as the design specification of its ideal controller

$$J(x, u, 0, N) = \sum_{k=0}^{N-1} \ell\big(x(k), u(k)\big), \tag{7.9}$$

where $N$ is the final time, and

$$\ell\big(x(k), u(k)\big) = \begin{bmatrix} x(k) \\ u(k) \end{bmatrix}^T \begin{bmatrix} Q_1 & Q_{12} \\ Q_{12}^T & Q_2 \end{bmatrix} \begin{bmatrix} x(k) \\ u(k) \end{bmatrix},$$

and $Q_1$, $Q_2$ and $Q_{12}$ are respectively, $n \times n$, $m \times m$ and $n \times m$ matrices. Let

$$Q = \begin{bmatrix} Q_1 & Q_{12} \\ Q_{12}^T & Q_2 \end{bmatrix}.$$

In the sequel, we assume that $Q$ is positive definite.

## 7.2 Static Strategy

### 7.2.1 Algorithm Description

Let $s^{T-1}$ be an admissible $T$-periodic quantization sequence. Let $\Lambda$ be the function defined by

$$\Lambda : \{0, \dots, M\}^m \longrightarrow \{0, 1\}^m,$$
$$s \longmapsto \delta,$$

such that

$$\begin{cases} \delta_i = 1 & \text{if } s_i \neq 0, \\ \delta_i = 0 & \text{if } s_i = 0. \end{cases}$$

Let $L_s^{T-1} = (L_s(0), \ldots, L_s(T-1))$ be the periodic sequence of control gains characterizing the optimal $T$-periodic controller of system (7.1) if the communication constraints are described by the periodic communication sequence $\delta^{T-1} = (\delta(0), \ldots, \delta(T-1)) = (\Lambda(s(0)), \ldots, \Lambda(s(T-1)))$. In this situation, the optimal controller, taking into account the medium access constraints modeled by $\delta^{T-1}$, is the state feedback control law:

$$u(k) = L_s(k)\tilde{x}(k), \tag{7.10}$$

where

$$\tilde{x}(k) = \begin{bmatrix} x(k) \\ u(k-1) \end{bmatrix}.$$

The control gains sequence $L_s^{T-1}$ may be derived by using the approach described in [27].

*Remark 7.5* (Control gains and communication constraints) The control gains $L_s(k)$ are designed to take into account the communication constraints $\delta_i(k) = 0 \Longrightarrow u_i(k) = u_i(k-1)$ (i.e., if $\delta_i(k) = 0$, then the $i$th line of $L_s(k)$ verifies $L_{s_{i,n+i}}(k) = 1$ and $L_{s_{i,j}}(k) = 0$ for $j \neq n+i$).

To each admissible $T$-periodic quantization sequence $s^{T-1}$, a periodic control gains sequence $L_s^{T-1}$, derived as mentioned previously, may be associated. A simple approach for controlling the system (7.1) is to use the following control algorithm:

$$\begin{aligned} p(k) &= s(k), \\ v(k) &= L_s(k)\tilde{x}(k), \\ u_i(k) &= \mathcal{Q}_{(p_i(k),U_i)}\big(v_i(k)\big) \quad \text{if } p_i(k) \neq 0, \\ u_i(k) &= u_i(k-1) \quad \text{if } p_i(k) = 0. \end{aligned} \tag{7.11}$$

This algorithm is called *static strategy*, because it is based on a fixed periodic quantization sequence, in opposition to the adaptive strategy, that will be described in the Sect. 7.3.

### *7.2.2 Practical Stabilization Using the Static Strategy*

In this subsection, we are interested in the practical stability properties of the static strategy. Since the control inputs can only take a finite number of values, achieving the asymptotic stability is practically impossible. In fact, in its seminal paper [72], Delchamps has proved that when the input of the system passes through a quantizer having a finite number of quantization levels, then the set of trajectories that correspond to the asymptotic stability has Lebesgue measure zero. This fundamental result motivates the use of practical stability for the studied problem. In the following, the main definitions that are related to the stability notions that will be considered are reviewed.

**Definition 7.3** ($(W, V)$-stability [82]) Let $V$ and $W$ be two compact sets of $\mathbb{R}^{n+m}$ containing the origin in their interiors and such that $V \subseteq W$. System (7.1), (7.2), (7.11) is called $(W, V)$-*stable* if

- $W$ is positively invariant for system (7.1), (7.2), (7.11),
- for all $\tilde{x}(0) \in W$, there exists $k_0$ (function of $\tilde{x}(0)$) such that any state trajectory of (7.1), (7.2), (7.11) with initial condition $\tilde{x}(0)$ satisfies $\tilde{x}(k) \in V$ for all $k \geq k_0$.

Furthermore, system (7.1), (7.2), (7.11) is called $(W, V)$-*stable in* $\mathscr{N}$-*steps* if $k_0 \leq \mathscr{N}$.

Assume that system (7.1), is uniformly exponentially stable using the control law (7.10). Let $\eta > 0$ and $\mathscr{B}_\eta$ the ball defined by

$$\mathscr{B}_\eta = \left\{\tilde{x} \in \mathbb{R}^{n+m}, \text{ such that } \|\tilde{x}\| \leq \eta\right\}. \tag{7.12}$$

When system (7.1), (7.10) is uniformly exponentially stable, a question that arises is whether it is possible to guarantee the ultimate boundedness of the trajectories of system (7.1), (7.2), (7.11) to any desired final set $\mathscr{B}_\eta$, if its inputs are quantized with a sufficiently high precision. However, due to the possibility of saturation of the control inputs, this ultimate boundedness property may not be verified if the initial extended state is situated in some regions of $\mathbb{R}^{n+m}$, where even the non-quantized system (7.1), (7.2), (7.10) subject to control inputs saturations may not be stabilizable. For that reason, and due to these saturation constraints that are considered in the proposed problem formulation, the latter question is reformulated in terms of $(W, V)$-stability. Consider the following perturbed system

$$x(k+1) = Ax(k) + Bu(k) + Bw(k). \tag{7.13}$$

Assume that the disturbance $w(k)$ belongs to a convex and compact set $\mathscr{W}$. The disturbance $w(k)$ may be seen as a model of the quantization error and its effect on the system.

Several methods have been proposed in the literature for computing robustly positively invariant sets, for example, [36, 140, 197]. Let $\mathscr{F}$ be the set of states that do not lead to the saturation of the control commands using the control law (7.10). The set $\mathscr{F}$ is defined by

$$\mathscr{F} = \left\{\tilde{x} \in \mathbb{R}^{n+m} \text{ such that } \forall k \in \{0, \ldots, T-1\},\ L_s(k)\tilde{x} \in \prod_{i=1}^{m}[-U_i, U_i]\right\}. \tag{7.14}$$

Let $\mathscr{G}_\gamma$ the greatest convex and compact set in $\mathscr{F}$ that is robustly positively invariant for the perturbed system (7.13), (7.2), (7.10) and for bounded disturbances $w(k)$ belonging to $\mathscr{W}_\gamma$ defined by

$$\mathscr{W}_\gamma = \{w(k) \in \mathbb{R}^m \text{ such that for all } k \in \mathbb{N}, \|w(k)\| \leq \gamma\}. \tag{7.15}$$

Under these assumptions, the set $\mathscr{G}_\gamma$ is positively invariant for system (7.1), (7.2), (7.11) if the "quantization error" is less than or equal to $\gamma$. The $(\mathscr{G}_\gamma, \mathscr{B}_\eta)$-stability of system (7.1), (7.2), (7.11) is addressed in the following theorem.

**Theorem 7.1** (($W, V$)-stability of the static strategy [24]) *For all $\varepsilon > 0$, if system* (7.1), (7.10) *is uniformly exponentially stable, and the set $\mathscr{G}_{\frac{\varepsilon}{\varphi}}$ (as previously defined) is robustly positively invariant for the system* (7.13), (7.2), (7.10), *where $\varphi$ is a constant that depends on the plant model* (7.1), *the controller* (7.10) *and $\varepsilon$, then there exists $p_0 \in \mathbb{N}$ such that system* (7.1), (7.2), *controlled by the control law* (7.11) *that is based on the periodic quantization sequence $p_0 \times s^{T-1}$ and the periodic control gains sequence $L_s^{T-1}$ is $(\mathscr{G}_{\frac{\varepsilon}{\varphi}}, \mathscr{B}_\varepsilon)$-stable.*

*Proof* Let $\gamma > 0$. Let $\tilde{x}(0) \in \mathscr{G}_\gamma$ be the initial condition of system (7.1), (7.2), (7.11) and $\tilde{x}(k)$ its extended state at instant $k$. Assume first that $\mathscr{G}_\gamma$ is robustly positively invariant for the system (7.13), (7.2), (7.10). Let $p \in \mathbb{N}$ and $v(k) = L_s(k)\tilde{x}(k)$. The quantization error $e^{(p)}(k)$ is defined for $i \in \{1, \ldots, m\}$ by

$$e_i^{(p)}(k) = \begin{cases} \mathscr{Q}_{(p \times s_i(k), U_i)}(v_i(k)) - v_i(k) & \text{if } s_i(k) \neq 0, \\ 0 & \text{if } s_i(k) = 0. \end{cases}$$

Let

$$\bar{A}(k) = \begin{bmatrix} A & 0_{n,m} \\ 0_{m,n} & 0_{m,m} \end{bmatrix} + \begin{bmatrix} BL_s(k) \\ L_s(k) \end{bmatrix}.$$

At a given discrete instant $k$, the extended state $\tilde{x}(k)$ of system (7.1), (7.11) verifies

$$\tilde{x}(k) = \Phi(k, 0)\tilde{x}(0) + \sum_{i=0}^{k-1} \Gamma(k, i)e^{(p)}(i), \tag{7.16}$$

where

$$\Phi(k, i) = \begin{cases} \prod_{j=1}^{k-i} \bar{A}(k-j) & \text{if } k > i, \\ I_{n+m} & \text{if } k = i, \end{cases} \tag{7.17}$$

and

$$\Gamma(k, i) = \begin{cases} (\prod_{j=1}^{k-i-1} \bar{A}(k-j))B & \text{if } i < k-1, \\ B & \text{if } i = k-1, \\ 0_{n,m} & \text{if } i = k. \end{cases} \tag{7.18}$$

The norm $\|.\|$ being a matrix norm, we may write

$$\left\|\tilde{x}(k)\right\| \leq \left\|\Phi(k,0)\right\| \left\|\tilde{x}(0)\right\| + \sum_{i=0}^{k-1} \left\|\Gamma(k,i)\right\| \left\|e^{(p)}(i)\right\|. \tag{7.19}$$

Since $\mathscr{G}_\gamma \subseteq \mathscr{F}$ and $\mathscr{G}_\gamma$ is positively invariant for system (7.1), (7.2), (7.11), then the control inputs corresponding to any state trajectory starting from $\tilde{x}(0) \in \mathscr{G}_\gamma$ never saturate. Consequently, the quantization error converges to zero as $p$ tends to infinity. Hence, for any desired worst-case quantization error $\gamma$, there exists $p_0 \in \mathbb{N}$ such that

$$\forall p \geq p_0,\ \forall k \in \mathbb{N}, \quad \left\|e^{(p)}(k)\right\| \leq \gamma.$$

For $p \geq p_0$,

$$\left\|\tilde{x}(k)\right\| \leq \left\|\Phi(k,0)\right\| \left\|\tilde{x}(0)\right\| + \sum_{i=0}^{k-1} \left\|\Gamma(k,i)\right\| \gamma. \tag{7.20}$$

Let $\varepsilon'$ sufficiently small and verifying $0 < \varepsilon' < \varepsilon$. Let $\mathscr{M} = \max_{\tilde{x} \in \mathscr{F}} \|\tilde{x}\|$. Since the system (7.1), (7.10) is uniformly exponentially stable, then $\Phi(k,0)$ converges exponentially to zero as $k$ tends to infinity. Consequently, there exists $k'_0 \in \mathbb{N}$ such that $\forall k \geq k'_0$, $\|\Phi(k,0)\| \leq \frac{\varepsilon'}{2.\mathscr{M}}$. Furthermore, we have

$$\begin{aligned}\sum_{i=0}^{k} \left\|\Gamma(k,i)\right\| &= \left\|\Phi(k,1)B\right\| + \left\|\Phi(k,2)B\right\| + \cdots + \left\|\Phi(k,k)B\right\| + \|B\| \\ &\leq \left(\left\|\Phi(k,1)\right\| + \left\|\Phi(k,2)\right\| + \cdots + \left\|\Phi(k,k)\right\| + 1\right)\|B\|. \end{aligned} \tag{7.21}$$

The uniform exponential stability of system (7.1), (7.10) implies that there exists two positive constants $c$ and $\beta < 1$ such that for all $i$

$$\left\|\Phi(k,i)\right\| \leq c\beta^{k-i}, \quad \text{for } k \geq i.$$

Consequently,

$$\begin{aligned}\sum_{i=0}^{k} \left\|\Gamma(k,i)\right\| &\leq \|B\|\left(1 + c\sum_{i=0}^{k} \beta^{k-i}\right) = \|B\|\left(1 + c\sum_{i=0}^{k} \beta^{i}\right) \\ &= \|B\|\left(1 + c\frac{1-\beta^{k+1}}{1-\beta}\right).\end{aligned}$$

Since the sequence $d_k = \sum_{i=0}^{k} \|\Gamma(k,i)\|$ is increasing and upper bounded by the convergent sequence $\|B\|(1 + c\frac{1-\beta^{k+1}}{1-\beta})$, then it is convergent too. Let

$$\varphi' = \lim_{k \to +\infty} \sum_{i=0}^{k} \left\|\Gamma(k,i)\right\|.$$

It follows that there exists $k_0'' \in \mathbb{N}$ such that $\forall k \geq k_0'', |\sum_{i=0}^{k} \|\Gamma(k,i)\| - \varphi'| \leq \frac{\varepsilon'}{2\gamma}$. Let $k_0 = \max(k_0', k_0'')$. Consequently,

$$\forall k \geq k_0, \quad \|\tilde{x}(k)\| \leq \frac{\varepsilon'}{2\mathscr{M}} \|\tilde{x}(0)\| + \left(\frac{\varepsilon'}{2\gamma} + \varphi'\right)\gamma \leq \varepsilon' + \varphi'\gamma. \tag{7.22}$$

Choosing $\varphi = \frac{\varphi'\varepsilon}{\varepsilon - \varepsilon'}$ and $\gamma = \frac{\varepsilon}{\varphi}$, the theorem is proved. □

Theorem 7.1 states appropriate conditions guaranteeing the positive invariance of the extended state as well its the ultimate boundedness to a given ball including the origin, using the static strategy. The next *Corollary* is a direct consequence, which shows that it is possible to refine this neighborhood of the origin if the quantization precision is increased.

**Corollary 7.1** (Quantization precision and ultimate boundedness near the origin) *Under the same hypothesis of Theorem* 7.1, *if the quantization errors are upper bounded by two constants* $\gamma_1$ *and* $\gamma_2$ *such that* $\gamma_1 \geq \gamma_2$, *then* $\mathscr{B}_{\varepsilon_1} \supseteq \mathscr{B}_{\varepsilon_2}$, *where* $\varepsilon_1 = \gamma_1\varphi$ *and* $\varepsilon_2 = \gamma_2\varphi$.

*Proof* Since the constants $\mathscr{M}$ and $\varphi$ are positive, then the function $f(\gamma) = \gamma\mathscr{M} + (\gamma + \varphi)\gamma$ is monotonically increasing. Consequently, if $\gamma_1 \geq \gamma_2$, then $\varepsilon_1 \geq \varepsilon_2$. □

## 7.3 Model Predictive Control *MPC*

### *7.3.1 Algorithm Description*

Assume that a set $\mathscr{C}$ of admissible periodic quantization sequences is defined (a method for the construction of this sequence will be presented in Sect. 7.3.2. Then, the problem of the integrated control and communication may be solved on-line by using the model predictive control algorithm. The model predictive control is an elegant solution to handle the hybrid aspect of the considered model, where both the control inputs and the quantization frequency decisions need to be determined at each sampling period. Model predictive control was successfully applied to the control of hybrid systems [21] and to the problems of integrated control and medium access allocation [27].

By using the *MPC* approach, an optimization problem is solved at each sampling period, in order to determine both the control inputs of the plant $u(k)$ and their quantization precision $p(k)$. This problem is formulated as described in the set of

Eqs. (7.23) below.

$$
\begin{cases}
\min\limits_{s^{T-1}\in\mathscr{C}} \quad \sum\limits_{h=0}^{N-1} \ell\big(\hat{x}(k+h|k), \hat{u}(k+h|k)\big) \\
\text{subject to} \\
\quad \hat{x}(k|k) = x(k), \\
\quad \hat{u}(k-1|k) = u(k-1), \\
\quad \text{and for all } h \in \{0,\ldots,N-1\}, \ \text{for all } i \in \{1,\ldots,m\}, \\
\quad \hat{p}(k+h|k) = s(h), \\
\quad \hat{v}(k+h|k) = L_{s^{T-1}}(h)\big[\hat{x}^T(k+h|k)\hat{u}^T(k+h-1|k)\big]^T, \\
\quad \hat{u_i}(k+h|k) = \mathscr{Q}_{(\hat{p}_i(k+h|k),U_i)}\big(\hat{v}_i(k+h|k)\big) \quad \text{if } \hat{p}_i(k+h|k) \neq 0, \\
\quad \hat{u_i}(k+h|k) = \hat{u_i}(k+h-1|k) \quad \text{if } \hat{p}_i(k+h|k) = 0, \\
\quad \hat{x}(k+h+1|k) = A\hat{x}(k+h|k) + B\hat{u}(k+h|k).
\end{cases}
\tag{7.23}
$$

In this context, the control performance $J(\hat{x}, \hat{u}, k, N+k)$ over a horizon of $N$ (which is assumed to be a multiple of $T$) is predicted, for the $|\mathscr{C}|$ communication sequences of $\mathscr{C}$. Here, $\hat{x}(k+h|k)$ and $\hat{u}(k+h|k)$ constitute the predicted values of $x(k+h)$ and $u(k+h)$, for a given quantization sequence $s^{T-1}$ and associated control gains sequence $L_{s^{T-1}}^{T-1}$. Notice that $\hat{p}(k+h|k)$ represents the quantization precision of the predicted inputs $\hat{u}(k+h|k)$. If $\hat{p}_i(k+h|k) = 0$, then the predicted control input $\hat{u}(k+h|k)$ cannot be updated. Consequently, its previous value $\hat{u}(k+h-1|k)$ will be maintained. Control inputs $u_i(k+h|k)$ which have to be updated (i.e., whose precision vectors satisfy $\hat{p}_i(k+h|k) \neq 0$) are computed by the application of the quantization map $\mathscr{Q}$ to the $i$th element of the result of the state feedback operation $\hat{v}(k+h|k) = L_{s^{T-1}}(h)[\hat{x}^T(k+h|k)\hat{u}^T(k+h-1|k)]^T$. The solution of this optimization problem is the admissible periodic quantization sequence $s^{T-1^*} = (s^*(0), \ldots, s^*(T-1))$ that minimizes the cost function $J(\hat{x}, \hat{u}, k, k+N)$ subject to the communication constraints. According to the *MPC* philosophy, the precision vector and control inputs at instant $k$ are respectively, given by:

$$
p(k) = \hat{p}^*(k|k) = s^*(0), \tag{7.24}
$$

and

$$
\begin{cases}
u_i(k) = \hat{u}_i^*(k|k) = \mathscr{Q}_{(p_i(k),U_i)}\big(\hat{v}_i^*(k|k)\big) & \text{if } p_i(k) \neq 0, \\
u_i(k) = u_i(k-1) & \text{if } p_i(k) = 0,
\end{cases}
\tag{7.25}
$$

where

$$
\hat{v}^*(k|k) = L_{s^{T-1*}}(0)\tilde{x}(k). \tag{7.26}
$$

In the sequel, we will represent the model predictive control law (7.23), (7.24), (7.26), (7.25), when based on a given set of admissible periodic quantization sequences $\mathscr{C}$, by the function $\kappa_{\mathscr{C}}(\tilde{x}(k))$ of the extended state $\tilde{x}(k)$.

The choice of the set $\mathscr{C}$ plays an important role in ensuring the practical stability and performance improvements. In the following, a heuristic method for choosing the elements of the set $\mathscr{C}$ is proposed. To simplify the notation, the cost function $J(\hat{x}, \hat{u}, k, k+N)$ will be simply denoted by $J(\tilde{x}, s)$, where $s$ is the quantization sequence that is used (as well as its associated control gains sequence) to evaluate the predicted values $\hat{x}$ and $\hat{u}$. It may be also denoted as $J(\tilde{x}(k), \mathscr{U}(k))$, where $\mathscr{U}(k) = (\hat{u}(k|k), \ldots, \hat{u}(k+N-1|k))$ is a sequence of control inputs that determine the evolution of the system from state $\tilde{x}(k)$ and the associated cost.

## 7.3.2 A Heuristic Approach for the Choice of Quantization Sequences

### 7.3.2.1 Action Domains

In order to simplify the presentation, and without loss of generality, we assume that

$$R_I = bM, \tag{7.27}$$

where $M$ represents the maximal precision of the digital-to-analog converters, generally considered as the "infinite precision", and $b$ is the maximal number of control inputs that may be sent with the maximal precision $M$, during the sampling period. Let $a^{1,Q} = (a_1, \ldots, a_Q)$ be a sequence of increasing integers such that $a_1 = 1$ and $\forall i \in \{1, \ldots, Q\}$, $a_i$ divides $M$. Let $\mathscr{S}(m, M)$ be the set of periodic quantization sequences of width $m$ and maximal precision $M$. Based on the sequence $a^{1,Q}$, it is possible to characterize $Q$ remarkable sub-sets of $\mathscr{S}(m, M)$. These particular sub-sets are called *action domains*. Formally, the $l$th action domain $\mathscr{A}_l, l \in \{1, \ldots, Q\}$ is the set of all quantization sequences $s^{T_l-1} = (s(o), \ldots, s(T_l-1))$ such that $\forall i \in \{1, \ldots, m\}$, $\forall k \in \{0, \ldots, T_l-1\}$, $s_i(k) \in \{\frac{M}{a_l}, 0\}$. This means that in the $l$th action domain, $a_l b$ inputs may be updated with the precision $\frac{M}{a_l}$. The first action domain $\mathscr{A}_1$ represents the quantization sequences that update the minimal number of plant inputs with the maximal precision $M$. The last action domain $\mathscr{A}_Q$ represents the set of quantization sequences that update the maximal number of plant inputs with the minimal precision $\frac{M}{a_Q}$.

### 7.3.2.2 Basic Sequences

Let $\Gamma^*(r)$ be an optimal off-line periodic communication sequence that minimizes the $\mathscr{H}_2$ norm of the system assuming that at most $r$ control inputs may be updated during the sampling period with an infinite quantization precision. An efficient method for obtaining optimal and suboptimal solutions (with a guaranteed error bounds) to this optimization problem and by making use of the branch and bound algorithm, was proposed in [29]. In this case, the generated optimal off-line communication sequences, are only dependent on the intrinsic characteristics of the system. In order to further reduce the computational requirements of the algorithm, the search is restricted to the most relevant quantization sequences. To this end, to each action domain, a particular quantization sequence, called *basic sequence*, is assigned. More formally, to action domain $A_l$, a basic sequence $s_B^l$ is associated according to the following relation:

$$s_B^l = \frac{M}{a_l} \Gamma^*(ba_l). \tag{7.28}$$

Although this assignment is suboptimal since the optimization is performed by assuming an infinite precision of the control inputs; it allows assigning the update rate of the different control signals according to the systems's dynamics. In fact, solving off-line the problem to find the optimal quantization sequence in the sense of the $\mathcal{H}_2$ performance index is very complex, since it suffers from the curse of dimensionality. The assumption of an infinite quantization precision considerably reduces this complexity, since convex optimization problems may be used inside the optimization algorithm (i.e., bounding phase of the branch and bound algorithm), instead of searching over all the possible discrete values of the control inputs, which, unfortunately, increase exponentially with the quantization precision. In the following, we will denote by $\mathscr{C}_l$ the set of admissible periodic quantization sequences that are obtained by the circular permutation of the basic sequence $s_B^l$. Unless stated otherwise, the set $\mathscr{C}$ that will be used by the proposed *MPC* algorithm is defined by

$$\mathscr{C} = \bigcup_{l=1}^{Q} \mathscr{C}_l.$$

### 7.3.3 *Attraction Properties of the* MPC

Theorem 7.1 states the conditions allowing to guarantee the ultimate boundedness to a given ball including the origin by using the static strategy. *Corollary* 7.1 shows that it is possible to refine this neighborhood of the origin if the quantization precision is increased. When the *MPC* algorithm is used, an interesting question is to determine the conditions under which the *MPC* will allow to perform the trade-off between quantization precision and update rates and to "attract" the system to the smallest ball around the origin in steady state, while improving the convergence rate by changing action domains in transient states. The notion of attraction is formalized in the following definition:

**Definition 7.4** (($W$, $V$)-attraction) Let $W$ a compact set and $V$ a set containing the origin in its interior and verifying $W \cap V = \emptyset$. System (7.1), (7.2) controlled by a control law $\kappa(\tilde{x}(k))$ is called *(W,V)-attractive* if for all trajectory starting in $W$ (i.e., verifying $\tilde{x}(0) \in W$), there exists a time instant $k_0$ such that $\tilde{x}(k_0) \in V$.

A system is called $(W, V)$-*attractive* when any trajectory starting in $W$ is driven in a finite time to $V$. A $(W, V)$-stable system is $(W, V)$-attractive, but the reciprocal is not necessarily true. In the sequel, sufficient conditions under which these attraction properties are verified, are stated. But before stating these results, the following technical lemma is necessary:

**Lemma 7.1** (Characterization of the $(W, V)$-attraction [24]) *Assume that there exists a control law $\kappa(\tilde{x}(k))$, a positive definite function $P(\tilde{x})$ of the extended state of the closed-loop system* (7.1), (7.2) *(controlled by $\kappa$), and a compact set $W$ that does not contain the origin and such that for all $\tilde{x}(k) \in W$, $\Delta P(\tilde{x}(k)) = P(\tilde{x}(k+1)) - P(\tilde{x}(k)) < 0$. Let $V = \{\tilde{x} \in \mathbb{R}^{n+m} - W$ for which there exists $\tilde{y} \in W$ such that $P(\tilde{x}) < P(\tilde{y})\}$. If $\tilde{x}(0) \in W$, then there exists an instant $k_0$ such that $\tilde{x}(k_0) \in V$.*

*Proof* The proof is performed by contradiction. Let $\tilde{x}(0)$ a given initial state in $W$. Assume that for all $k \in \mathbb{N}$, $\tilde{x}(k) \in W$. Since $W$ is compact (and consequently closed) and does not contain the origin, then there exists $\delta < 0$ such that $\max_{\tilde{x}\in W} \Delta P(\tilde{x}) = \delta$. Therefore, at any instant $k$,

$$P\big(\tilde{x}(k)\big) = P\big(\tilde{x}(0)\big) + \sum_{i=1}^{k}\Big(P\big(\tilde{x}(i)\big) - P\big(\tilde{x}(i-1)\big)\Big) \leq P\big(\tilde{x}(0)\big) + \delta k.$$

Consequently, when $k \longrightarrow +\infty$, $P(\tilde{x}(k)) \longrightarrow -\infty$, which contradicts the fact that $P(\tilde{x}(k))$ is a positive definite. For that reason, we conclude that there exists a time instant $k_0$ such that $\tilde{x}(k_0) \notin W$. Since $P(\tilde{x}(k))$ is strictly decreasing for $k \leq k_0$, then necessarily $P(\tilde{x}(k_0)) < P(\tilde{x}(0))$. Thus, $\tilde{x}(k_0) \in V$. □

Lemma 7.1 simply shows that when a positive definite function of the state is strictly decreasing along the trajectories of the system in a region $W$ that does not contain the origin; the state cannot stay indefinitely in this region and must approach the origin. This lemma will be the basis for proving the attraction properties of the *MPC* algorithm.

Let $P(\tilde{x}(k)) = J(\tilde{x}(k), \mathscr{U}^*(k))$. The function $P(\tilde{x}(k))$ represents the cost function that is minimized by the *MPC* algorithm and will play the role of a "Lyapunov function". By construction, $P$ is a positive definite function. We have the following result.

**Theorem 7.2** ($(W, V)$-attraction of the *MPC* law $\kappa_{\mathscr{C}_l}$ [24]) *Let $\xi > 0$. Let*

$$W = \big\{\tilde{x} \in \mathbb{R}^{n+m} \text{ such that } P(\tilde{x}) \leq \xi\big\}$$

*and $V$ two compact sets of $\mathbb{R}^{n+m}$ containing the origin in their interiors and such that $V \subseteq W$. Let $\varepsilon$ be a small positive number, $\bar{W} = \{\tilde{x} \in W$ such that $\ell(x, \kappa_{\mathscr{C}_l}(\tilde{x})) \geq \max_{\tilde{x}_v \in V} \ell(x_v, \kappa_{\mathscr{C}_l}(\tilde{x}_v)) + \varepsilon\}$ and $\bar{V} = W - \bar{W}$. If the $(W, \bar{V})$-stability in $\mathscr{N}$-steps of the system* (7.1), (7.2) *is ensured by the static strategy based on any quantization sequence $s^l \in \mathscr{C}_l$, and if the horizon $N$ of the* MPC *algorithm is chosen such that $N > \mathscr{N}$, then the* MPC *algorithm $\kappa_{\mathscr{C}_l}$ is $(\bar{W}, \bar{V})$-attractive.*

*Proof* Let $\bar{\eta} = \max_{\tilde{x}_v \in V} \ell(x_v, \kappa_{\mathscr{C}_l}(\tilde{x}_v))$. At time step $k = 0$, the solutions of the open-loop optimal control and communication problem are $\mathscr{U}^*(0) = (\hat{u}^*(0|0), \ldots, \hat{u}^*(N-1|0))$ and $s^*$. The control input that will be applied to the plant, according to the receding horizon philosophy will be $u(0) = \hat{u}^*(0|0)$. At time step $k = 1$, the sequence $\check{\mathscr{U}}(1) = (\hat{u}^*(1|0), \ldots, \hat{u}^*(N|0))$ is a feasible sequence, where $\hat{u}^*(N|0)$ is defined for $i \in \{1, \ldots, m\}$ by

$$\hat{u}_i^*(N|0) = \begin{cases} \mathscr{Q}_{(s_i^*(N), U_i)}(\hat{v}_i^*(N|0)) & \text{if } s_i^*(N) \neq 0, \\ \hat{u}_i^*(N-1|0) & \text{if } s_i^*(N) = 0, \end{cases}$$

and

$$\hat{v}(N|0) = L_{s^*}(N)\big[\hat{x}^*(N|0) \quad \hat{u}^*(N-1|0)\big]^T.$$

The associated cost is

$$J\big(\tilde{x}(1), \check{\mathscr{U}}(1)\big) = J\big(\tilde{x}(0), \mathscr{U}^*(0)\big) - \ell\big(x(0), u(0)\big) + \ell\big(\hat{x}^*(N|0), \hat{u}^*(N|0)\big). \quad (7.29)$$

Since the static strategy based on the quantization sequence $s^* \in \mathscr{C}_l$ ensures $(W, \bar{V})$-stability in $\mathscr{N}$-steps of the system (7.1), (7.2), and knowing that $N > \mathscr{N}$, then $\hat{x}^*(N|0) \in \bar{V}$. Consequently, by taking into account the construction of $\bar{V}$, $\ell(\hat{x}^*(N|0), \hat{u}^*(N|0)) < \bar{\eta}$. We may write

$$\begin{aligned} P\big(\tilde{x}(1)\big) - P\big(\tilde{x}(0)\big) &= J\big(\tilde{x}(1), \mathscr{U}^*(1)\big) - J\big(\tilde{x}(0), \mathscr{U}^*(0)\big) \\ &\leq J\big(\tilde{x}(1), \check{\mathscr{U}}(1)\big) - J\big(\tilde{x}(0), \mathscr{U}^*(0)\big) \\ &< -\ell\big(x(0), \kappa_{\mathscr{C}_l}\big(\tilde{x}(0)\big)\big) + \bar{\eta}. \end{aligned} \quad (7.30)$$

This reasoning remains true for all other instants $k > 0$. The positive-definite function $P$ associated to the *MPC* control law is then strictly decreasing when $\tilde{x}(0) \in \bar{W} \subseteq \{\tilde{x} \in \mathbb{R}^{n+m}$ such that $\ell(x, \kappa_{\mathscr{C}_l}(\tilde{x})) > \bar{\eta}\}$. By construction, $\bar{W}$ is compact, does not contain the origin and verifies $\bar{W} \cap \bar{V} = \emptyset$. Consequently, by applying the previous lemma, there exists an instant $k_0$ such that $\tilde{x}(k_0) \notin \bar{W}$ and $P(\tilde{x}(k_0)) < P(\tilde{x}(0)) \leq \xi$, thus, $\tilde{x}(k_0) \in \bar{V}$. □

Theorem 7.2 states the conditions under which any trajectory starting in $\bar{W}$ is driven in finite time to $\bar{V}$, using the *MPC* law $\kappa_{\mathscr{C}_l}$. Let $\mathscr{G}$ the set of extended states that can be practically stabilized by at least one static strategy (as defined by the previous heuristic approach). More precisely, $\mathcal{G}$ is formally defined by

$$\mathcal{G} = \bigcup_{i=1}^{Q} \mathcal{G}_{\frac{M}{a_i}}.$$

In the sequel, to simplify the notion, let $J(\tilde{x}, l_i) = \min_{s \in \mathcal{C}_i} J(\tilde{x}, s)$. Let $\mathcal{R}_i$ be the region defined by

$$\mathcal{R}_i = \{\tilde{x} \in \mathcal{G} \text{ such that for all } j > i,\ J(\tilde{x}, l_i) < J(\tilde{x}, l_j)\}.$$

It is easy to see that the regions $\mathcal{R}_i$ ($i \in \{1, \dots, Q\}$) form a partition of $\mathcal{G}$. An interesting question is to see if the *MPC* algorithm $\kappa_{\mathcal{C}}$ is able to guarantee the attraction of any trajectory starting in $\mathcal{G}$ to a desired ball $X_f$. Theorem 7.3 provides sufficient conditions to ensure this property. To simplify the notation, let $\mathcal{R}_0 = X_f$.

**Theorem 7.3** (Convergent switching between action domains in the *MPC* law $\kappa_{\mathcal{C}}$ [24]) *If $\mathcal{R}_1 \subset \mathcal{R}_2 \subset \cdots \subset \mathcal{R}_Q$ and if the* MPC *algorithm $\kappa_{\mathcal{C}_l}$ ensures $(\mathcal{R}_l, \mathcal{R}_{l-1})$-attraction of system* (7.1), (7.2) *for $1 \le l \le Q$, then system* (7.1), (7.2), *controlled by the* MPC *control law $\kappa_{\mathcal{C}}$ is $(\mathcal{G}, X_f)$-attractive.*

*Proof* As mentioned earlier, $(\mathcal{R}_1, \dots, \mathcal{R}_Q)$ form a partition of $\mathcal{G}$. For that reason, there exists $l \in \{1, \dots, Q\}$ such that the extended state $\tilde{x}$ belongs to the region $\mathcal{R}_l$. Consequently, by the construction of the *MPC* algorithm $\kappa_{\mathcal{C}}$, as long as the extended state $\tilde{x}$ belongs to $\mathcal{R}_l$, the *MPC* algorithm $\kappa_{\mathcal{C}_l}$ will be applied, until the extended state reaches a region $\mathcal{R}_j$ with $j \ne l$. The $(\mathcal{R}_l, \mathcal{R}_{l-1})$-attraction implies that this region is necessarily included in $\mathcal{R}_{l-1}$. Consequently, $j < l$. Next, by applying recursively the same argument to the region $\mathcal{R}_j$, the theorem is proved. □

In simulation, it is observed that when the prediction horizon is sufficiently large, adding more elements to $\mathcal{C}$ leads to a better control performance. However, if the set $\mathcal{C}$ is reduced to $\mathcal{C} = \{s_B^1, \dots, s_B^Q\}$, proving the $(W, V)$-stability becomes easy, as stated in the following:

**Corollary 7.2** (Convergent switching between action domains in $\kappa_{\mathcal{C}}$ for $\mathcal{C} = \{s_B^1, \dots, s_B^Q\}$) *If $\mathcal{R}_1 \subset \mathcal{R}_2 \subset \cdots \subset \mathcal{R}_Q$ and if the static strategy $s_B^l$ ensures $(\mathcal{R}_l, \mathcal{R}_{l-1})$-stability of system* (7.1), (7.2) *for $1 \le l \le Q$, then the system* (7.1), (7.2), *controlled by the* MPC *control law $\kappa_{\mathcal{C}}$ for $\mathcal{C} = \{s_B^1, \dots, s_B^Q\}$ is $(\mathcal{G}, X_f)$-stable.*

The proof may be easily established following the same lines as to the one proposed for *Corollary* 7.1.

## 7.4 Simulation Results

Consider the continuous-time LTI system defined by the state and input matrices:

$$A_c = \begin{bmatrix} A_{c_s} & 0 & 0 & 0 \\ 0 & A_{c_s} & 0 & 0 \\ 0 & 0 & A_{c_s} & 0 \\ 0 & 0 & 0 & A_{c_s} \end{bmatrix} \quad \text{and} \quad B_c = \begin{bmatrix} B_{c_s} & 0 & 0 & 0 \\ 0 & B_{c_s} & 0 & 0 \\ 0 & 0 & B_{c_s} & 0 \\ 0 & 0 & 0 & B_{c_s} \end{bmatrix},$$

where

$$A_{c_s} = \begin{bmatrix} 0 & 110 \\ -900 & 10 \end{bmatrix} \quad \text{and} \quad B_{c_s} = \begin{bmatrix} 0 \\ 210 \end{bmatrix}.$$

The system is composed of four independent and identical second-order open-loop unstable subsystems, with eigenvalues $5 \pm 314.6j$. The design criteria of the ideal controller for the closed-loop of each subsystem are defined by the matrices $Q_{c_s} = \mathrm{Diag}(30, 10)$ and $R_{c_s} = 0.01$. Assume now that the communication channel linking the controller to the four distant actuators has a bandwidth of 10 kbps, which means that every 2 ms, at most 20 bits of information can be sent to the actuators.

Based on these bandwidth constraints, choosing $M = 20$, and using the heuristic approach discussed in the previous section, three action domains $\mathcal{A}_1$, $\mathcal{A}_2$ and $\mathcal{A}_3$ are associated to the global system, and correspond to the sequence $a^{1,3} = (a_1, a_2, a_3) = (1, 2, 4)$. Since the four subsystems are identical, the basic sequences that are associated to each action domain are defined by

$$s_B^1 = \left( \begin{bmatrix} 20 \\ 0 \\ 0 \\ 0 \end{bmatrix}, \begin{bmatrix} 0 \\ 20 \\ 0 \\ 0 \end{bmatrix}, \begin{bmatrix} 0 \\ 0 \\ 20 \\ 0 \end{bmatrix}, \begin{bmatrix} 0 \\ 0 \\ 0 \\ 20 \end{bmatrix}, \ldots \right),$$

$$s_B^2 = \left( \begin{bmatrix} 0 \\ 10 \\ 0 \\ 10 \end{bmatrix}, \begin{bmatrix} 10 \\ 0 \\ 10 \\ 0 \end{bmatrix}, \begin{bmatrix} 0 \\ 10 \\ 0 \\ 10 \end{bmatrix}, \begin{bmatrix} 10 \\ 0 \\ 10 \\ 0 \end{bmatrix}, \ldots \right),$$

and

$$s_B^3 = \left( \begin{bmatrix} 5 \\ 5 \\ 5 \\ 5 \end{bmatrix}, \begin{bmatrix} 5 \\ 5 \\ 5 \\ 5 \end{bmatrix}, \begin{bmatrix} 5 \\ 5 \\ 5 \\ 5 \end{bmatrix}, \begin{bmatrix} 5 \\ 5 \\ 5 \\ 5 \end{bmatrix}, \ldots \right).$$

The static strategy is first evaluated. In the simulation results (depicted in Figs. 7.3, 7.4 and 7.5), the control performance corresponding to the application of the basic sequences $s_B^1$, $s_B^2$ and $s_B^3$ (of action domains $\mathcal{A}_1$ $\mathcal{A}_2$ and $\mathcal{A}_3$) is evaluated and compared. In these simulations, it is assumed that the global system is started from the initial condition $[0.4\ 0\ 0.4\ 0\ 0.4\ 0\ 0.4\ 0]^T$ and disturbed at $t = 59$ ms. It can be

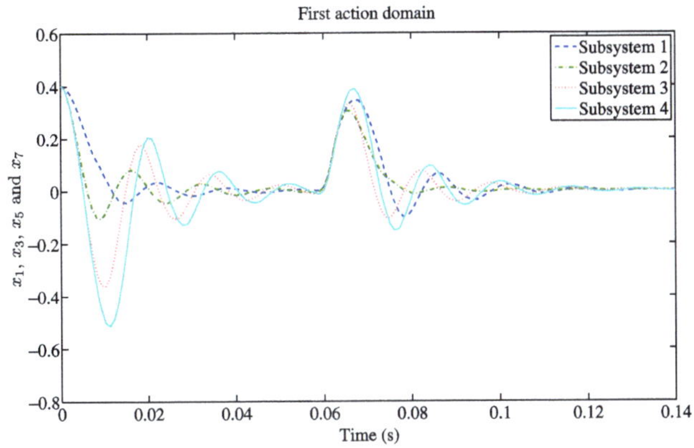


**Fig. 7.3** Global system responses obtained for the basic sequence of action domain $\mathscr{A}_1$

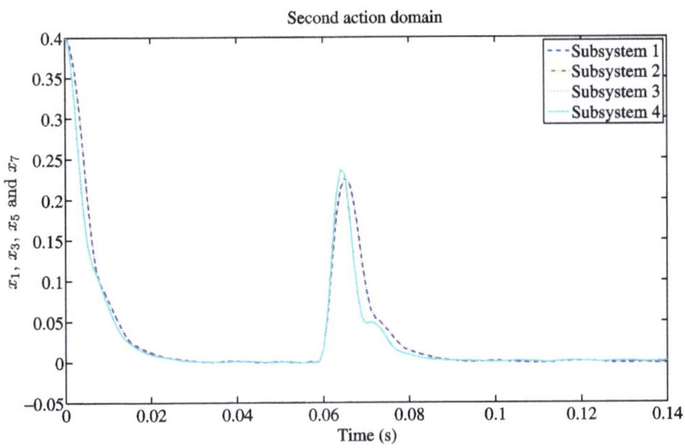


**Fig. 7.4** Global system responses obtained for the basic sequence of action domain $\mathscr{A}_2$

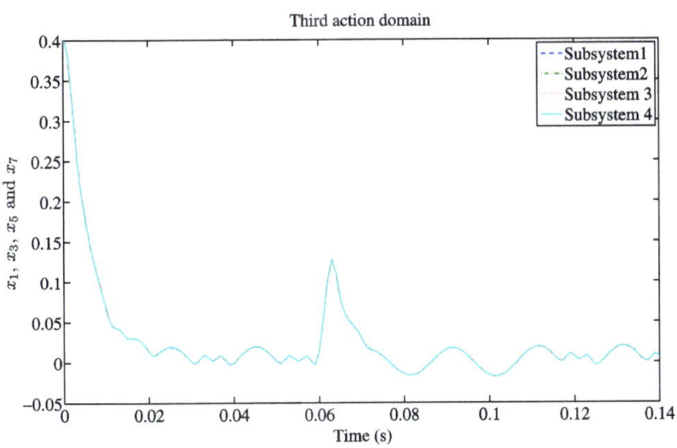


**Fig. 7.5** Global system responses obtained for the basic sequence of action domain $\mathscr{A}_3$

observed that the response corresponding to the basic sequence $s_B^1$ presents an oscillatory behavior, resulting from the effect of the long effective update period (which is equal to 8 ms). These oscillations disappear when the basic sequence $s_B^2$ is used. Now, the use of the basic sequence $s_B^3$ further improves the disturbance rejection capabilities as well as the response time of the systems, but, unfortunately, leads to a degradation of the steady state performance (chaotic behavior near the origin). The main observation from this example can be resumed as follows: the use of a high action domain improves the transient behavior whereas the use of a low action domain improves the steady state precision. Performing a switching between these actions domains in an automatic and clairvoyant way will allow to take advantage of the benefits of each action domain and to improve the control performance by trading quantization precision for update rates.

The *MPC* algorithm is next evaluated. The proposed *MPC* algorithm was designed by using the heuristic approach discussed in the previous section. According to this approach, the set $\mathscr{C}$ will contain the basic sequences $s_B^1 \times \frac{R_I}{M}$, $s_B^2 \times \frac{R_I}{M}$ and $s_B^3 \times \frac{R_I}{M}$ as well as their circular permutations. The cardinality of $\mathscr{C}$ (i.e. the number of used quantization sequences) is equal to 7. For that reason, 3 bits must be reserved to the encoding of the decoding key. Consequently, the parameters $R_I$ and $R_D$ are chosen such that

$$R_I = 16 \quad \text{and} \quad R_D = 3.$$

Next, the set $\mathscr{P}$ containing the possible precision vectors (corresponding to this choice of $\mathscr{C}$) is defined by

$$\mathscr{P} = \left\{ \begin{bmatrix} 16 \\ 0 \\ 0 \\ 0 \end{bmatrix}, \begin{bmatrix} 0 \\ 16 \\ 0 \\ 0 \end{bmatrix}, \begin{bmatrix} 0 \\ 0 \\ 16 \\ 0 \end{bmatrix}, \begin{bmatrix} 0 \\ 0 \\ 0 \\ 16 \end{bmatrix}, \begin{bmatrix} 0 \\ 8 \\ 0 \\ 8 \end{bmatrix}, \begin{bmatrix} 8 \\ 0 \\ 8 \\ 0 \end{bmatrix}, \begin{bmatrix} 4 \\ 4 \\ 4 \\ 4 \end{bmatrix} \right\}.$$

The control gains that are associated to each quantization sequence are derived in the same way as for the static strategy.

The responses of the global system corresponding to the use of the *MPC* algorithm are depicted in Fig. 7.6 (top). It may be observed that when using the *MPC* approach, some tradeoff between precision and system reaction rapidity is operated. After the system disturbance, the quantization sequence of control signals is changed and switch from that of action domain 1 (one control signal is coded with maximal precision) to that of action domain 3 (four control signals are coded with the same precision). The advantages resulting from the use of each basic sequence are achieved. The oscillations caused by the long update period as well as the chaotic behavior near the practical stability region are eliminated and the response to the unpredictable disturbances improved. These improvements are due to a more "intelligent" choice of the number of quantization levels of the control signals, which are allocated according to the state value of the system. The time evolution of the used action domains is depicted in Fig. 7.6 (right). It may be observed that when the four systems are disturbed at the same time, the *MPC* algorithm chooses to send at

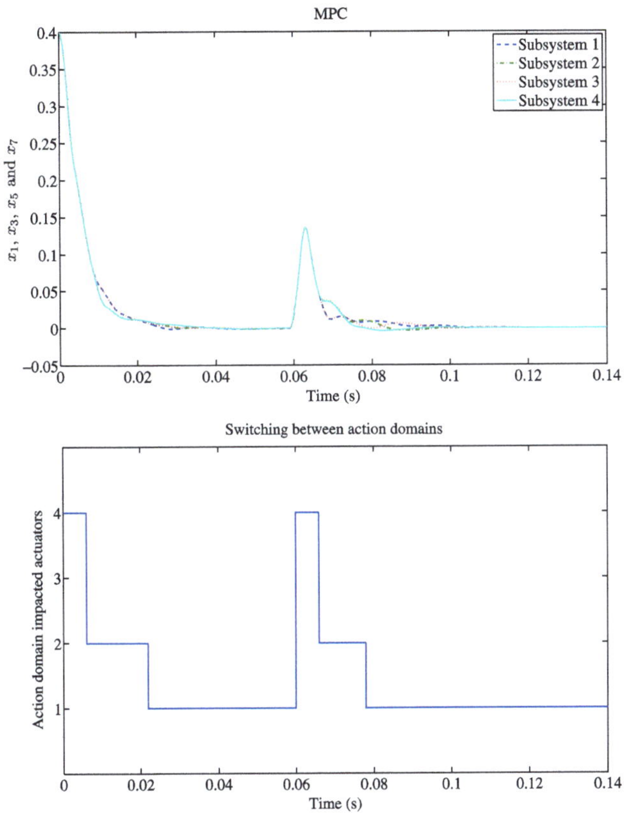


**Fig. 7.6** Global system responses obtained for the *MPC* algorithm (*top*) and used action domains (*bottom*)

the same time the control inputs to the four systems, using a precision vector from the first action domain. As long as the systems are stabilized, precision vectors from the second and finally from the third action domain are chosen. The corresponding quantization precision of the different control signals is depicted in Fig. 7.7.

## 7.5 Notes and Comments

In this chapter, the problem of the control over limited bandwidth communication channels was studied. A finely grained model was adopted, ensuring the respect of the bandwidth constraints and allowing the influence characterization of update frequency and quantization precision on the control performance. A simple static strategy was first proposed, and its $(W, V)$-stability properties studied. An efficient approach for the improvement of disturbance rejection capabilities and steady state precision was then proposed. This approach dynamically assigns the quantization

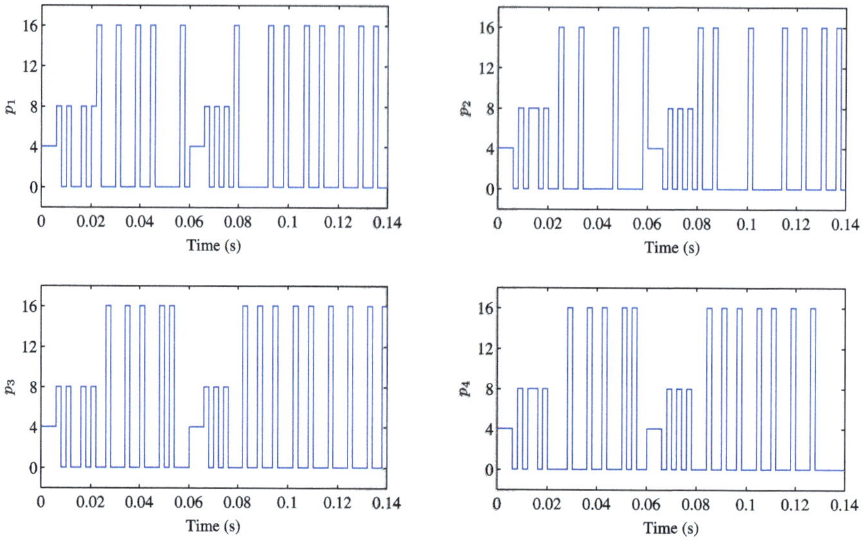


**Fig. 7.7** Quantization precision of control signals

precision of the control signals in order to improve the control performance, and takes into account the communication and computation requirements of the introduced dynamic protocol. It naturally allows handling *LTI* systems with multiple inputs. Sufficient conditions for ensuring the practical stability of this approach were stated and proved. The proposed approach was evaluated and illustrated through a numerical example.

Allocation of communication resources: the "per symbol" paradigm,[1] the information exchange is modeled at the symbol level. The quantization of measurements and control commands is thus implicitly taken into account. The general model is given in Fig. 2.4. In this model, the communication channel can transmit at most $R$ bits per time unit. Due to these resource limitations, measurements and control commands must be encoded (as a flow of symbols) before their transmission and decoded at their reception. Various coding techniques may be employed. A fundamental question is to determine the necessary and/or sufficient data-rate allowing the existence of a coder, a decoder as well as a controller able to achieve the stabilization of the system. Many contributions have tried to bring more insight into this fundamental question, by treating various models of resource limitations.

In [72], Delchamps has shown that it is impossible to asymptotically stabilize a discrete-time unstable LTI system, whose output passes through a quantizer having a finite number of quantization levels. In this setting, it is necessary to introduce and use other stability concepts, like practical stability.

The problem of state estimation, in the presence of state and measurement noise, was studied in [253]. In the considered model, the state observer is situated at the

[1] For more details please refer to Chap. 2, Sect. 2.2.2.

same location as the plant. However, the controller is located at a distant place. Consequently, the observations must be sent to the controller through a finite bandwidth communication channel. This problem was shown to be different from the classic estimation and vector quantization problems. The concept of *finitely recursive coder-estimator sequence* was then introduced. Necessary conditions as well as sufficient conditions, which are related to the stability and the convergence of various coding-estimation algorithms, have been proposed. These conditions connect the network data-rate to the dynamical characteristics of the plant.

In [254], *Wong and Brockett* introduced the concept of *containability*, as a weaker stability condition, to tackle the problem of the stabilization of networked systems through limited capacity communication networks, where the values of the measurements (which are received by the controller) and the controls (which are sent to the plant) belong to a finite set of values $\mathscr{J}$ (because of the quantization which is induced by the limited data-rate communication channel). In the case of continuous-time LTI systems, which are impulsively controlled, they proved that a necessary condition to ensure the containability is $e^{\frac{2}{R}\mathrm{tr}(A_c)} \leq |\mathscr{J}|$ where $|\mathscr{J}|$ is the size of the alphabet, $\frac{1}{R}$ is the transmission duration of one bit and $A_c$ is the state matrix. They also proved that if the initial condition of the system lies in a bounded set, then a memoryless coding and control is sufficient to ensure the containability, if some conditions affecting the data-rate are met.

In [182], *Nair and Evens* considered a class of discrete-time, linear and time-varying plants. The initial state is the realization of a random variable. Communications constraints only affect the sensors-to-controller link. The controller is directly connected to the actuators. The considered problem is the explicit synthesis of a coder (on the sensors-to-controller link) and of a controller that minimize a cost function of the state over finite and infinite horizons. The cost function over the finite horizon is the $m$th output moment. A coder/controller scheme was proposed. Under some technical assumptions, which are related to the probability density function of the initial state, to some conditions depending on the size of the alphabet, to the data-rate and to the plant dynamics, a necessary and sufficient optimality condition of the proposed coder/controller was established. A necessary and sufficient condition for the existence of a coder/controller that asymptotically stabilizes the system (in the sense that the $m$th output moment converges to zero over an infinite horizon) has been proposed. In the special case when the plant is invariant, unstable and finite-dimensional, this last condition simplifies to $R > \log_2 |\lambda|$ where $\lambda$ is the unstable open-loop pole with the largest magnitude.

In [42], Brockett and Liberzon proposed the idea of "zooming" as a means for ensuring the asymptotic stability of continuous-time and discrete time system whose control loops are closed through a finite bandwidth communication network. The zooming technique consists in changing the sensitivity of the quantizer over the time, based on the available quantized measurements. The relationship between performance and complexity of the quantized stabilization using the zooming technique has been studied in [82].

In [80], *Elia and Mitter* addressed the problem of the stabilization of single-input linear systems whose measurements and control commands are quantized. By first considering quantizers with a countable number of levels, and supposing the exact knowledge of the state, they proved that the coarsest quantizer that allows the quadratic stabilization of a discrete-time single-input LTI system, is logarithmic, and may be computed by solving an appropriate LQR problem. The state-feedback control problem as well as the state observation problem have been solved within this theoretical framework. Furthermore, these results have been thereafter extended to the continuous-time single-input and periodically sampled linear systems. Next, the expression of the optimal sampling period (for the suggested quantizers) has been explicitly established. It only depends on the sum of the unstable eigenvalues of the continuous system. This approach has been finally extended to address quantizers with a finite number of levels.

In [227], Tatikonda and Mitter considered discrete-time LTI systems. The control loop is closed through a limited capacity communication channel. Consequently, before their transmission, the measurements are quantized and encoded in symbols by a coder. At their reception, a decoder reconstructs a state estimate that will be used by the controller, which is directly connected to the plant. Two types of coders have been studied:

- class 1 coders, which know past measurements, past controls and past transmitted symbols that were sent over the channel,
- class 2 coders, which only know past measurements.

Stabilization and asymptotic observability properties have been addressed. It was shown that a necessary conditions for the existence of coders and decoders making it possible to guarantee these two properties is given by $R > \sum_{\lambda(A)} \max\{0, \log|\lambda(A)|\}$, the sum is over the eigenvalues of the state matrix $A$. This necessary condition is independent from coder classes and becomes sufficient if the whole state is measured and if class 1 coders are used.

More recently, many researchers have extended the model of networked control systems in order to include other aspects of constrained bandwidth such as packet drop-out, distributed nature of calculation and communication nodes, communication protocols as well as delay induced models. In *Tsumura et al.* [237], the stabilization problem of a linear system via quantized feedback with stochastic packet losses is considered. It is shown that this upper bound of the coarseness is strictly given by the packet loss probability and the unstable poles of the plants. In *Chaillet and Bicchi* [59], the problem of stabilizing sufficiently smooth nonlinear time-invariant plants over a network with limited bandwidth is addressed. Reliable packet switching networks are explicitly considered, for which both the time between consecutive accesses to each node and the delay by which each data packet is received, processed, and fed back to the plant are unknown but bounded. In order to compensate the unpredictably varying delays, a model-based strategy is proposed and a bound on the tolerable delays and access frequency is explicitly provided. Next, *Franci and Chaillet* [87] propose a dynamic quantization strategy, able to cope with time-varying model uncertainties. *Nesiç and Liberzon* [184] propose a unified design framework for networked and quantized control systems (*NQCS*). The designed quantization/time-scheduling protocols are proved to be stable and a detailed methodology explaining how this can be done for several representative protocols is given. In [69], *De Persis and Mazenc* used appropriate *Lyapunov–Krasowskii* functionals to design quantized control laws for nonlinear continuous-time systems in the presence of input constant delays. The resulting quantized feedback is parameterized with respect to the quantization density and maximal allowable delay tolerated by the system is characterized as a function of the quantization density.

# Chapter 8
# Optimal State-Feedback Resource Allocation

In computer systems, the scheduler is the entity that is responsible of the allocation of the computational resources. Consequently, using efficiently these resources amounts to design appropriate scheduling algorithms. The need for an efficient use of the computational resources comes from cost pressure, which requires the realization of system functionalities with minimum resources. Traditionally, control design and real-time scheduling design have been decoupled in the design processes of embedded control applications. This separation of concerns allowed each community to focus on specific problems, and led to the spectacular development we are familiar with today [12]. However, this separation of concerns relies on the fact that these two domains use very simplified models of their interface with each other. In fact, as pointed out in [53], control designers assume that the implementation is able to provide an equidistant sampling and actuation, and to ensure a fixed input output latency as well as an infinite quantization precision. Real-time designers on the other hand see the control loop as a periodic task with a hard deadline.

In both control and real-time scheduling theories, the notion of "instantaneous computational needs" of a control application is not defined. Computation and communication resources are consequently allocated according to the "worst case needs" of these applications. This corresponds to the use of periodic sampling, which on one hand simplifies the control design problem, but on the other hand leads to an unnecessary usage of some computational resources. In fact, a control task that is close to the equilibrium needs less computing resources than another one which is severely disturbed [23, 26, 167]. Similarly, the real-time scheduling design of control tasks is based on their worst-case execution time. As illustrated in [55], a resource saving may be obtained if these hard real-time constraints are "conveniently" relaxed, since control systems may be situated in between hard and soft real-time systems. Our opinion is that a significant saving in computing resources may be performed if more elaborate models are used by the two communities, which amounts to the co-design and the scheduling [12, 55, 212, 255].

A. Çela et al., *Optimal Design of Distributed Control and Embedded Systems*,
Communications and Control Engineering, DOI 10.1007/978-3-319-02729-6_8,

In this chapter, a new approach for the co-design of control and real-time scheduling is proposed. In the first part, we motivate the use of the $\mathcal{H}_2$ performance index for the problem of the optimal integrated control and non-preemptive off-line scheduling of control tasks. A new approach for solving this problem is proposed. This approach decomposes the problem into two sub-problems, which may be solved separately. The first sub-problem amounts in finding the optimal non-preemptive off-line schedule, and may be solved by using the branch and bound method. The second sub-problem resolution uses the lifting technique to determine the optimal control gains, based on the solution of the first sub-problem. We illustrate how the existing knowledge from the considered control systems models may be used to considerably reduce the time needed to find the optimal solution, taking full advantage of the use of the branch and bound method. In the second part, a plant state feedback scheduling algorithm, called reactive pointer placement (*RPP*) scheduling algorithm is proposed. Its objective is to improve the control performance by reacting as fast as possible to unexpected disturbances. Performance improvements as well as stability guarantees using the *RPP* algorithm are formally proven and then illustrated on a comprehensive implementation model, which was simulated using the tool TRUETIME [5]. Finally, the *RPP* algorithm is implemented on an embedded processor in order to achieve the concurrent real-time speed regulation of two DC motors.

The rest of this chapter is organized as follows. After a brief overview of the related work in the second section, the formulation and solving of the optimal control and non-preemptive off-line scheduling according to the $\mathcal{H}_2$ performance criterion is addressed in the third section. The *RPP* scheduling algorithm is introduced, described and evaluated using simulation in the fourth section. Finally, some remarks on the experimental evaluation of the proposed approach are presented in the fifth section. Some notes and comments end the chapter.

## 8.1 Optimal Off-line Scheduling

### *8.1.1 Problem Formulation*

Consider a collection of $N$ continuous-time linear time-invariant (LTI) systems $(S^{(j)})_{1\leq j\leq N}$. Assume that each system $S^{(j)}$ is controlled by a task $\tau^{(j)}$, which is characterized by its worst-case execution time $d^{(j)}$. Since we are interested in deriving an optimal off-line schedule, the scheduling decisions (i.e., releasing and starting of control tasks) must be made periodically (i.e., at instants that are multiple of an elementary *tick period*), rather than at arbitrary time instants, as illustrated in ([157], Chap. "Clock-Driven Scheduling"). Let $T_p$ be this elementary tick period. The time line is then partitioned into intervals of length $T_p$ called *time slots*. Control tasks may be started only at the beginning of the time slots. In some cases, the control task may need more than one time slot to execute. In the proposed model, we assume that the control tasks are executed *non-preemptively*. This assumption has two

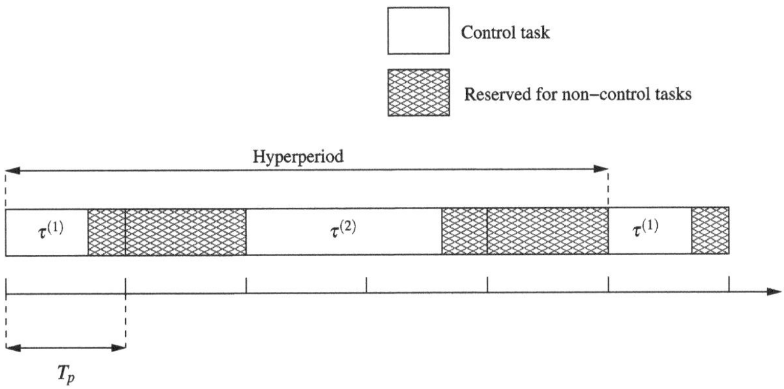


**Fig. 8.1** An example of an off-line scheduling of control and non-control tasks

essential benefits. First, it simplifies the problem formulation as well as its solution. Second, non-preemptive scheduling ensures a minimal and constant input/output latency for control tasks. In fact, if the input and output operations are performed respectively at the beginning and at the end of a control task, preemption causes variations in the input/output latency, which may degrade the control performance, as illustrated in [56, 166].

Usually, control tasks share the processor with other sporadic or aperiodic tasks, which do not perform the computations of the control laws, and called *non-control tasks*. Examples of such tasks include signal processing, monitoring or communication tasks. The total computational load that these tasks induce may be described by their processor utilization rate, noted $\mathscr{R}$. If these non-control tasks do not have any real-time constraints, then they may be scheduled in background, after the completion of the control tasks jobs (i.e., in the slices of the time slots where control tasks do not execute), as shown in ([157], Chap. "Clock-Driven Scheduling"). Otherwise, their schedulability and real-time constraints have to be checked by using some standard methods from real-time system theory. This issue is particularly tackled in the previously cited reference. By using off-line scheduling, the scheduling pattern (starting time instants of control tasks or non-control tasks reserved slots) is repeated identically every *major cycle*, or *hyperperiod*. In the following, we will denote by $T \times T_p$ the hyperperiod of the static schedule. The hyperperiod is equal to $T$ when expressed in terms of numbers of elementary time slots $T_p$.

*Example 8.1* An example of an off-line schedule of two control tasks $\tau^{(1)}$ and $\tau^{(2)}$ and a set of sporadic tasks using at most 44 % of the *CPU* (i.e., whose utilization rate $\mathscr{R} = 0.44$) is given in Fig. 8.1.

Task scheduling may be described by associating *Boolean scheduling functions* $\gamma^{(j)}$ to control tasks $\tau^{(j)}$ such that

$$\gamma^{(j)}(k) = \begin{cases} 1 & \text{if the execution of task } \tau^{(j)} \text{ finishes in the interval} \\ & [(k-1)T_p, kT_p), \\ 0 & \text{otherwise.} \end{cases} \tag{8.1}$$

We call $\gamma^{(j)}(k)$ the *execution end indicator* of the jobs of task $\tau^{(j)}$. Due to the use of non-preemptive scheduling, the $\lceil \frac{d^{(j)}}{T_p} \rceil$ slots containing or preceding the end of a job of task $\tau^{(j)}$ are allocated to its execution. Using this observation, the processor time slots utilization by a given control task $\tau^{(j)}$ may be described by

$$e^{(j)}(k) = \sum_{l=k}^{k+\lceil \frac{d^{(j)}}{T_p} \rceil - 1} \gamma^{(j)}(l). \tag{8.2}$$

The variable $e^{(j)}(k) \in \{0, 1\}$ is the *task!execution indicator* corresponding to the jobs of task $\tau^{(j)}$ and verifies

$$e^{(j)}(k) = 1 \iff \begin{array}{l} \text{the processor is allocated to task } \tau^{(j)} \text{ during a sub-interval} \\ \text{of } [(k-1)T_p, kT_p). \end{array} \tag{8.3}$$

During the interval $[(k-1)T_p, kT_p)$, the processor can execute only one control task. This constraint may be modeled by the following inequality

$$\sum_{j=1}^{N} e^{(j)}(k) \leq 1. \tag{8.4}$$

In order to guarantee the computational needs of non-control tasks, described by their processor utilization rate $\mathscr{R}$, the scheduling decisions of control tasks must satisfy

$$\mathscr{R} + \frac{1}{TT_p} \sum_{k=0}^{T-1} \sum_{j=1}^{N} d^{(j)} \gamma^{(j)}(k) \leq 1. \tag{8.5}$$

Each system $S^{(j)}$ is characterized by its sampled-data model, derived at the sampling period $T_p$ according to the approach of [137], and described by

$$x^{(j)}(k+1) = A^{(j)} x^{(j)}(k) + B_1^{(j)} w^{(j)}(k) + B_2^{(j)} u^{(j)}(k), \tag{8.6a}$$

$$z^{(j)}(k) = C_1^{(j)} x^{(j)}(k) + D_{11}^{(j)} w^{(j)}(k) + D_{12}^{(j)} u^{(j)}(k), \tag{8.6b}$$

where $x^{(j)}(k) \in \mathbb{R}^{n_j}$ is the state vector, $w^{(j)}(k) \in \mathbb{R}^{q_j}$ is the disturbance input, $u^{(j)}(k) \in \mathbb{R}^{m_j}$ is the control input and $z^{(j)}(k) \in \mathbb{R}^{p_j}$ is the controlled output.

In the sequel, we will assume that for all $j \in \{1, \ldots, N\}$:

1. the pair $(A^{(j)}, B_2^{(j)})$ is controllable,
2. $Q^{(j)} = \begin{bmatrix} (C_1^{(j)})^T \\ (D_{12}^{(j)})^T \end{bmatrix} \begin{bmatrix} C_1^{(j)} & D_{12}^{(j)} \end{bmatrix} = \begin{bmatrix} Q_{xx}^{(j)} & Q_{xu}^{(j)} \\ (Q_{xu}^{(j)})^T & Q_{uu}^{(j)} \end{bmatrix} \geq 0$, with $Q_{uu}^{(j)} > 0$.

Let $\mathscr{S}$ be the global system composed by the systems $(S^{(j)})_{1\leq j\leq N}$. Assumption 1 [131, 262] is a necessary condition for the existence of a hyperperiod $T \times T_p$ and an off-line schedule (with hyperperiod $T \times T_p$) ensuring the reachability of the global system $\mathscr{S}$. When the scheduling of system $\mathscr{S}$ is performed according to a given off-line schedule ensuring its reachability, Assumption 2 [62, 78] guarantees the existence of a stabilizing controller (and also an optimal controller) and is usually made in optimal control problems.

Straightforward algebraic manipulations lead to the following extended state model representation of the global system $\mathscr{S}$:

$$x(k+1) = Ax(k) + B_1 w(k) + B_2 u(k), \tag{8.7a}$$

$$z(k) = C_1 x(k) + D_{11} w(k) + D_{12} u(k). \tag{8.7b}$$

Let $n = \sum_{j=1}^N n_j$, $m = \sum_{j=1}^N m_j$, $q = \sum_{j=1}^N q_j$ and $Q$ the matrix defined by

$$Q = \begin{bmatrix} C_1^T \\ D_{12}^T \end{bmatrix} \begin{bmatrix} C_1 & D_{12} \end{bmatrix}. \tag{8.8}$$

In the considered modeling, when a control task finishes its execution, then it immediately updates the plant, which means that

$$u^{(j)}(k) \text{ is updated during interval } [(k-1)T_p, kT_p) \iff \gamma^{(j)}(k) = 1. \tag{8.9}$$

The digital-to-analog converters use zero-order-holders to maintain the last received control commands constant until new control values are updated. Consequently, if a control command is not updated during the time slot $[(k-1)T_p, kT_p)$, then it is held constant. This assertion may be modeled by

$$\gamma^{(j)}(k) = 0 \implies u^{(j)}(k) = u^{(j)}(k-1). \tag{8.10}$$

*Remark 8.1* The introduction of the scheduling variables allows modeling the computational delays in some abstract way. The computational delay of a given task is thus approximated in terms of numbers of elementary time slots.

In order to formulate the joint problem of optimal control and scheduling, in addition to the modeling of the tasks and the representation of the system's dynamics, it is necessary to chose an adequate criterion of performance. The previous studies that were carried out on the joint problem of the optimal control and scheduling, starting from a given initial condition, have shown that the optimal schedule is dependent on the chosen initial states of the systems [26, 27]. This dependence may be exploited by the on-line scheduling algorithms in order to improve the control performance. But when only a fixed schedule is desired, it is necessary to use performance criteria that depend on the intrinsic characteristics of the system, not on a particular evolution or initial state. The use of the well-known $\mathscr{H}_2$ performance criterion provides a solution to meet these objectives. In fact, by using this performance index, the obtained off-line schedules will be independent of any initial conditions and disturbances. Moreover, the results may be easily transposed to a linear quadratic Gaussian (*LQG*)context, as illustrated in ([215], pp. 365–366).

### 8.1.2 Decomposability of the Optimal Integrated Control and Off-line Scheduling Problem

An off-line schedule is called *admissible* if it satisfies the constraints (8.2), (8.4) and (8.5). In theory, for a fixed $T$, the number of feasible off-line schedules with hyperperiod $T \times T_p$ is finite. Consequently, finding an optimal off-line schedule with hyperperiod $T \times T_p$ that minimizes the $\mathscr{H}_2$ norm of the global system amounts to the exploration of the set of feasible off-line schedules (with hyperperiod $T \times T_p$) and to the computation of the $\mathscr{H}_2$ norm of each feasible off-line schedule, in order to find an optimal one. However, in practice, *explicit search* methods like exhaustive search will rapidly suffer from the curse of dimensionality when used for solving problems with relatively large values of $H$, $T$ or $N$. That's why we propose the use of the branch and bound method, which is an *implicit search* method, allowing a more intelligent exploration of the set of admissible off-line schedules.

When the scheduling of the control tasks is performed according to an admissible fixed off-line schedule with hyperperiod $T \times T_p$, the scheduling functions $\gamma^{(j)}(k)$ and $e^{(j)}(k)$ that are associated to this off-line schedule will be both periodic with period $T$ (i.e., $\gamma^{(j)}(k) = \gamma^{(j)}(k+T)$ and $e^{(j)}(k) = e^{(j)}(k+T)$). It may be easily shown [29] that the model of the global system $\mathscr{S}$ may be described by a $T$-periodic discrete-time state representation. This remains true when the tasks are scheduled using the optimal off-line schedule that was previously determined. Consequently, using the well-known results of the optimal periodic control theory, we know that the optimal control of each system $S^{(j)}$ is a state-feedback control law. Since $\mathscr{H}_2$ optimal control problems are solved over an infinite horizon, the optimal controller of the global system will be $T$-periodic.

These observations show that the optimal integrated control and off-line scheduling problem may be decomposed into two sub-problems, which may be solved successively: the optimal scheduling sub-problem (which has to be solved first) and the optimal control sub-problem (whose solution requires the knowledge of the used scheduling).

### 8.1.3 Formal Definition of the $\mathscr{H}_2$ Norm

Assume now that the control tasks are scheduled according to an admissible off-line schedule with hyperperiod $T \times T_p$ and that ensure the reachability of system $\mathscr{S}$.

As previously mentioned, the off-line schedule as well as the optimal controller are $T$-periodic. For those reasons, the system $\mathscr{S}$ will be $T$-periodic. Consequently, in order to determine the $\mathscr{H}_2$ norm of the global system $\mathscr{S}$, we adopt a definition of this norm which is based on the impulse response of linear discrete-time periodic systems [256]. This definition generalizes the well-known definition of the $\mathscr{H}_2$ norm of discrete-time LTI systems. Let $(e_i)_{1\leq i\leq q}$ be the canonical basis vectors in $\mathbb{R}^q$ and $\delta_k$ the Dirac impulse applied at instant $k$. With these notations, $\delta_k e_i$ is a Dirac impulse applied to the $i$th disturbance input at instant $k$. Let $g_{ik}$ be the resulting controlled output of the global system $\mathscr{S}$ (in closed-loop) by assuming zero initial conditions and that the control input $u$ to the global system $\mathscr{S}$ satisfies the constraints (8.10). More formally, the response $g_{ik}$ to the Dirac impulse $\delta_k e_i$ assuming that the global system is controlled by the stabilizing control input $u$ that satisfies the constraints (8.10), is completely defined by the following equations:

$$\begin{aligned}
&x(0) = 0,\\
&w(k) = e_i,\\
&\text{and for all } l \in \mathbb{N},\\
&w(l) = 0 \quad \text{if } l \neq k,\\
&\gamma^{(j)}(k) = 0 \quad \Longrightarrow \quad u^{(j)}(k) = u^{(j)}(k-1) \quad \text{for all } j \in \{1,\ldots,N\},\\
&x(l+1) = Ax(l) + B_1 w(l) + B_2 u(l),\\
&z(l) = C_1 x(l) + D_{11} w(l) + D_{12} u(l),\\
&g_{ik}(l) = z(l).
\end{aligned} \tag{8.11}$$

The $\mathscr{H}_2$ norm of the periodic system $\mathscr{S}$ is defined as

$$\|\mathscr{S}\|_2 = \sqrt{\frac{1}{T}\sum_{k=0}^{T-1}\sum_{i=1}^{q}\|g_{ik}\|_{l_2}^2}. \tag{8.12}$$

The use of this definition to compute the $\mathscr{H}_2$ norm involves the computation of $\|g_{ik}\|_{l_2}$, which requires the observation of the system's response over an infinite time horizon. In this work, a very close approximation of $\|g_{ik}\|_{l_2}$ is obtained through a finite horizon $H$ from the instant where the impulse is applied. For that reason, it is necessary to choose $H$ greater than the response time of the global system $\mathscr{S}$. In the practical implementation of the algorithm, it is possible to visualize the responses to the different Dirac impulses, and to evaluate how much these responses (which correspond to exponentially stable trajectories) are close to zero. The $\mathscr{H}_2$ norm of the system is proportional to the square root of the sum of the squares of the $l_2$ norms corresponding to these responses. As a result, the "finite-horizon computed $\mathscr{H}_2$ norm" $\|\mathscr{S}\|_2(H)$ converges asymptotically to the true $\mathscr{H}_2$ norm $\|\mathscr{S}\|_2(\infty)$ computed over an infinite horizon, when $H \longrightarrow +\infty$. Consequently, for a desired precision, specified by the maximal absolute error $\varepsilon_{H_2}$ between the true $\mathscr{H}_2$ norm

$\|\mathscr{S}\|_2(\infty)$ computed over an infinite horizon and the "finite-horizon computed $\mathscr{H}_2$ norm" $\|\mathscr{S}\|_2(H)$, there will exist a horizon $H_{\min}$, such that, for all $H \geq H_{\min}$, $|\|\mathscr{S}\|_2(\infty) - \|\mathscr{S}\|_2(H)| \leq \varepsilon_{H_2}$. Based on this remark, and on the visualization tools which were implemented, the horizon $H_{\min}$ may be determined iteratively.

### *8.1.4 Solving of the Optimal Scheduling Sub-problem*

In this section, the translation of the optimal scheduling sub-problem in the sense of $\mathscr{H}_2$ into the mixed integer quadratic formulation is described. This translation requires the transformation of the involved constraints into linear equalities and/or inequalities. The constraints (8.10) may be translated into an equivalent conjunction of linear inequalities and equalities if extra variables are introduced, as illustrated in [21]. Since (8.10) is equivalent to

$$u^{(j)}(k) - u^{(j)}(k-1) = \gamma^{(j)}(k)u^{(j)}(k) - \gamma^{(j)}(k)u^{(j)}(k-1), \tag{8.13}$$

then, by introducing the extra variables

$$v^{(j)}(k) = \gamma^{(j)}(k)u^{(j)}(k), \tag{8.14}$$

the constraints (8.10) may be rewritten in the equivalent form

$$\begin{aligned}
v^{(j)}(k) &\leq U^{(j)}\gamma^{(j)}(k), \\
v^{(j)}(k) &\geq L^{(j)}\gamma^{(j)}(k), \\
v^{(j)}(k) &\leq u^{(j)}(k) - L^{(j)}\big(1 - \gamma^{(j)}(k)\big), \\
v^{(j)}(k) &\geq u^{(j)}(k) - U^{(j)}\big(1 - \gamma^{(j)}(k)\big),
\end{aligned} \tag{8.15}$$

where $U^{(j)}$ and $L^{(j)}$ are respectively the upper and the lower bounds of the control commands $u^{(j)}$ of the system $S^{(j)}$. In practice, these bounds correspond to the saturation thresholds of the actuators. If such a saturation does not occur, then $U^{(j)}$ (respectively $L^{(j)}$) may be set large enough (respectively, small enough) with respect to the values that the control signals may take during the evolution of the system. Note that the product $o^{(j)}(k) = \gamma^{(j)}(k)u^{(j)}(k-1)$ can be also translated by using the same procedure.

Thus, the constraints involved in this problem may be classified into two groups. The first group is related to the scheduling constraints (8.2), (8.4) and (8.5). Let

$$\bar{\Gamma}^{(j)} = \begin{bmatrix} \gamma^{(j)}(0) \\ \vdots \\ \gamma^{(j)}(H-1) \end{bmatrix} \quad \text{and} \quad \bar{\Gamma} = \begin{bmatrix} \bar{\Gamma}^{(1)} \\ \vdots \\ \bar{\Gamma}^{(N)} \end{bmatrix}.$$

Then the constraints belonging to this group may be described by

$$\mathscr{A}_s \bar{\Gamma} \leq \mathscr{B}_s \tag{8.16}$$

where $\mathscr{A}_s$ and $\mathscr{B}_s$ are respectively, a Boolean matrix and a Boolean vector of appropriate dimensions. For more details, please refer to Chap. 4.

The second group is related to the computation of the impulsive responses $g_{ik}$, for $0 \le k \le T-1$ and $1 \le i \le q$, over the horizon $H$. Let $u^{ik}$, $x^{ik}$, $z^{ik}$, $v^{ik}$ and $o^{ik}$ be respectively the values of the control, the state, the controlled output and the auxiliary variables corresponding to a Dirac impulse applied at instant $k$ to the $i$th disturbance input of the global system. Let $S^{ik}$ be the set of the involved constraints, for a given response $g_{ik}$. Here, $S^{ik}$ includes the state model (8.7a), (8.7b), control updates constraints (8.10), the Dirac impulse verifying $w_i^{ik}(k)=1$ and $w_r^{ik}(l)=0$ for $r \neq i$ and $l \neq k$, in addition to the constraints that must be added to the problem to ensure the causality of the response (i.e., $u^{ik}(l)=0$ for $l<k$).

Let

$$\bar{U}^{ik}=\begin{bmatrix} u^{ik}(0) \\ \vdots \\ u^{ik}(H-1) \end{bmatrix}, \qquad \bar{X}^{ik}=\begin{bmatrix} x^{ik}(0) \\ \vdots \\ x^{ik}(H-1) \end{bmatrix}, \qquad \bar{Z}^{ik}=\begin{bmatrix} z^{ik}(0) \\ \vdots \\ z^{ik}(H-1) \end{bmatrix},$$

$$\bar{V}^{ik}=\begin{bmatrix} v^{ik}(0) \\ \vdots \\ v^{ik}(H-1) \end{bmatrix}, \qquad \bar{O}^{ik}=\begin{bmatrix} o^{ik}(0) \\ \vdots \\ o^{ik}(H-1) \end{bmatrix} \quad \text{and} \quad \mathscr{V}^{ik}=\begin{bmatrix} \bar{U}^{ik} \\ \bar{X}^{ik} \\ \bar{Z}^{ik} \\ \bar{V}^{ik} \\ \bar{O}^{ik} \end{bmatrix}.$$

Then the set of constraints $S^{ik}$ may be described by

$$\mathscr{A}^{ik}\begin{bmatrix} \bar{\Gamma} \\ \mathscr{V}^{ik} \end{bmatrix} \le \mathscr{B}^{ik}, \tag{8.17}$$

where $\mathscr{A}^{ik}$ and $\mathscr{B}^{ik}$ are matrices of appropriate dimensions. Consequently, the optimal scheduling sub-problem in the sense of the $\mathscr{H}_2$ norm may be written in the form

$$\begin{cases} \displaystyle\min_{\bar{\Gamma},(\mathscr{V}^{ik})_{1\le i\le q, 0\le k\le T-1}} \sum_{k=0}^{T-1}\sum_{i=1}^{q} (\mathscr{V}^{ik})^T \mathscr{H} \mathscr{V}^{ik} \\ \mathscr{A}_s\bar{\Gamma} \le \mathscr{B}_s, \\ \mathscr{A}^{ik}\begin{bmatrix} \bar{\Gamma} \\ \mathscr{V}^{ik} \end{bmatrix} \le \mathscr{B}^{ik}, \quad \text{for } 1\le i\le q, 0\le k\le T-1, \end{cases} \tag{8.18}$$

where $\mathscr{H}$ is a matrix of appropriate dimensions, whose elements are functions of $C_1$, $D_{11}$, $D_{12}$ and $T$. For more details please refer to Chap. 4.

The problem (8.18) is a mixed integer quadratic problem (*MIQP*), and its resolution leads to an optimal off-line schedule (computed over the horizon $H$) $\bar{\Gamma}^*$. This resolution may be performed using the branch and bound algorithm.

*Remark 8.2* In *operation research* literature, the branch and bound method is considered among the most efficient methods for solving *MIQP* problems. As its name indicates, this method is based on two complementary mechanisms: *branching* and *bounding*.

- Branching makes possible decomposing a given problem into subproblems (by adding additional constraints), such that the union of the feasible solutions of these subproblems forms a partition (in the worst case a covering) of the feasible solutions of the original problem. In this manner, the resolution of the original problem is reduced to the resolution of the subproblems obtained by its branching.
- Bounding consists on computing an upper and a lower bound of the optimal solution of a given node. The bounding stage allows the branch and bound to avoid exploring the nodes where it is possible to certify that they do not contain any optimal solution. In fact, if the upper bound of a subproblem **A** is larger than the lower bound of another subproblem **B**, then the optimal solution cannot lie in the feasible set of solutions of subproblem **A**. For that reason, it becomes useless to branch subproblem **A**. In this way, the subproblem **A** is then *pruned*.

An intrinsic advantage of this method is that it allows finding sub-optimal solutions with a guaranteed distance to the true optimal solutions. The optimality distance is the difference between the cost of the best obtained solution and the lower bound on the optimal cost. The fact that the branch and bound algorithm allows finding sub-optimal solutions with a guaranteed distance to the true optimal ones comes from the fact that at each stage, this algorithm is able to compute a lower bound on the cost of the true optimal solutions (using continuous relaxation, i.e., replacing integer variables in $\{0, 1\}$ by continuous ones in $[0, 1]$ and solving an efficient quadratic program). For a description of the application of the branch and bound method for solving MIQP problems, the reader may consult [22] (Chap. 4, pp. 49–52) and references therein.

### 8.1.5 Solving the Optimal Control Sub-problem

Given an admissible off-line schedule, the ordered execution of control tasks during the major cycle $[0, T \times T_p)$ may be described by the sequence $(s(0), \ldots, s(\mathscr{T} - 1))$, where $\mathscr{T}$ is the number of control task executions during the hyperperiod $[0, T \times T_p)$. For example, the sequence $(s(0), s(1), s(2), s(3)) = (1, 2, 2, 3)$ indicates that, during the hyperperiod, the processor begins by executing task $\tau^{(1)}$, followed by two consecutive executions of task $\tau^{(2)}$, which are followed by the execution of task $\tau^{(3)}$. The total number of control task executions during the hyperperiod is $\mathscr{T} = 4$. Knowing the optimal off-line schedule $\bar{\Gamma}^*$, which is a solution of the optimization problem (8.18), it is then possible to derive the optimal control gains, according to the $\mathscr{H}_2$ performance criterion. In opposition to problem (8.18), the determination of the optimal control gains can be performed on each system separately.

In the practical implementation of the control tasks, it is impossible to directly measure the disturbances acting on the system. It is only possible to detect their effects on the state. When the controller has only access to the plant state, the optimal $\mathcal{H}_2$ controller becomes identical to the optimal *LQR* controller (for a complete proof, see [63], p. 143). For that reason, in the sequel, we will consider that $D_{11} = 0$. The expression of $z^{(j)}(k)$ reduces to

$$z^{(j)}(k) = C_1^{(j)}(k)x^{(j)}(k) + D_{12}^{(j)}(k)u^{(j)}(k).$$

Let $k_q^{(j)}$ be the index of the time slot corresponding to the $(q+1)$th update of the control commands of task $\tau^{(j)}$. More formally, $k_q^{(j)}$ are the discrete instants verifying

$$\exists k \in \{1, \ldots, T\}, \exists i \in \mathbb{N} \quad \text{such that } k_q^{(j)} = k + iT \text{ and } \gamma^{(j)^*}(k) = 1.$$

Based on this definition, the optimal control sub-problem can be stated as follows:

$$\begin{cases} \min\limits_{u^{(j)}} \displaystyle\sum_{k=0}^{+\infty} z^{(j)^T}(k) z^{(j)}(k) \\ \text{subject to} \\ \quad x^{(j)}(k+1) = A^{(j)}x^{(j)}(k) + B_2^{(j)}u^{(j)}(k), \\ \quad u^{(j)}(k) \text{ may be updated} \quad \text{if and only if} \quad \exists q \in \mathbb{N} \quad \text{such that } k_q^{(j)} = k, \\ \quad u^{(j)}(k) = u^{(j)}(k-1) \quad \text{if and only if} \quad \forall q \in \mathbb{N}, \quad k_q^{(j)} \neq k. \end{cases} \tag{8.19}$$

Since zero order holders are used to maintain the last received control inputs constant, an equivalent down-sampled discrete-time representation of system $S^{(j)}$ can be deduced. Let

$$\mathbf{x}^{(j)}(q) = x^{(j)}\left(k_q^{(j)}\right),$$
$$\mathbf{u}^{(j)}(q) = u^{(j)}\left(k_q^{(j)}\right).$$

Then the following relation is obtained:

$$\mathbf{x}^{(j)}(q+1) = \mathbf{A}^{(j)}(q)\mathbf{x}^{(j)}(q) + \mathbf{B}_2^{(j)}(q)\mathbf{u}^{(j)}(q) \tag{8.20}$$

with

$$\mathbf{A}^{(j)}(q) = A^{(j)^{k_{q+1}^{(j)} - k_q^{(j)}}},$$
$$\mathbf{B}_2^{(j)}(q) = \sum_{i=0}^{k_{q+1}^{(j)} - k_q^{(j)} - 1} A^{(j)^i} B_2^{(j)}.$$

Equation (8.20) describes the evolution of the state of the system $S^{(j)}$ at the time slots where its control inputs are updated. With the remark that for $k$ such that

$k_q^{(j)} \leq k < k_{q+1}^{(j)}$,

$$\begin{bmatrix} x^{(j)}(k) \\ u^{(j)}(k) \end{bmatrix} = \begin{bmatrix} A^{(j)^{k-k_q^{(j)}}} & \sum_{i=0}^{k-k_q^{(j)}-1} A^{(j)^i} B_2^{(j)} \\ 0_{m_j,n_j} & I_{m_j} \end{bmatrix} \begin{bmatrix} \mathbf{x}^{(j)}(q) \\ \mathbf{u}^{(j)}(q) \end{bmatrix},$$

it follows that

$$\sum_{k=k_q^{(j)}}^{k_{q+1}^{(j)}-1} z^{(j)^T}(k) z^{(j)}(k) = \begin{bmatrix} \mathbf{x}^{(j)}(q) \\ \mathbf{u}^{(j)}(q) \end{bmatrix}^T \mathbf{Q}^{(j)}(q) \begin{bmatrix} \mathbf{x}^{(j)}(q) \\ \mathbf{u}^{(j)}(q) \end{bmatrix},$$

where

$$\begin{aligned} \mathbf{Q}^{(j)}(q) = & \sum_{k=k_q^{(j)}}^{k_{q+1}^{(j)}-1} \begin{bmatrix} A^{(j)^{k-k_q^{(j)}}} & \sum_{i=0}^{k-k_q^{(j)}-1} A^{(j)^i} B_2^{(j)} \\ 0_{m_j,n_j} & I_{m_j} \end{bmatrix}^T \\ & \times Q^{(j)} \begin{bmatrix} A^{(j)^{k-k_q^{(j)}}} & \sum_{i=0}^{k-k_q^{(j)}-1} A^{(j)^i} B_2^{(j)} \\ 0_{m_j,n_j} & I_{m_j} \end{bmatrix} \\ = & \begin{bmatrix} \mathbf{Q}_1^{(j)}(q) & \mathbf{Q}_{12}^{(j)}(q) \\ \mathbf{Q}_{12}^{(j)^T}(q) & \mathbf{Q}_2^{(j)}(q) \end{bmatrix}. \end{aligned}$$

Let $\mathscr{T}^{(j)}$ be the number of executions of task $\tau^{(j)}$ during the hyperperiod. Since the optimal schedule is periodic, then the matrices $\mathbf{A}^{(j)}(q)$, $\mathbf{B}^{(j)}(q)$ and $\mathbf{Q}^{(j)}(q)$ are $\mathscr{T}^{(j)}$-periodic. Based on these notations, it is easy to see that the problem (8.19) is equivalent to the following periodic optimal control problem:

$$\begin{cases} \min_{\mathbf{u}^{(j)}} \sum_{q=0}^{\infty} \begin{bmatrix} \mathbf{x}^{(j)}(q) \\ \mathbf{u}^{(j)}(q) \end{bmatrix}^T \mathbf{Q}^{(j)}(q) \begin{bmatrix} \mathbf{x}^{(j)}(q) \\ \mathbf{u}^{(j)}(q) \end{bmatrix} \\ \text{subject to} \\ \quad \mathbf{x}^{(j)}(q+1) = \mathbf{A}^{(j)}(q)\mathbf{x}^{(j)}(q) + \mathbf{B}_2^{(j)}(q)\mathbf{u}^{(j)}(q). \end{cases} \tag{8.21}$$

It is worth mentioning that the problem (8.21) is a periodic optimal control problem over an infinite horizon. Let $\mathbf{S}^{(j)}(q)$ be the solution of the Riccati equation associated to the optimal control problem (8.21), which is described by

$$\begin{aligned} \mathbf{S}^{(j)}(q) = & \mathbf{A}^{(j)^T}(q)\mathbf{S}^{(j)}(q+1)\mathbf{A}^{(j)}(q) + \mathbf{Q}_1^{(j)}(q) \\ & - \left(\mathbf{A}^{(j)^T}(q)\mathbf{S}^{(j)}(q+1)\mathbf{B}^{(j)}(q) + \mathbf{Q}_{12}^{(j)}(q)\right) \\ & \times \left(\mathbf{B}^{(j)^T}(q)\mathbf{S}^{(j)}(q+1)\mathbf{B}^{(j)}(q) + \mathbf{Q}_2^{(j)}(q)\right)^{-1} \\ & \times \left(\mathbf{B}^{(j)^T}(q)\mathbf{S}^{(j)}(q+1)\mathbf{A}^{(j)}(q) + \mathbf{Q}_{12}^{(j)^T}(q)\right). \end{aligned} \tag{8.22}$$

Furthermore, the problem above (8.21) admits a unique solution $\mathbf{u}^{(j)^*}(q)$ defined by

$$\mathbf{u}^{(j)^*}(q) = -\mathbf{K}^{(j)}(q)\mathbf{x}^{(j)}(q)$$

with

$$\begin{aligned}\mathbf{K}^{(j)}(q) = &\left(\mathbf{Q}_2^{(j)}(q) + \mathbf{B}^{(j)^T}(q)\mathbf{S}^{(j)}(q+1)\mathbf{B}^{(j)}(q)\right)^{-1} \\ &\times \left(\mathbf{B}^{(j)^T}(q)\mathbf{S}^{(j)}(q+1)\mathbf{A}^{(j)}(q) + \mathbf{Q}_{12}^{(j)^T}(q)\right).\end{aligned} \tag{8.23}$$

By using the lifting approach described in [27], it may be proved that the matrices $\mathbf{S}^{(j)}(q)$ and $\mathbf{K}^{(j)}(q)$ are both $\mathscr{T}^{(j)}$-periodic. The optimal cost corresponding to an evolution of system $S^{(j)}$ from the instant $k_\iota^{(j)}$ to $+\infty$ is given by

$$\sum_{k=k_\iota^{(j)}}^{+\infty} z^{(j)^T}(k)z^{(j)}(k) = \mathbf{x}^{(j)^T}(\iota)\mathbf{S}^{(j)}(\iota)\mathbf{x}^{(j)}(\iota). \tag{8.24}$$

*Remark 8.3* Note that the optimal control gains $\mathbf{K}^{(j)}(q)$, which are $\mathscr{T}^{(j)}$-periodic, are independent of $\iota$, which represents the initial time of the optimal control problem. This considerably simplifies the implementation of the controller, and justifies their use in the on-line scheduling approach, which will be described in the next section.

*Remark 8.4* In equation (8.24), the cost corresponding to an evolution of the system $S^{(j)}$ from the instant $k_\iota^{(j)}$ to $+\infty$ is expressed as a quadratic function of the state $\mathbf{x}^{(j)}(\iota) = x^{(j)}(k_\iota^{(j)})$. However, Eq. (8.24) may only be used at instants where the control inputs of system $S^{(j)}$ are updated (i.e., belonging to the set $\{k_q^{(j)}, q \in \mathbb{N}\}$). In the following, we prove how the cost corresponding to an evolution of system $S^{(j)}$ from an arbitrary time instant $k$ to $+\infty$ may be written as an appropriate quadratic function of an extended state $\tilde{x}^{(j)}(k)$ such that

$$\tilde{x}^{(j)}(k) = \begin{bmatrix} x^{(j)}(k) \\ u^{(j)}(k-1) \end{bmatrix}.$$

Let $\mathscr{Y}^{(j)} = [\, 0_{m_j,n_j} \; I_{m_j} \,]$, $\bar{\mathscr{Y}}^{(j)} = [\, I_{n_j} \; 0_{n_j,m_j} \,]$ and $\tilde{\Psi}^{(j)}$ the matrix defined by:

$$\tilde{\Psi}^{(j)} = \begin{bmatrix} [\, A^{(j)} \; 0_{n_j,m_j} \,] + B^{(j)}\mathscr{Y}^{(j)} \\ \mathscr{Y}^{(j)} \end{bmatrix}.$$

Consider now an appropriate integer $k$ such that $k_{q-1}^{(j)} < k < k_q^{(j)}$. Since for the (positive) integer $l$ such that $k \leq l < k_q^{(j)}$, $u^{(j)}(l) = u^{(j)}(k) = u^{(j)}(k_{q-1}^{(j)})$, then for $k \leq l \leq k_q^{(j)}$

$$\tilde{x}^{(j)}(l) = \tilde{\Psi}^{(j)^{l-k}}\tilde{x}^{(j)}(k).$$

Consequently, the cost corresponding to an evolution of the system $S^{(j)}$ from an arbitrary time instant $k$ to $+\infty$ is given by

$$\sum_{l=k}^{+\infty} z^{(j)^T}(l) z^{(j)}(l) = \tilde{x}^{(j)^T}(k) \tilde{S}^{(j)}(k) \tilde{x}^{(j)}(k),$$

where

$$\tilde{S}^{(j)}(k) = \begin{cases} [\tilde{\Psi}^{(j)^{k_q^{(j)}-k}}]^T \check{S}^{(j)}(q) \tilde{\Psi}^{(j)^{k_q^{(j)}-k}} + \sum_{l=k}^{k_q^{(j)}-1} [\tilde{\Psi}^{(j)^{l-k}}]^T Q^{(j)} \tilde{\Psi}^{(j)^{l-k}} \\ \text{if } k_{q-1}^{(j)} < k < k_q^{(j)}, \\ \mathbf{S}^{(j)}(q) \quad \text{if } k = k_q^{(j)} \end{cases} \tag{8.25}$$

and

$$\check{S}^{(j)}(q) = \bar{\mathscr{Y}}^{(j)^T} \mathbf{S}^{(j)}(q) \bar{\mathscr{Y}}^{(j)}.$$

### 8.1.6 An Illustrative Numerical Example

In order to allow the evaluation of the systems dynamic characteristics influence on the obtained optimal off-line schedules, we consider a collection of 3 sampled-data *LTI* systems $S^{(1)}$, $S^{(2)}$ and $S^{(3)}$. Assume further that $S^{(1)}$ and $S^{(2)}$ are two second-order systems such that the first one is open-loop unstable whereas the second is open-loop stable. Assume also that the system $S^{(3)}$ is a fourth-order open-loop unstable system and corresponds to a linearized model of a fast inverted pendulum. This benchmark was obtained through the discretization of three continuous-time systems having different response times and stability properties. These three systems are given by:

$$x^{(1)}(k+1) = \begin{bmatrix} 0.9967 & 0.0266 \\ -0.2500 & 0.9987 \end{bmatrix} x^{(1)}(k) + \begin{bmatrix} 0.0132 \\ 0.9998 \end{bmatrix} w_1^{(1)}(k) + \begin{bmatrix} 0.0133 \\ 0.9999 \end{bmatrix} u^{(1)}(k),$$

$$z^{(1)}(k) = \begin{bmatrix} \begin{bmatrix} 89.4427 & 0 \\ 0 & 3.1623 \end{bmatrix} x^{(1)}(k) \\ 10^{-3} \begin{bmatrix} 0 & 15.9630 & 14.7321 \\ 0 & 14.7321 & 69.1671 \end{bmatrix} w^{(1)}(k) \\ u^{(1)}(k) \end{bmatrix},$$

$$x^{(2)}(k+1) = \begin{bmatrix} 0.9937 & 0.3458 \\ -0.0250 & 1.0007 \end{bmatrix} x^{(2)}(k) + \begin{bmatrix} 0.1243 \\ 0.7911 \end{bmatrix} w_1^{(2)}(k) + \begin{bmatrix} 0.1384 \\ 0.8008 \end{bmatrix} u^{(2)}(k),$$

$$z^{(2)}(k) = \begin{bmatrix} \begin{bmatrix} 2.2361 & 0 \\ 0 & 1 \end{bmatrix} x^{(2)}(k) \\ 10^{-3} \begin{bmatrix} 0 & 4.1674 & 3.8133 \\ 0 & 3.8133 & 17.4947 \end{bmatrix} w^{(2)}(k) \\ 3.1623\, u^{(2)}(k) \end{bmatrix},$$

$$x^{(3)}(k+1) = \begin{bmatrix} 1.1180 & 0 & 0.0025 & 0.5531 \\ 0 & 1 & 0.2129 & 0 \\ 0 & 0 & 0.7613 & 0 \\ 0.4518 & 0 & 0.0093 & 1.1180 \end{bmatrix} x^{(3)}(k) + 10^{-4} \begin{bmatrix} 23 \\ 9 \\ 75 \\ 81 \end{bmatrix} w_1^{(3)}(k)$$

$$+ 10^{-4} \begin{bmatrix} 28 \\ 11 \\ 88 \\ 106 \end{bmatrix} u^{(3)}(k),$$

$$z^{(3)}(k) = \begin{bmatrix} \begin{bmatrix} 10 & 0 & 0 & 0 \\ 0 & 31.62 & 0 & 0 \\ 0 & 0 & 1 & 0 \\ 0 & 0 & 0 & 1 \end{bmatrix} x^{(3)}(k) \\ 10^{-3} \begin{bmatrix} 0 & 0.3088 & 0.3809 & 0.0712 & 0.0922 \\ 0 & 0.3809 & 0.4984 & 0.1031 & 0.1278 \\ 0 & 0.0712 & 0.1031 & 0.1118 & 0.1176 \\ 0 & 0.0922 & 0.1278 & 0.1176 & 0.1249 \end{bmatrix} w^{(3)}(k) \\ u^{(3)}(k) \end{bmatrix}.$$

Assume also that each system $S^{(j)}$ is controlled by an independent control task $\tau^{(j)}$, $j \in \{1, 2, 3\}$, and that the task $\tau^{(3)}$ is followed by some communication task $\tau^c$. Thus, a data dependency between tasks $\tau^{(3)}$ and $\tau^c$ exists. Consequently, the task $\tau^c$ has to be executed immediately after the task $\tau^{(3)}$.

We assume that the execution platform is an Allen-Bradley CompactLogix 5320 programmable logic controller (PLC) [3] from Rockwell Automation. This execution platform is characterized by the execution times of its basic assembler instructions. The complete list of the execution times of all the instructions can be found in [3]. Based on this assembler language, a handwritten assembler code of the control tasks was derived. The estimated worst-case execution times (WCET) of the different tasks are given in Table 8.1. We also assume that a set of sporadic and aperiodic tasks may need to be executed on the same processor. Based on this requirement as well as on the WCET of the tasks, the elementary time slot duration $T_p$ was chosen equal to 1 ms.

The execution of the tasks $\tau^{(1)}$ and $\tau^{(2)}$ require only one time slot. The consecutive execution of the tasks $\tau^{(3)}$ and $\tau^c$ requires two time slots. Sporadic and aperiodic tasks require at most 40 % of the *CPU* power.

The optimal solutions, corresponding to different choices of $T$, are illustrated in Table 8.2. The relative optimality gap of the used branch and bound algorithm

**Table 8.1** Worst-case execution times of the control tasks

| Task | WCET (μs) |
|---|---|
| $\tau^{(1)}$ | 282.94 |
| $\tau^{(2)}$ | 282.94 |
| $\tau^{(3)} + \tau^{c}$ | 779.82 + 240.19 |

is equal to $10^{-5}$, which means that the best-obtained solution will be considered as an *optimal* solution if the difference between its cost and the lower bound of the *true optimal* cost is less than 0.01 %. The computations were performed on a PC equipped with a 3.6 GHz Intel Pentium IV processor and 1 GB of RAM. The optimization problem was solved using the solver CPLEX (Release 9.1.0) from ILOG. In this particular implementation of the optimization algorithm, $H$ must be a multiple of $T$. It is sufficient to choose $H$ larger than 27 to guarantee a maximal absolute error $\varepsilon_{\mathscr{H}_2} = 10^{-4}$.

The optimization results are given in Table 8.2. In this table, the two columns *CPU Time* indicate the time needed by the optimization algorithm to end. The algorithm stops when it proves that the best obtained solution is *close enough* to the lower bound of the optimal solution (i.e., the relative difference between the best obtained solution and the true optimal one is less than the specified relative optimality gap). The optimization results indicate that the minimal optimal schedule is of length $T = 5$. In this schedule, the task $\tau^{(2)}$ is first executed, followed by the execution of the task $\tau^{(1)}$, which is followed by the execution of the task $\tau^{(3)}$, which is finally followed by the execution of the task $\tau^{(1)}$. The length of the optimal schedules leading to the best $\mathscr{H}_2$ norm ($\mathscr{H}_2 = 9.7463$) is a multiple of 5. The resource allocation depends on the dynamics of the systems as well as on their sensitivity to the Dirac impulse disturbance of the $\mathscr{H}_2$ performance evaluation. It may be proved (using the formal definition of the $\mathscr{H}_2$ norm of the system $\mathscr{S}$) that a circular permutation of an optimal schedule still remains optimal.

The column *CPU Time Default* indicates the required *CPU* time in the case we use the default parameters of the solver. The column *CPU time with initial solu-*

**Table 8.2** Optimal $\mathscr{H}_2$ norm as a function of $T$

| $T$ | $H$ | $\mathscr{H}_2$ norm | Optimal schedule | *CPU* time default (s) | *CPU* time with initial solution (s) |
|---|---|---|---|---|---|
| 4 | 28 | 13.2472 | 321... | 86 | 13 |
| 5 | 30 | 9.7463 | 2131... | 58 | 40 |
| 6 | 30 | 11.7966 | 11213... | 243 | 185 |
| 7 | 28 | 13.0122 | 13213... | 533 | 482 |
| 8 | 32 | 12.1146 | 312131... | 1482 | 1151 |
| 9 | 27 | 10.6545 | 1312312... | 1796 | 1244 |
| 10 | 30 | 9.7463 | 21312131... | 4288 | 2062 |

*tion* indicates the required *CPU* time when the cost of a feasible initial solution is taken into account by the algorithm. In fact, it is always possible for the designer to derive a feasible schedule ad-hoc, by employing rules of thumb like those described in [14] (for example, choosing the sampling period $T_s$ of a first-order system such that $\frac{T_r}{T_s} \approx 4 \cdots 10$, where $T_r$ is the rising time of the system). The advantage of the branch and bound method is that it is able to use these feasible solutions to considerably reduce the search space by pruning the regions that are proved to be worst than those given by the initial solutions. This approach allows reducing the number of regions to be explored by the algorithm. Finally, note that the required time to find the optimal solution is lower than the time needed to prove that it is optimal. For example, for $T = 6$, the optimal solution was found after 69 seconds but 185 seconds were necessary to prove that it is optimal.

*Remark 8.5* When the number of systems (and the corresponding control tasks) increases, the potential for improving the global performance is more important but the complexity of the decision increases also, due to the "curse of dimensionality". In such a case, finding the true optimal off-line schedule becomes difficult. The advantage of the proposed approach lies in its ability to provide sub-optimal solutions with an upper-bounded distance from the optimality, along the execution of the branch and bound algorithm. For problems including a large number of tasks, the algorithm has to be run during a predetermined amount of time by starting from a feasible off-line schedule that the designer might determine ad-hoc using appropriate methods, in order to find solutions that are better than the given initial schedule, which was derived by the designer.

*Remark 8.6* In the obtained off-line schedule, the task $\tau^{(1)}$ was executed twice in the hyperperiod. Consequently, the optimal control gains for its two instances may be different. For that reason, a subscript will be added to distinguish two instances of the same task that may use different control gains. Thus, the two instances of task $\tau^{(1)}$ will be noted $\tau_1^{(1)}$ and $\tau_2^{(1)}$.

## 8.2 On-line Scheduling of Control Tasks

Assume that an off-line schedule, specifying how the different control and non-control tasks should be executed, was designed. This off-line schedule, which may be stored in some table, describes when each control task should be started and possibly the starting time instants of the time slots that are pre-allocated to the non-control tasks.

At runtime, the execution of the periodic off-line schedule may be described by using the notion of *pointer*. The pointer may be seen as a variable $p$, which contains the index of the control task to execute. The pointer is incremented after each control task execution. If it reaches the end of the sequence, its position is reset. After each

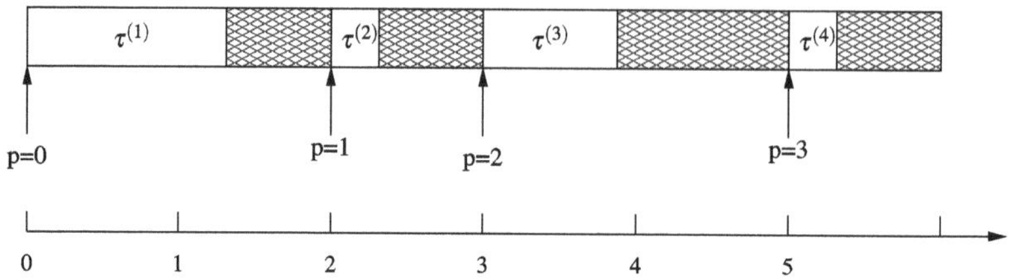


**Fig. 8.2** Absolute positions of the pointer

**Table 8.3** The map $t$

| Pointer position $p$ | Absolute position $t(p)$ |
|---|---|
| 0 | 0 |
| 1 | 2 |
| 2 | 3 |
| 3 | 5 |

task execution, the position of the pointer is updated according to the rule

$$p := (p+1) \mod \mathscr{T}. \tag{8.26}$$

By knowing the pointer position $p$, the control task to execute is $s(p)$. It is convenient to define the map $t$, which associates to the pointer position its absolute position (expressed as multiples of $T_p$ in the schedule). The map $t$ linking the different pointer positions to their absolute positions in the example of Fig. 8.2 is defined in Table 8.3.

Since the non-control tasks are sporadic or aperiodic, they do not necessarily use all the pre-allocated time slots for their execution. Consequently, it appears useful and perspicacious to employ this unused *CPU* power for improving the quality of control. Such improvements may be possible by executing more frequently the control tasks of the systems that have more need for additional computing resources. This requires specific scheduling algorithms, because the control tasks are used to control dynamical systems, where the timing of the input and output operations is capital for guaranteeing appropriately the stability and the performance requirements.

The idea behind the proposed scheduling strategy is to use the free available computing resources (i.e., where there are no sporadic nor aperiodic tasks to execute for example) in order to execute a feedback scheduler, which is responsible of determining the best scheduling of the control tasks in the future. In the sequel, the issues that are related to the execution of the feedback scheduler are discussed as well as the optimal pointer placement scheduling strategy, which represents the cornerstone of the used adaptive scheduling strategy. Finally, the method for reduction of the computational complexity of the feedback scheduler is described.

### 8.2.1 Execution of the Feedback Scheduler

As previously mentioned, the feedback scheduler is executed in the non-control tasks reserved time frames, when they are "free". Depending on the possibilities of the execution platform (whether it supports preemption or not), the feedback scheduler can be executed as:

- A preemptive task, with a given priority and deadline, and which will be discarded by the scheduler if it misses its deadline (in this case, the schedule execution is not modified).
- A non-preemptive task which is executed on predefined sub-slots of the non-control tasks slots.

In both cases, the possible real-time constraints of the sporadic or aperiodic tasks should be taken into account, when the feedback scheduler task is added.

*Remark 8.7* In real-time scheduling theory, static scheduling is mostly used in safety critical systems, in order to ensure a strong temporal determinism. The first motivation behind the use of a basic static schedule in the on-line scheduling heuristic is the need for predictability. In fact, the adaptive feedback scheduler, which will be described thereafter, needs to compute a predicted cost function in order to determine the scheduling decision. When employing the basic sequence, the computation of the predicted cost is considerably reduced since it boils down to the computation of a reduced number of quadratic functions. Furthermore, the use of the basic static schedule simplifies also the computations of the feedback gains. Finally, the implementation of the feedback scheduler as an extension of the basic sequencer may be easier than handling dynamic priorities in priority-based schedulers. In fact, only a few real-time operating systems support the run-time change of priorities.

### 8.2.2 Adaptive Scheduling of Control Tasks

Choosing the pointer replacement as an on-line decision allows exploiting more efficiently the available computational resources in order to improve the control performance. These resources are thus allocated according to the "instantaneous needs" of the controlled systems. In the proposed adaptive scheduling strategy, instead of a systematic use of (8.26), the position of the pointer is placed according to the knowledge of the controlled plants states, in order to ensure the improvement of the control performance as measured by a quadratic cost function. This strategy relies on computing $\mathscr{T}$ predicted cost functions corresponding to an evolution of the global system for $\mathscr{T}$ different positions of the pointer. Let

$$\tilde{x}(k) = \begin{bmatrix} \tilde{x}^{(1)}(k) \\ \vdots \\ \tilde{x}^{(N)}(k) \end{bmatrix}.$$

If the pointer is placed at position $p$ at instant $k$, then the cost function corresponding to an evolution of system $S^{(j)}$ over an infinite horizon starting from the state $\tilde{x}^{(j)}(k)$ at instant $k$ and using the static scheduling algorithm is

$$J^{(j)}(k, p) = \sum_{i=0}^{\infty} z^{(j)^T}(k+i) z^{(j)}(k+i) = \tilde{x}^{(j)}(k)^T \tilde{S}^{(j)}\big(t(p)\big)\tilde{x}^{(j)}(k). \quad (8.27)$$

If the pointer is placed at position $p$ at instant $k$, then the cost function corresponding to an evolution of the global system $\mathscr{S}$ over an infinite horizon starting from the state $\tilde{x}(k)$ at instant $k$ and using the static scheduling algorithm is

$$J(k, p) = J^{\text{sts}}\big(\tilde{x}(k), k, +\infty, p\big) = \sum_{j=1}^{N} J^{(j)}(k, p). \quad (8.28)$$

This strategy (called optimal pointer placement scheduling) was employed in [26] in the context of networked control systems in order to reduce the considerable computational complexity, which is required to find the true optimal control and scheduling decisions (this latter problem was investigated in [23]).

In the sequel, we describe how this concept may be deployed for monoprocessor scheduling. As mentioned previously, the feedback scheduler is triggered in some non-control task reserved time frame, if this time frame is free (i.e., there are no sporadic nor aperiodic tasks to execute, for example). It first acquires the state of all the controlled plants at instant $k_a$ (the instant where the feedback scheduler begins its execution), and then computes $\mathscr{T}$ predicted cost functions corresponding to an evolution over an infinite horizon, starting at instant $k_x$ (the instant where the next control task will begin its execution) for the $\mathscr{T}$ possible pointer positions. The task that will be executed at instant $k_x$ is the one corresponding to the pointer position that gives the minimal predicted cost. Figure 8.3 illustrates the method (in the case of the adaptive scheduling of the numerical example of Sect. 8.1.6).

Note that the feedback scheduler makes use of the knowledge of the state at instant $k_a$ to compute the predicted cost functions $J(k_x, p)$, for $p \in \{0, \ldots, \mathscr{T} - 1\}$, corresponding to an evolution starting at $k_x$. If the plant model is employed to predict the state at instant $k_x$ when knowing the state at instant $k_a$, then it is easy to establish that

$$J(k_x, p) = \tilde{x}^T(k_x)\tilde{S}\big(t(p)\big)\tilde{x}(k_x) = \tilde{x}^T(k_a)\hat{S}\big(t(p)\big)\tilde{x}(k_a).$$

Thus, the expression of $\hat{S}(t(p))$ can be easily deduced from the plant model, the expression of $\tilde{S}(t(p))$ and the control gains.

### 8.2.3 Reduction of the Feedback Scheduler Overhead

In the following, a method of reduction of the computational complexity of the feedback scheduler is proposed. This method relies on some intuitive observations,

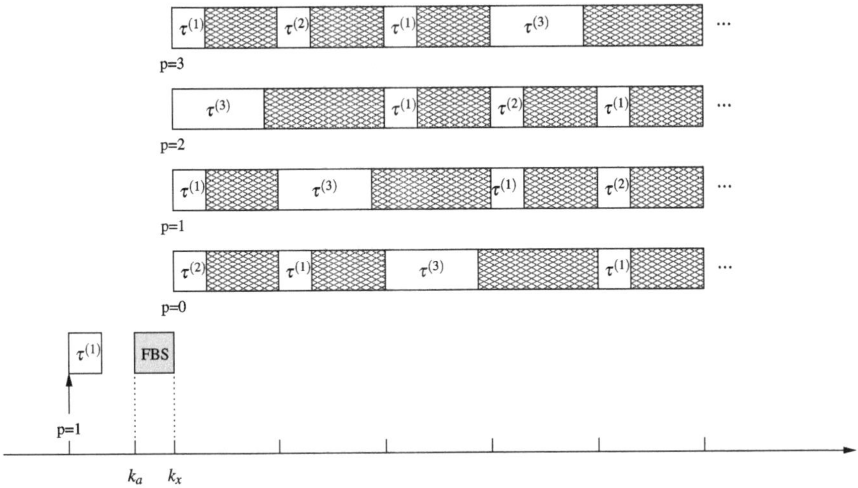


**Fig. 8.3** Optimal pointer placement scheduling of tasks $\tau^{(1)}$, $\tau^{(2)}$ and $\tau^{(3)}$. FBS is the abbreviation of feedback scheduler

which may be related to the concept of practical stability. The method is motivated by the fact that when a system $S^{(j)}$ is at the equilibrium (i.e., $x^{(j)} = 0$), its computational resources can be used to perform other tasks. If the system $S^{(j)}$ is close to the equilibrium, it may be less frequently updated than other systems that experience severe disturbances. This "proximity" to the equilibrium can be formalized by defining a practical equilibrium region $R^{(i)}$ for each system $S^{(i)}$. To define how much a system $S^{(i)}$ is close to the equilibrium, positive constants $\varepsilon_x^{(i)}$ are introduced. Here, $\varepsilon_x^{(i)}$ have to be chosen small enough such that when $\|\tilde{x}^{(i)}(k)\|_\infty \leq \varepsilon_x^{(i)}$, the system $S^{(i)}$ is considered to be "practically" at the equilibrium. Each practical equilibrium region can be formally defined by

$$R^{(i)} = \left\{\tilde{x}^{(i)}(k) \text{ such that } \left\|\tilde{x}^{(i)}(k)\right\|_\infty \leq \varepsilon_x^{(i)}\right\}.$$

In other words, systems (located) "in" the practical equilibrium region are systems whose affected computational resources may be released for the profit of other tasks. In order to ensure stability and performance improvements, this dynamic resource allocation has to be done by using a well-defined methodology, which will be developed in the sequel.

The basic ideas behind the proposed method for the reduction of the computational complexity of the feedback scheduler relies on changing the adaptive scheduling paradigm from

*find the best pointer position*

to

*find a pointer position that is better than the cyclic schedule.*

The answer to the latter problem requires less computational resources than the answer to the first one, and can take advantage from the knowledge that a given system is practically stable. For a given system $S^{(j)}$, let

$$\bar{J}^{(j)}_{\min}(p) = \min_{\tilde{x}^{(j)}(k)\in R^{(j)}} \left(J^{(j)}(k,p)\right)$$

and

$$\bar{J}^{(j)}_{\max}(p) = \max_{\tilde{x}^{(j)}(k)\in R^{(j)}} \left(J^{(j)}(k,p)\right),$$

then we have the following proposition.

**Proposition 8.1** *Let $p_1$ and $p_2$ be two pointer positions and $I$ a subset of $\{1,\ldots,N\}$. If*

$$\forall i \in I, \quad \left\|\tilde{x}^{(i)}(k)\right\|_\infty \le \varepsilon_x^{(i)}$$

*and*

$$\sum_{j\in\{1,\ldots,N\}-I} J^{(j)}(k,p_2) + \sum_{j\in I} \bar{J}^{(j)}_{\max}(p_2) < \sum_{j\in\{1,\ldots,N\}-I} J^{(j)}(k,p_1) + \sum_{j\in I} \bar{J}^{(j)}_{\min}(p_1)$$

*then*

$$J(k,p_2) < J(k,p_1).$$

*Proof* This result follows straightforwardly from the fact that

$$\begin{aligned} J(k,p_1) &= \sum_{j\in\{1,\ldots,N\}-I} J^{(j)}(k,p_1) + \sum_{j\in I} J^{(j)}(k,p_1) \\ &\ge \sum_{j\in\{1,\ldots,N\}-I} J^{(j)}(k,p_1) + \sum_{j\in I} \bar{J}^{(j)}_{\min}(p_1) \end{aligned}$$

and

$$\begin{aligned} J(k,p_2) &= \sum_{j\in\{1,\ldots,N\}-I} J^{(j)}(k,p_2) + \sum_{j\in I} J^{(j)}(k,p_2) \\ &\le \sum_{j\in\{1,\ldots,N\}-I} J^{(j)}(k,p_2) + \sum_{j\in I} \bar{J}^{(j)}_{\max}(p_2). \end{aligned}$$

□

The constants $\bar{J}^{(j)}_{\min}(p)$ and $\bar{J}^{(j)}_{\max}(p)$ above may be easily pre-computed off-line by using a standard QP solver.

*Example 8.2* In order to illustrate the gains in terms of computational complexity, we reconsider the example proposed in the Sect. 8.1.6. The computation of the true pointer position requires $\mathscr{T}(n+m-1)(n+m+1) = 480$ additions and

**Algorithm 8.1:** Pseudocode of the feedback scheduling algorithm, called Reactive Pointer Placement (*RPP*) scheduling algorithm

```
Read x(k);
p := p + 1  mod 𝒯;
if ∀i ∈ I_p, ‖x̃^(i)(k)‖_∞ ≤ ε_x^(i) or I_p = ∅ then
    if ∃π ∈ 𝒫_p / Σ_{j∈Ī_p} J^(j)(k, π) + Σ_{j∈I_p} J̄_max^(j)(π) <
    Σ_{j∈Ī_p} J^(j)(k, p) + Σ_{j∈I_p} J̄_min^(j)(p) then
        p := π;
    end if
end if
execute task s(p);
```

$\mathscr{T}(n+m)(n+m+1) = 528$ multiplications. If the systems $S^{(2)}$ and $S^{(3)}$ are practically stable, searching for a pointer position that may be better than the next pointer position require, in the worst case, $\mathscr{T}(n_1 + m_1 - 1)(n_1 + m_1 + 1) = 32$ additions and $\mathscr{T}(n_1 + m_1)(n_1 + m_1 + 1) = 48$ multiplications, thus a substantial reduction of respectively 95 % and 90 % of the computational complexity.

In the case when input operations are performed by independent hardware devices and that their computational overhead may be neglected, a possible pseudocode of the feedback scheduler can be used, and its description is given in the listing below. This feedback scheduling algorithm is called, reactive pointer placement (*RPP*) scheduling algorithm. In this listing, for a given $p \in \{0, \ldots, \mathscr{T} - 1\}$, $I_p$ is a set of plant indices and $\bar{I}_p = \{1, \ldots, N\} - I_p$. For a given $p \in \{0, \ldots, \mathscr{T} - 1\}$, $\mathscr{P}_p$ is a set of pointer positions that does not contain $p$. Typically, $I_p$ is chosen such that $\bar{I}_p = \{s(\pi), \pi \in \mathscr{P}_p\}$. When this choice is performed, $I_p$ contains $s(p)$. Note that $I_p$ may also be chosen such that $I_p = \emptyset$ (when more computational resources are available to the feedback scheduler). The choice of the elements of $\mathscr{P}$ and $I_p$ depends on the available computing resources that may be dedicated to the feedback-scheduler. More resources are available, more potential pointer positions may be tested. It is important to point out that the computational resources needed are time-varying and are in the worst-case proportional to the cardinality of the sets $\bar{I}_p$ (which is the complementary of the set $I_p$ in $\{1, \ldots, N\}$) and $\mathscr{P}_p$. In fact, these resources are essentially needed for the on-line computation of the functions $J^{(j)}(k, \pi) - J^{(j)}(k, p)$, for $j \in \bar{I}_p$, as illustrated in the *RPP* algorithm listing (see Algorithm 8.1).

*Remark 8.8* When using the proposed complexity reduction methodology, it is possible to bound the computational requirements of the feedback scheduler even if the number of control tasks increases. In general, the situations where all the independent systems are severely disturbed at the same time appear less frequently than the situations where some systems are disturbed and some others are in normal operation. The *RPP* algorithm allocates the available resources according to the needs of the controlled systems. Therefore, intuitively, when the number of tasks increases, the potential to improve the control performance may be more important because the quantity of unneeded resources that may be allocated to the disturbed systems will also increase.

### 8.2.4 Stability and Performance Improvements

Let $J^{\text{rpp}}(\tilde{x}(i), i, f)$ be the cost function corresponding to an evolution from instant $k = i$ to instant $k = f$ starting from the extended state $\tilde{x}(i)$ where the *RPP* scheduling algorithm is applied. The performance improvements of the *RPP* scheduling algorithm are stated in the following theorem.

**Theorem 8.1** [28] *Let $\tilde{x}(0)$ be a given initial extended state of the global system $\mathscr{S}$ (composed of the systems $(S^{(j)})_{1\leq j\leq N}$) and $p_0$ an initial pointer position of the static scheduling algorithm. Then*

$$J^{\text{rpp}}\big(\tilde{x}(0), 0, +\infty\big) \leq J^{\text{sts}}\big(\tilde{x}(0), 0, +\infty, p_0\big).$$

*Proof* Let $\tilde{x}^{\text{rpp}}$ be the extended state trajectory of the system $\mathscr{S}$ when scheduled by using *RPP*, $p(l)$ the pointer position at the $(l+1)$th execution of *RPP* and $k_l$ the index of the time slot corresponding to the end of this $(l+1)$th execution. Without any loss of generality, assume that $k_0 = 0$. Let $J^{\text{rpp-sts}}(l)$ be the cost function corresponding to an evolution starting from the initial state $\tilde{x}(0)$ where *RPP* is applied from the instant $k_0 = 0$ to the instant $k_l$ and then followed by the application of the static scheduling algorithm (which is applied from instant $k_{l+1}$ to $+\infty$). By construction of the *RPP* scheduling strategy (as described in listing Algorithm 8.1), and by an appropriate use of Proposition 8.1, the pointer position at the $(l+1)$th execution is set to position $p(l)$ such that

$$\begin{aligned} &J^{\text{sts}}\big(\tilde{x}^{\text{rpp}}(k_0), k_0, +\infty, p(0)\big) \\ &\quad \leq J^{\text{sts}}\big(\tilde{x}(0), 0, +\infty, p_0\big), \quad \text{if } l = 0, \end{aligned} \tag{8.29a}$$

$$\begin{aligned} &J^{\text{sts}}\big(\tilde{x}^{\text{rpp}}(k_l), k_l, +\infty, p(l)\big) \\ &\quad \leq J^{\text{sts}}\big(\tilde{x}^{\text{rpp}}(k_l), k_l, +\infty, \big(p(l-1)+1\big) \bmod \mathscr{T}\big), \quad \text{if } l > 0, \end{aligned} \tag{8.29b}$$

where $J^{\text{sts}}(\tilde{x}^{\text{rpp}}(k_l), k_l, +\infty, p(l))$ (respectively, $J^{\text{sts}}(\tilde{x}^{\text{rpp}}(k_l), k_l, +\infty, (p(l-1) + 1) \bmod \mathscr{T}))$ represents the predicted cost corresponding to an evolution over an infinite horizon of the global system from the state $\tilde{x}^{\text{rpp}}(k_l)$ where the pointer at instant $k_l$ is placed at position $p(l)$ (respectively $(p(l-1)+1) \bmod \mathscr{T}$). In fact, as previously mentioned, the scheduling decisions preformed by the *RPP* algorithm are based on an appropriate prediction of the evolution of the system, for selected pointer positions, and under the assumption that the static scheduling algorithm is used during these predicted evolutions. Note that $(p(l-1)+1) \bmod \mathcal{T})$ represents the pointer position obtained by incrementing the pointer according to relation (8.26) (i.e open-loop static scheduling).

Adding $J^{\text{rpp}}(\tilde{x}(0), 0, k_{l-1})$ to both left and right terms of inequality (8.29b), for $l > 0$, and by taking into account the fact that

$$J^{\text{rpp-sts}}(l) = J^{\text{rpp}}\big(\tilde{x}(0), 0, k_{l-1}\big) + J^{\text{sts}}\big(\tilde{x}^{\text{rpp}}(k_l), k_l, +\infty, p(l)\big),$$

and

$$J^{\text{rpp-sts}}(l-1) = J^{\text{rpp}}\big(\tilde{x}(0), 0, k_{l-1}\big) + J^{\text{sts}}\big(\tilde{x}^{\text{rpp}}(k_l), k_l, +\infty, \big(p(l-1)+1\big) \bmod \mathcal{T}\big),$$

we get

$$J^{\text{rpp-sts}}(l) \le J^{\text{rpp-sts}}(l-1), \quad \text{for } l > 0. \tag{8.30}$$

Recall that

$$J^{\text{sts}}\big(\tilde{x}^{\text{rpp}}(k_0), k_0, +\infty, p(0)\big) = J^{\text{rpp-sts}}(0).$$

Thus, it is ease to see that

$$\lim_{l \to +\infty} J^{\text{rpp-sts}}(l) = J^{\text{rpp}}\big(\tilde{x}(0), 0, +\infty\big).$$

Finally, the use of the inequalities (8.29a) and (8.30) leads to

$$J^{\text{rpp}}\big(\tilde{x}(0), 0, +\infty\big) \le J^{\text{sts}}\big(\tilde{x}(0), 0, +\infty, p_0\big). \tag{8.31}$$

□

Theorem 8.1 simply indicates that the *RPP* strategy guarantees the performance improvements with respect to the static scheduling algorithm. It is worth mentioning that the stability of the *RPP* scheduling algorithm follows straightforwardly from Theorem 8.1 as stated in the following corollary.

**Corollary 8.1** [28] *If $Q$ is positive definite and if the asymptotic stability of the global system $\mathcal{S}$ (composed of systems $(S^{(j)})_{1 \le j \le N}$) is guaranteed by the static scheduling algorithm, then it is also ensured by the* RPP *scheduling algorithm.*

*Proof* When $Q$ is positive definite, then $J^{\text{sts}}(\tilde{x}(0), 0, \infty, p_0)$ (resp. $J^{\text{rpp}}(\tilde{x}(0), 0, \infty)$) is finite if and only if system $\mathcal{S}$ is asymptotically stable. Now, by knowing the asymptotic stability of the system scheduled when the static scheduling algorithm is employed, the use of the relation (8.31) allows to conclude. □

*Remark 8.9* When the *RPP* strategy is used, the effective frequency with which the control tasks are executed is not necessarily constant. In particular, some tasks may be executed more frequently during some periods of time. Based on Theorem 8.1, we proved that this irregular execution performed according to some well defined methodology (on-line scheduling that minimizes an infinite horizon cost) provides a better (and in the worst-case situation a similar) control performance than (to) the basic static schedule. The performance improvements may be seen as a direct application of the Bellman optimality principle. In fact, the *RPP* algorithm has more degrees of freedom than the static scheduling algorithm. Its worst-case performance corresponds to the static scheduling strategy. The control coefficients (gains) that will be used by each control task are taken into account when the infinite horizon cost is computed. For that reason, there is no need to recompute the control coefficients of the tasks that will be executed at some irregular rate. This considerably reduces the complexity of the on-line scheduling algorithm.

### *8.2.5 Reduction of Input Readings Overhead*

In some situations, the input operations are performed directly by the processor. The control application may be also networked: the control inputs from the plant and outputs to the plant are transmitted over the network. Consequently, reading the inputs from the plant at each execution of the feedback scheduler may result in a computational or bandwidth overhead. In order to reduce this overhead, the feedback scheduler has to read at most $n_s$ inputs, such that $n_s < n$. The most efficient way is to read these inputs according to the optimal off-line sampling periods of the controlled systems. In the following, a heuristic approach for selecting the inputs that are read by the feedback scheduler is proposed. Since the feedback scheduler is executed in the non-control slots which follow the execution of the control tasks, we may associate to each pointer position of set of $n_s$ control inputs that will be read. Let $\Pi$ be the set of all the permutations of the $\mathscr{T}$-tuple $(0, 1, \ldots, \mathscr{T}-1)$. Let $\pi \in \Pi$ a given permutation and $\zeta_\pi$ the sequence defined by

$$\zeta_\pi = \big(\zeta_\pi(0), \zeta_\pi(1), \ldots, \zeta_\pi(\mathscr{T}-1)\big) = \big(s\big(\pi(0)\big), s\big(\pi(1)\big), \ldots, s\big(\pi(\mathscr{T}-1)\big)\big). \tag{8.32}$$

Assume that a $n_s$-tuple of permutations $\pi = (\pi_1, \ldots, \pi_{n_s})$ is defined. If the pointer is at position $p$, then the inputs $\{\zeta_{\pi_1}(p), \ldots, \zeta_{\pi_{n_s}}(p)\}$ are read. The *binary detection indicators* associated to each system $S^{(j)}$ are defined by:

$$\begin{cases} \mathscr{Z}_\pi^{(j)}(p) = 1 & \text{if } \exists k \in \{1, \ldots, n_s\} \text{ such that } \zeta_{\pi_k}(p) = j, \\ \mathscr{Z}_\pi^{(j)}(p) = 0 & \text{otherwise.} \end{cases} \tag{8.33}$$

The binary detection indicators simply show whether or not the outputs of system $S^{(j)}$ are read, for a given position $p$ of the sequence pointer. Let $\bar{\mathscr{Z}}_\pi^{(j)}(k) = \mathscr{Z}_\pi^{(j)}(k \bmod \mathscr{T})$. The inputs that are read at position $p$ are given by the detection sequence

$$\zeta = \big(\big\{\zeta_{\pi_1^*}(0), \ldots, \zeta_{\pi_{ns}^*}(0)\big\}, \ldots, \big\{\zeta_{\pi_1^*}(\mathscr{T}-1), \ldots, \zeta_{\pi_{ns}^*}(\mathscr{T}-1)\big\}\big)$$

that is determined by the solution $(\pi_1^*, \ldots, \pi_{n_s}^*)$ of the following optimization problem

$$\begin{cases} (\pi_1^*, \ldots, \pi_{n_s}^*) \\ \quad = \min_{(\pi_1,\ldots,\pi_{n_s})\in\Pi^{n_s}} \sum_{j=1}^{N} \max_{k_1,k_2,k_1\leq k_2} \big\{k_2 - k_1, \text{ such that: } \bar{\mathscr{Z}}_\pi^{(j)}(k_1) = 1, \ \bar{\mathscr{Z}}_\pi^{(j)}(k_2) = 1 \\ \qquad\qquad \text{and for all } k_1 < k < k_2, \ \bar{\mathscr{Z}}_\pi^{(j)}(k) = 0\big\}. \end{cases} \tag{8.34}$$

This optimization problems aims in minimizing the sum over $j$ of the "maximal distances" between two successive output reading of system $S^{(j)}$.

### 8.2.6 A Numerical Example

In order to evaluate the proposed approach, consider the real-time implementation of the example already presented in the Sect. 8.1.6. Based on the assembler language of the CompactLogix 5320 PLC, a handwritten assembler code of the feedback scheduler was written. The tasks execution was simulated using the toolbox TRUETIME [5, 56], which allows the co-simulation of distributed real-time control systems by appropriately taking into account the effects of the execution of the control tasks and the data transmission on the controlled systems dynamics.

Based on Table 8.1, the processor utilization of the control tasks, when scheduled by using a basic sequencer, is equal to 32.57 %. This means that aperiodic tasks as well as the feedback scheduler may have at most a utilization rate of 67.43 %. Note that this implementation assumes that input operations are performed by independent hardware devices. An important issue concerns the execution overhead of the feedback scheduler. In order to be able to implement the proposed feedback scheduling algorithm, the worst case execution time of the feedback scheduler must be less or equal to 717.06 µs or 980.08 µs, depending on the non-control tasks slots lengths. To solve this problem, an implementation of the feedback scheduler (as described in Sect. 8.2.3) was performed. When the pointer is at position $p$, the feedback scheduler tries to answer to the question:

*Is position $\pi^*(p)$ better than position $p$?*

where $\pi^* = (\pi^*(0), \pi^*(1), \pi^*(2), \pi^*(3)) = (1, 0, 3, 2)$. The parameters $\varepsilon_x^{(1)} = \varepsilon_x^{(2)} = \varepsilon_x^{(3)} = 0.001$, $I_0 = \{2, 3\}$, $I_1 = \{1, 3\}$, $I_2 = \{2, 3\}$, $I_3 = \{1, 2\}$ and $\mathscr{P}_p = \{\pi^*(p)\}$, for $p \in \{0, \ldots, 3\}$ were chosen. These parameters are simply a consequence of the available computational resources to execute the feedback scheduler. In order to improve the responsiveness of the feedback scheduler, $\pi^*(p)$ was computed using the optimization heuristic (8.34). Based on these choices, the worst-case execution times of the feedback scheduler (for a given pointer position) are given in Table 8.4. Note that, in this example, testing over all the pointer positions based

**Table 8.4** Worst-case execution time of the feedback scheduler (for each pointer position)

| Pointer position | WCET (μs) |
|---|---|
| 0 | 455.27 |
| 1 | 455.27 |
| 2 | 968.60 |
| 3 | 455.27 |

on the global system model (which corresponds to the use of the OPP algorithm) requires an execution time of 3393.8 μs.

The global system responses corresponding to the states $x_1$, $x_3$, $x_5$ and $x_6$ of the global system (i.e., states $x_1^{(1)}$, $x_1^{(2)}$, $x_1^{(3)}$ and $x_2^{(3)}$) as well as the associated accumulated global cost are depicted in Fig. 8.4. It is assumed that the global system is started from the initial state $[1\ 0\ 1\ 0\ 1\ 0\ 0\ 0]^T$. The three systems $S^{(1)}$, $S^{(2)}$ and $S^{(3)}$ reach the practical stability region at $t = 0.019$ s, $t = 0.026$ s and $t = 0.057$ s, respectively. Assume now that at $t = 0.0663$ s, the system $S^{(1)}$ is severely disturbed. The conditions of the "reactive pointer change" are fulfilled. The *RPP* algorithm reacts then at the instant $t = 0.0665$ s to execute the task $\tau_2^{(1)}$ instead of the task $\tau^{(3)}$ (as illustrated in Fig. 8.5). These changes allowed to better react to this disturbance and to improve the quality of control (as illustrated in Fig. 8.4). It may be seen that when these disturbances occur, the execution time of the feedback scheduler increases (as illustrated in Fig. 8.5). This increase is due to the additional

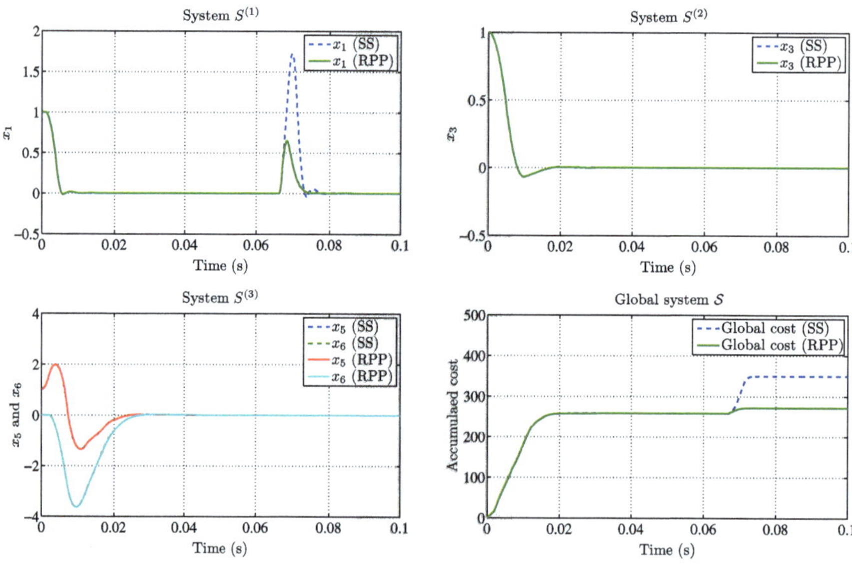


**Fig. 8.4** Global system responses and accumulated cost using the static scheduling (StS) and the *RPP* scheduling algorithms

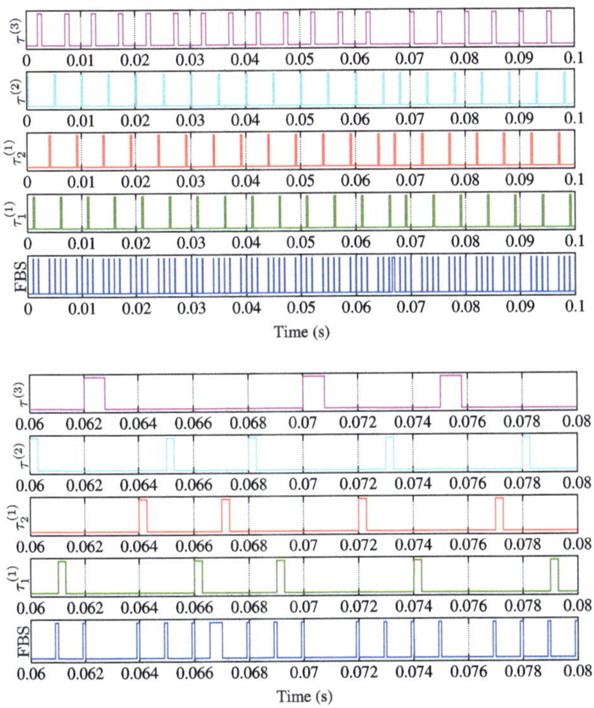


**Fig. 8.5** *RPP* schedule (*top*) and zoom on the *RPP* schedule between instants 60 ms and 80 ms (*bottom*)

test which is needed to guarantee that the "reactive pointer change" will improve the performances while maintaining the stability.

Finally, the *RPP* scheduling algorithm is tested in the case when all the systems are in the practical equilibrium region, except one, which is perpetually disturbed. In the simulations depicted in Fig. 8.6 correspond to the following situation: the systems $S^{(1)}$ and $S^{(2)}$ remain in the equilibrium region whereas the system $S^{(3)}$ is continuously disturbed with a band limited white noise characterized by a noise power of 0.1 and a correlation time of $1 \times 10^{-4}$. The simulation results indicate significant improvements in control performance (Fig. 8.6, left). These improvements are due to the fact that the unused computing resources are allocated by the feedback scheduler to the control of the system $S^{(3)}$, as illustrated in (Fig. 8.6, right).

*Remark 8.10* In all the situations, the *RPP* scheduling algorithm is able to maintain the stability of all the plants and to provide a performance that is better (and in the worst-case similar) to the one obtained by the static scheduling algorithm, even in the case when all the plants are disturbed at the same time. In this example, the restriction concerning the fact that *RPP* is able to react to only one disturbance is only due to the limitations of the computational resources that are allocated to its execution. If more resources are available to the execution of the feedback scheduler,

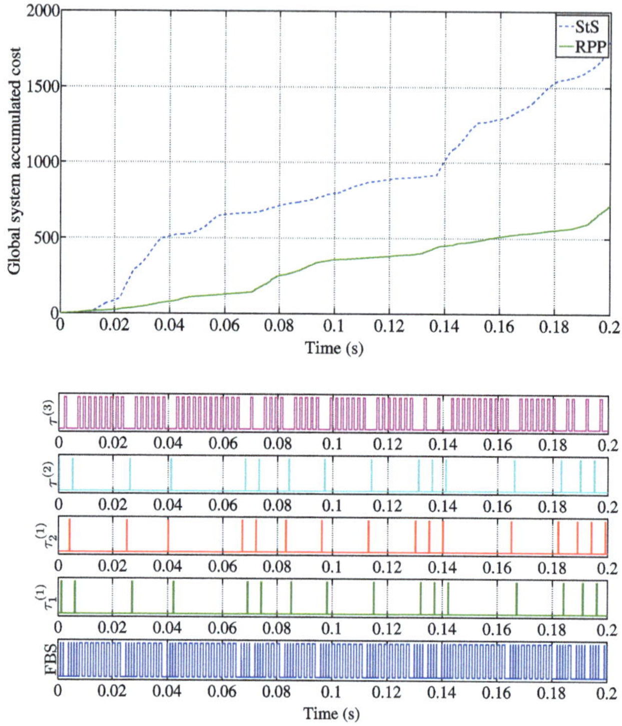


**Fig. 8.6** Accumulated cost functions (*top*) and tasks schedules (*bottom*) when the systems $S^{(1)}$ and $S^{(2)}$ are in the equilibrium regions whereas the system $S^{(3)}$ is continuously disturbed

then the parameter $I_p$ may be set to $\emptyset$. In this situation, the reactive pointer change may be performed even when all the plants are disturbed at the same time. This situation will be illustrated in the experimental study proposed in the next section.

## 8.3 Application to a *DC* Motor Associated to an Embedded Processor

In order to evaluate the extent of the considered theoretical assumptions, to study experimentally the robustness of the proposed on-line scheduling algorithm and to compare its performance to state of the art methods (i.e fixed priority preemptive scheduling), the proposed methodology is applied to a real world application. The considered application is the concurrent speed control of two DC motors when using an embedded processor. The control objective is to impose desired some appropriate angular velocities to each DC motor. The considered motors present some non-linearities at zero crossing velocities (dry friction). However, they may be approximated by first-order LTI models. The control tasks are executed on an aJile aJ-PC104 board, equipped with a JEM2 processor, clocked at 40 MHz. A PC

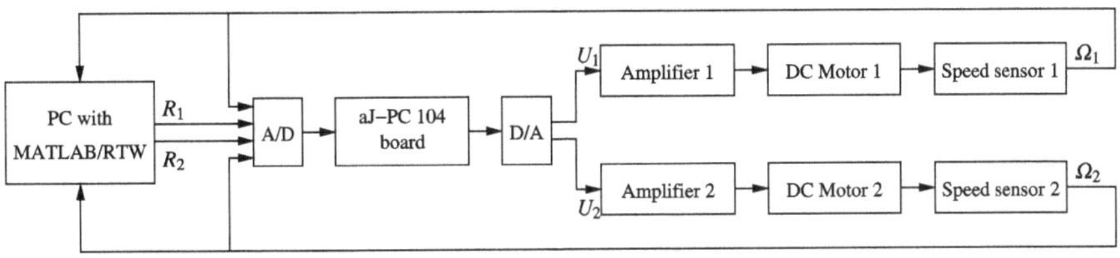


**Fig. 8.7** Experimental setup

**Table 8.5** Measured execution times of the segments of a motor control task and of the *RPP* algorithm

| Motor control task | | *RPP* algorithm | |
|---|---|---|---|
| Segment | Execution time (μs) | Segment | Execution time (μs) |
| Inputs reading | 363 | Inputs reading | 371 |
| Control computation | 58 | Scheduling computation | 230 |
| Output application | 322 | | |
| Total | 743 | Total | 601 |

with a Matlab/RTW (Real-Time Workshop) board generates the desired set points ($R_1$ and $R_2$) and measures the voltages that are the image of the motors speed (i.e. proportional to the motors angular velocities). The experimental setup is described in Fig. 8.7. The model of each DC motor (viewed between the input of the amplifier and the output of the speed sensor) may be respectively approximated by the following first-order models:

$$\Omega_1(s) = \frac{0.94}{1+6.2s}U_1(s),$$

and

$$\Omega_2(s) = \frac{1.17}{1+2.9s}U_2(s),$$

where $\Omega_i$ and $U_i$ are respectively, the image of the angular velocity (output of the speed sensor) and the control voltage (input to the amplifier) of the $i$th DC motor, $i \in \{1,2\}$. Using the standard *LQR* method, two proportional-integral (*PI*) controllers were derived in order to impose similar closed-loop characteristics for the two motors.

Each *DC* motor is controlled by a control task $\tau^{(i)}$, $i \in \{1,2\}$. The processor is shared with other periodic and sporadic tasks. Assume that, during the experiments, a set of periodic tasks, with respective periods 4 ms, 40 ms and 200 ms, are executed in parallel (with the highest priorities, assigned according to the rate monotonic rule). Their *CPU* utilization load is equal to 60 %. The measured execution times of the different segments (i.e., parts) of the motors control tasks and the feedback scheduler are given in Table 8.5. The segments *Inputs reading* and

*Output application* represent the computations that are necessary for the acquisition of the controller inputs and the application of the control outputs, respectively. The segment *Control computation* represents the computation of the state feedback control, whereas the segment *Scheduling computation* represents the code of the *RPP* algorithm, as illustrated in the listing Algorithm 8.1 (evaluation and comparison of the quadratic cost functions and pointer placement decision).

The objective of the experiments is to compare the control performance when the two control tasks $\tau^{(1)}$ and $\tau^{(2)}$ are scheduled using:

- The rate monotonic scheduling algorithm (RM), where the preemption is authorized, and where each DC motor is controlled by an independent control task running at the period of 1 s.
- The *RPP* scheduling algorithm with the parameters $T_p = 0.5$ s, a basic optimal off-line schedule 12... (obtained by using the $\mathcal{H}_2$ optimization), $I_1 = \emptyset$, $I_2 = \emptyset$, $\mathcal{P}_0 = \{1\}$ and $\mathcal{P}_1 = \{0\}$.

Based on these parameters, the controllers that are used in both cases are identical (both derived at the sampling period of 1 s).

The experimentation scenario is the following: Initially, the two DC motors are stopped (i.e., their angular velocity is zero). The higher priority tasks use 60 % of the *CPU*. Assume that at the instant $t = 3$ s, the set point of the first motor is set to 40 rd/s, and, next, at the instant $t = 12$ s, the set point of the second motor is set to $-30$ rd/s. Assume further that at the instant $t = 23$ s, the set points of the two motors are respectively changed to 30 rd/s and $-50$ rd/s. Next, assume that at the instant $t = 30$ s, a set of sporadic tasks, using more than 40 % of the *CPU*, begin their execution. The priority of these tasks is lower than the priority of the motors control tasks but higher than the priority of the feedback scheduler, and at the instant $t = 35$ s, the set point of the second motor is set to $-30$ rd/s. Finally, at the instant $t = 43$ s, the set point of the first motor is set to 50 rd/s.

The experimental results (i.e., the measured angular velocities of DC motors 1 and 2 using the *RM*and the *RPP* scheduling algorithms) are depicted in Fig. 8.8. They show that a significant improvement in the control performance is achieved by the *RPP* scheduling algorithm with respect to the rate monotonic scheduling algorithm. These improvements consist in a considerable reduction of the overshoot and an improvement of the response time, manifesting themselves after set point changes. These improvements are due to a more efficient use of the available computing resources by the *RPP* algorithm. It is worth mentioning that the steady state behavior using the two algorithms is similar. At the instant $t = 23$ s, the set points are applied at the same time. The *RPP* algorithm chooses to allow the resources to the control task of the second motor, which experiences the largest deviations, in order to optimize the global cost. From the instant $t = 30$ s, the admitted sporadic tasks induce a situation of processor overload. The feedback scheduler cannot be executed. In fact, as previously indicated, the feedback scheduler has the lowest priority in the system. Consequently, in this situation of overload, motors control tasks are executed according to the optimal static $\mathcal{H}_2$ schedule. For that reason, the obtained control performance (obtained from instant $t = 30$ s) is similar to that obtained with the rate monotonic algorithm (i.e., very similar responses are observed).

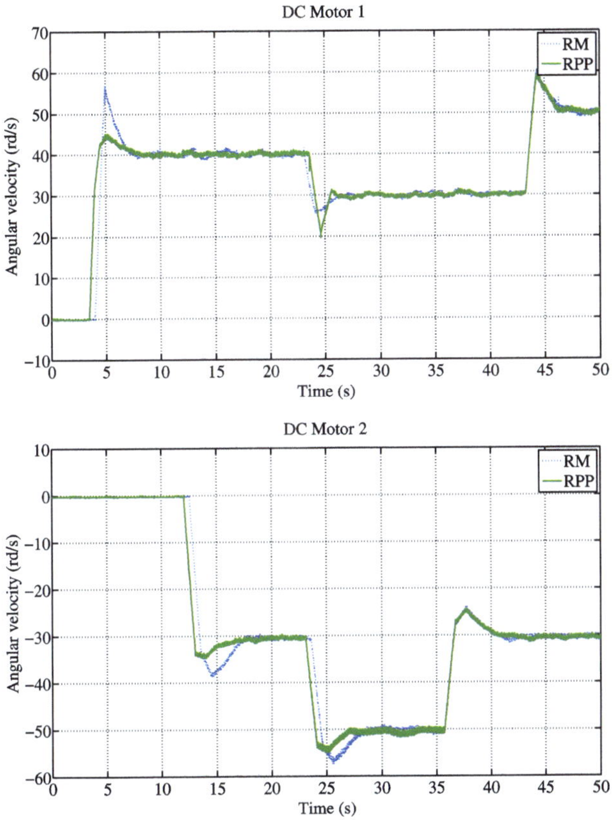


**Fig. 8.8** Speed responses corresponding to the *RPP* and RM scheduling algorithms

## 8.4 Notes and Comments

In this chapter, an approach for control tasks scheduling is proposed. This approach aims to improve the control performance through a more efficient use of the available computational resources. First, an optimal integrated control and non-preemptive off-line scheduling problem is formulated. This problem is based on the $\mathscr{H}_2$ performance criterion (which is closely approximated over a sufficiently large finite horizon) to statically allocate the computing resources according to the intrinsic characteristics of the controlled systems. When using such an approach, the *sampling periods* of the control tasks are optimally chosen. The proposed method is based on the decomposition of the optimal control and off-line scheduling problem into two independent sub-problems. The first sub-problem aims at finding the optimal cyclic schedule and is solved by using the branch and bound method. The second sub-problem takes into account the result of the first sub-problem to determine the optimal control gains by applying the lifting technique. A plant state-based feedback scheduling mechanism is then proposed, enabling to enhance the control performance with respect to the optimal off-line scheduling algorithm. This algorithm combines a more frequent "reading" of the systems' inputs with a state feedback based resource allocation in order to improve the control performance. The underlined ideas of the proposed approach opens interesting perspectives. and some of them have received a lot of attention in the open literature. We classified the related works and contribution in four categories, as follows:

- *Optimal control and monoprocessor scheduling*: The problem of the optimal selection of the control tasks periods subject to schedulability constraints was discussed in *Seto et al.* [208]. In *Rehbinder et al.* [199], the optimal off-line scheduling of control tasks in the sense of *LQG* was addressed in the case when all the control tasks have the same constant execution time. Its solution was performed by using an appropriate exhaustive search method. As a consequence the, application of this approach is limited to the systems having a limited number of tasks. Similarly, the problem of the optimal mono processor scheduling of control tasks in order to optimize an appropriate robustness metric (i.e., the stability radius) was treated in *Palopoli et al.* [192, 193]. The problem of the optimal control and scheduling in the sense of *LQG* was introduced and discussed in [155]. The proposed method allows pruning the search tree in order to reduce the combinatoric explosion.
- *Scheduling of control tasks in environments with variable computing workload*: It is well known that the worst-case analysis techniques proposed by *Sha et al.* [211] may be used in order to guarantee the deadlines of the tasks with variable but bounded execution times. However, when the average execution time is smaller than the worst-case execution time (*WCET*), these techniques lead, in general, to an oversized design. Recently, in order to handle variations in tasks execution times and system overload more efficiently, new approaches have been proposed. Among them, we may cite the feedback scheduling algorithms discussed in *Lu et al.* [160], *Cervin et al.* [55], *Robert et al.* [203], *Xia and Sun* [255] as well as the elastic task model given in *Buttazzo et al.* [49].

  In the elastic task model proposed by *Buttazzo et al.* [49], a periodic task set containing $\mathcal{N}$ tasks may be seen as a sequence of $\mathcal{N}$ linear springs. In this model, the utilization factor of a task is analogous to the spring's length. The tasks may change their utilization rate in order to handle overload conditions, which may occur, for example, if a new task is admitted to the computing system. In order to ensure the schedulability of the task set, the tasks are compressed or decompressed. For instance, in *Liu et al.* [158], the elastic task model was applied to the scheduling of control tasks with variable execution times. The use of this method allows the application of the approach of *Seto et al.* [208] in order to find the optimal tasks periods based on tasks average execution times (instead of their worst-case execution times), leading to an improvement of the control performance. Next, *Buttazzo et al.* [47] generalized this approach to take into account the degradations that may occur to the control system if its control task designed to work at a given rate runs at another one. The analytical expressions of the performance degradations were explicitly given. Furthermore, a compensation method was proposed. This method allows to trade-off the performance degradations and the required memory space (needed to store the parameters of the pre-computed control laws to be used by the compensation algorithm).

The work of *Bini and Cervin* [34] tackled the control/scheduling co-design problem of the optimal period assignment for multi-loop control systems. The key improvement consists in the derivation of closed-form formulae for optimal sampling periods based on approximate control delay enabling delay-aware period assignment under fixed-priority scheduling. In *Samii et al.* [205], a control/scheduling co-design method for distributed control systems integrating controller design is proposed. Both static and priority-based scheduling of the tasks and messages optimizing the overall control performance are obtained. Next, in *Samii et al.* [206], the control-performance optimization for embedded multi-mode control systems is addressed. The complexity reduction problem is solved by an appropriate selection of the system's modes to be implemented and an appropriate synthesis approach producing schedules and controllers for an efficient deployment of embedded multi-mode control systems is proposed.

Inside the class of on-line *FBS* methods there are two main trends. The first one mainly concerns the methods relaying on the instantaneous plant state information and metric in order to adapt the sampling period. In this class, one can find the works of *Martì et al.* [169] and *Ben Gaid et al.* [32], where quadratic approximation of the relation between task periods and control costs allowed to explicitly formulate the optimal state-based sampling period problem as a convex optimization problem. The second class includes the methods calculating the future sampling periods based on finite or infinite horizon metric. In this sense, one can cite the results proposed in *Henriksson and Cervin* [114], *Castane et al.* [52], *Cervin et al.* [57]. This type of approach was also adopted by *Ben Gaid et al.* [28, 29] where the sampling period is calculated as a function of the states of the controlled plants, a given static scheduling and a quadratic metric over an periodic infinite horizon.

- *Adaptive sampling*: As pointed out by [125], the methods of *Dorf et al.* [77] and [176] are closely related. In this context, *Hsia* proposed a generic approach allowing to derive adaptive sampling laws [126].

  More recently, an important direction of research on the *DCES* performance enhancement through adaptive sampling is offered by the use of the so-called *Event Driven Controllers* (*EDC*). The aim of such a controller is to reduce the calculation and the communication resource utilization in order to provide in priority those tasks which need more.

  As stated in *Årzèn* [11], event-driven control (*EDC*) appears to be closer in nature to the way a human behaves as a controller. The *EDC* controller are triggered by some *external* events or they are self-triggered. The works of *Heemels et al.* [110], *Åström and Bernhardsson* [15], *Tabuada and Wang* [226], *Suh et al.* [223], *Heemels et al.* [111], *Henningsson et al.* [113], *Lunze* [163], *Wang et al.* [246], *Marchand et al.* [165], and many others fall in the event-triggered class whereas the works of *Velasco et al.* [242], *Anta and Tabuada* [6–9], *Wang and Lemmon* [247–250],*Mazo and Tabbuada* [170, 172], *Araujo* [10] represent some of the contribution in the class of self triggered controller.

In a general distributed control architecture related to *DCES*, the reduction of sampling frequency is not always sufficient to enhance system performances. It is also needed to synchronize the decisions between subsystems sharing a given calculation or communication resources. The recent works of *Tabuada* [225], *Seyboth* [210], *Donkers and Heemels* [75], *Mazo and Cao* [171], *Wang and Lemmon* [251], *de Persis* [70] go in this direction with the objective to coordinate the subsystems' local decision.

# Part III
# Insight on Stability and Optimization of Distributed Control and Embedded Systems

# Chapter 9
# Insight in Delay System Modeling of *DCESs*

In the *Part* II of this document, the object of our study is the *DCES* analysis and design with a special emphasis on the scheduling of control signals and their real time update depending on the system state. This simply means that the systems resources are allocated with respect to the system's performance enhancement and robustness. Such a state dependent scheduling may be classified in the set of event driven scheduling algorithms which have attracted recently the attention of many researchers in control domain.

Another characteristic of the class of applications treated there, which is more related to the technology of communication, concerns the communication and calculation model. The network was supposed to posses real-time characteristics and the transfer time of messages packets from a node processor to another node processor of the *Hardware/Software* application architecture was considered as being negligible. The calculation as well as the communication models were supposed to be synchronous. Thus, such an assumption simply means that the delay induced model is uniquely given by the static scheduling hyperperiod. Meanwhile, for the class of this type of applications, in our opinion, it is impossible to neglect the delay induced by scheduling of messages on the network as well as of control and other tasks (actuation, sampling, ...) running on the node processors.

The object of our study in Chap. 10 will be the influence of the induced delay on the stability and on the performance of some special class of *DCESs* corresponding to those studied and discussed in Chap. 7. The stability analysis will allow us defining appropriate stability domain for task period variation and for *DCES-induced* delay. Such a study, done generally off-line, will help us to better understand their influence on the system's stability and performance as well as to propose new control switching algorithms in order to handle network or processor over-load state and control messages packets dropping. As it will be seen in the Chaps. 12 and 13, the knowledge of the induced delay influence will also allow slowing down the system's performance deterioration by switching from a given control law to the zero-control one.

In our opinion, it is important to operate this analysis for the case where the induced delay is inferior to a sampling period as well when it is composed by a se-

A. Çela et al., *Optimal Design of Distributed Control and Embedded Systems*,
Communications and Control Engineering, DOI 10.1007/978-3-319-02729-6_9,

quence of sampling period. The objective is double. *First*, it concerns the choice of sampling periods. We will see that we can increase it without degrading the system's performances. This means that, for the same *level of service or performances*, we use less communication and calculation resources. During this analysis, we observe that the generally accepted intuition consisting in the fact that more communication and calculation resources allow to better control a given system is not always verified. *Second*, it is helpful for proposing new control and/or recovery strategies in the case of messages dropouts and control task preemption. Based on this analysis, in Chap. 11, we also propose scheduling algorithms able to handle resources and/or performance optimization of *DCESs*.

Our main objective is to be as realistic as possible with respect to the representation of communication and calculation model in the presence of induced delays. We will see that this representation is sufficiently informative to address stability analysis as well as the performance optimization of *DCESs*.

In the sequel, we briefly introduce the existing delay models of *DCESs*, the corresponding technical background, different related problems, as well as appropriate methodologies and approaches adopted in handling them. We absolutely do not have the pretentiousness to cover the whole field of *DCESs* or to provide all the solutions proposed so far in the literature. Meanwhile, in Sect. 9.6, we give a very brief presentation of some of them which seems to be in close relation or complementary the ones proposed in the book.

## 9.1 Preliminaries

The objective of this section is to scan some communication and calculation model of *DCESs* in order to derive their related induced delay models. As in the previous chapters, the formalism adopted is the state-space representation and the control law structure is the standard state feedback. So, if we ignore the distributed aspects of *DCESs*, its open-loop LTI continuous model can be written in the following simple form:

$$\dot{x}(t) = Ax(t) + Bu(t), \tag{9.1}$$

where $x(t)$ and $u(t)$ represent the system state and control input, respectively, and $A$, $B$ are the corresponding state and the input matrices. An equivalent view of the continuous-time state space model (9.1) is that of an infinite-resource *DCES* application.[1]

For the clarity and consistency of the presentation, we suppose that the open-loop system is not stable, that is its state matrix $A$ is *not* Hurwitz. By focusing on the case of open-loop unstable systems, it allows operating a more detailed analysis of the communication and calculation model. Furthermore, it also helps to better understanding how the resource allocation influences the system's stability and system performances.

[1] In other words, calculation speed and communication bandwidth are assumed to be infinite.

Practically, the state $x(t)$ is sampled by the sensors at discrete time instants commonly called *sampling instants*. We denote these sampling instants by $s_k, k \in \mathbb{N}$. The corresponding variable sampling period is denoted by $T(k) \triangleq s_{k+1} - s_k$. Once a sampled-data state, $x(s_k)$, is obtained from the controller node, it will be used to calculate the control signal and to send it to the actuator in order to generate a new control input for the system (9.1) at time $a_k$ ($k \in \mathbb{N}$), called the actuation instants. In the case of a general architecture of *DCESs*, this process typically includes state sampling by the sensor node, its transition from the sensor node to the controller node, calculation of a new control value by the controller, its transition from the controller node to the actuator node, and, finally, the corresponding updating of the control input of the controlled plant. Thus, in general, $a_k \geq s_k$ and we call the *input/output time-latency* $a_k - s_k$ the *DCES*-induced delay, denoted by $\tau(k) \triangleq a_k - s_k$. It is important to point out the fact that, in a normal situation, the induced delay may be produced from the data packets scheduling on the network as well as from the scheduling of different calculation tasks on the controller, sensor and actuator node processors.

At the actuator node, we generally use a zero-order-hold (*ZOH*) device in order to reconstruct the control signal between two actuation instants, leading to the following continuous-time control input:

$$u(t) = u(s_k), \quad a_k \leq t < a_{k+1}, \tag{9.2}$$

where $u(s_k)$ is the control input derived from the system's state, $x(s_k)$, and updated by the actuator at instant $a_k$.

It is clear that the input/output latency depends on the type of communication network, scheduling algorithm, as well as on the communication and calculation model assigned to each node processor. For more details on some special cases, we may refer to the analysis operated in *Lian et al.* [153] and in *Ben Gaid et al.* [31]. We can not neglect the influence of network and processor load which generally induce jitter or variation in the value of induced delay.

Combining the expression of the control signal (9.2) with the continuous model (9.1), we obtain the sampled-data model of a linear time invariant (LTI) *DCES*:

$$\begin{cases} \dot{x}(t) = Ax(t) + Bu(t), \\ u(t) = Kx(s_k), \end{cases} \quad \forall t \in [a_k, a_{k+1}), \; k \in \mathbb{N}. \tag{9.3}$$

This general sampled-data *LTI* model of *DCES* has some interesting properties that will be particularly exploited in order to handle the related analysis and design problems. In this sense, we recall the following result:

**Theorem 9.1** [90] *For a given sampled-data* LTI *system* (9.3) *with bounded sampling intervals and a given initial state* $x(0)$, *the following conditions are equivalent*:

1. $\lim_{t\to\infty} x(t) = 0$,
2. $\lim_{k\to\infty} x(s_k) = 0$.

To complete the result above, some further analysis proposed by *Hetel et al.* [121] pointed out a stronger connection between these classes of systems (continuous-time and its corresponding discrete-time ones) in terms of stability. In other words, the above theorem simplifies the study of such sampled-data systems by suggesting to use a discrete-time approach in studying the stability of *DCESs*. For more details on the subject, we refer the reader to the recent *PhD* thesis of *Fiter* [84].

This general model encompassing the influence of all the components composing *DCESs* may be simplified with respect to real applications. If the communication and calculation delay are negligible (infinite communication and calculation capacity), we may neglect the input/output latency or the induced delay. In this case, we use $t_k$ to denote the sampling instants (instead of $s_k$), and so the corresponding sampled-data control is given by:

$$u(t) = u(t_k), \quad t_k \le t < t_{k+1}. \tag{9.4}$$

Such a case corresponds to $s_k = a_k$, i.e., $\tau(k) = 0$. The corresponding sampling period is hence denoted by $T(k) \triangleq t_{k+1} - t_k$. In particular, when the sampling period is constant $T$, (9.4) can be further simplified as:

$$u(t) = u(kT), \quad kT \le t < (k+1)T. \tag{9.5}$$

The expressions (9.2), (9.4), and (9.5) represent three types of sampled-data control temporal model which allow concluding on the *DCES* stability.

When we adopt the commonly-used state feedback control, the expressions of control signals given by (9.2), (9.4), and (9.5) can be written as:

$$u(t) = Kx(s_k), \quad a_k \le t < a_{k+1},\ k \in \mathbb{N}, \tag{9.6}$$

$$u(t) = Kx(t_k), \quad t_k \le t < t_{k+1},\ k \in \mathbb{N}, \tag{9.7}$$

$$u(t) = Kx(kT), \quad kT \le t < (k+1)T,\ k \in \mathbb{N}, \tag{9.8}$$

where $K$ is the feedback gain matrix.

The *DCES* given by the combination of plan model (9.1) and the sample controller model (9.6) has the following discrete-time expression:

$$z(k+1) = \Phi\big(\tau(k), T(k)\big)z(k), \tag{9.9}$$

where $z(k) = \binom{x(s_k)}{u(a_{k-1})}$ and the two-variable transition matrix function $\Phi(\alpha, \beta)$ is given by

$$\Phi(\alpha, \beta) = \begin{pmatrix} e^{A\beta} + \int_\alpha^\beta e^{A\theta}\, d\theta\, BK & \int_0^\alpha e^{A\theta}\, d\theta\, B \\ K & 0 \end{pmatrix}. \tag{9.10}$$

Similarly, from (9.1) and (9.7), we obtain the following discrete-time model of the corresponding *DCES*:

$$x(t_{k+1}) = \Phi\big(T(k)\big)x(t_k) \tag{9.11}$$

where the one-variable transition matrix $\Phi(\beta)$ is given by:

$$\Phi(\beta) = e^{A\beta} + \int_0^{\beta} e^{A\theta}\, d\theta\, BK. \tag{9.12}$$

Finally, the third considered model of a *DCES* derived by (9.1) and (9.8) has the following discrete-time expression:

$$x\big((k+1)T\big) = \Phi(T)x(kT). \tag{9.13}$$

For the sake of simplicity and with a little abuse of notation, we define by $\Phi$ the transition matrix function of a *DCES*. When it has two arguments, the induced delay and the sampling period, its definition refers to (9.10). Finally, in the case when it has only one argument corresponding to the sampling period, its definition refers to (9.12).

## 9.2 Hyper-Sampling Period and Induced Mathematical Model

As presented in the forgoing section, the induced delays depend mainly on the scheduling policy of the network, that of calculation tasks on each node processors, as well as on their respective resources capacities. If we look more closely to the calculation model of the node processor, we may easily see that the induced delay depends on the priorities assigned to each task running on the processor as well as on the scheduling policy. In Chap. 8, we derived a scheduling policy for a given set of tasks called *RPP* which falls in the class of synchronous scheduling policy.[2] This scheduling policy defines the instants of the task executions, their duration, as well as the number of its instances in a given hyper-period. The task execution model may be seen as a number of execution or task instances in each hyper-period which gives, in some sense, a *macroscopic* view. On the other hand, the number of executions of a task in a hyper-period reflects the relative importance of this task. Another macroscopic view of the task execution model may be given by the finite and ordered set of sampling periods. So, we may construct a relevant *DCES* model based on the model of calculation and communication implemented on a *DCES*, that can be modeled as delay systems. It is worth mentioning that, for the stability analysis as well as for the performance optimization of *DCESs*, it is extremely important to have a precise variation bounds time characteristics of the delay.

In order to illustrate our purposes, we will treat a simple and reduced model of a *DCES* given in Fig. 9.1 where the *Hardware/Software* architecture is composed of $p$ tasks and $q$ processors. Suppose that the tasks $\upsilon^1$, $\upsilon^2$, and $\upsilon^3$ are executed on *processor 1*. As stated before, some of control tasks may be executed multiple times during each hyper-period.

In this sense, an example of tasks execution by a hyper-sampling period is given in Fig. 9.2. We may easily observe that task $\upsilon^1$ is executed twice within each hyper-period. It is supposed that a control task includes state signal sampling, control

[2]Synchronous scheduling policy gives a fixed order of task executions.

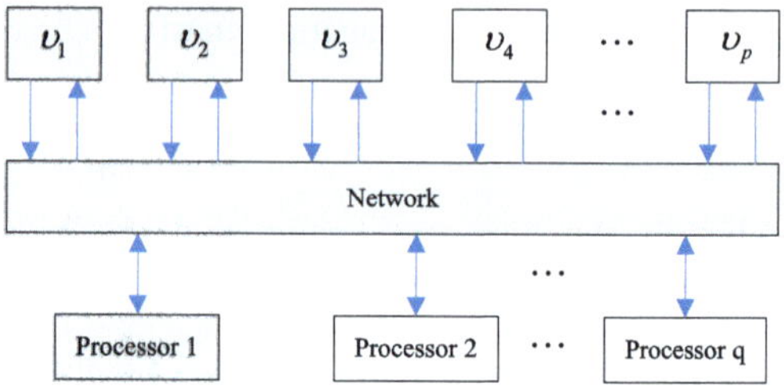


**Fig. 9.1** A general Hardware/Software architecture of a *DCES* composed of $q$ processors and $p$ real time tasks executed on these processors

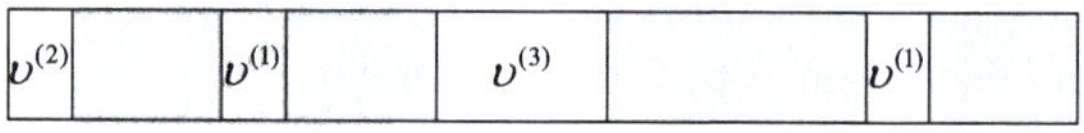


**Fig. 9.2** A hyper-period of three tasks executed by processor 1

signal computation, and control input actuation. As shown in Fig. 9.3, the task $\upsilon^1$, has two sub-sampling periods, $T_1$ and $T_2$ for each hyper-period and its timing is depicted in Fig. 9.4, where $s_k$ and $a_k$ ($k$ are non negative integers) denote sampling instants and actuation instants, respectively.

As stated before, communications between tasks induce also delays which are also dependent on the scheduling policy and messages priority handling. In more general terms, communication delays depend on the communication model between the *DCES* nodes. Without any loss of generality, we denote the induced delay by $\tau_1$ ($\tau_2$) if it corresponds to the sub-sampling period $T_1$ ($T_2$) as given in Fig. 9.4. It is assumed that $\tau_1, \tau_2 < \min\{T_1, T_2\}$.

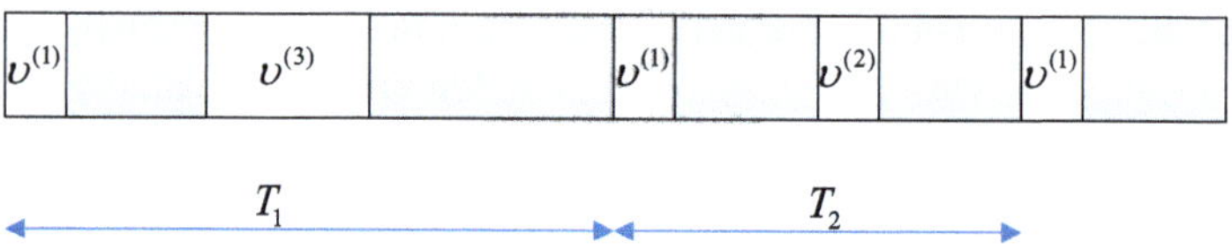


**Fig. 9.3** Two samplings of control task 1 executed by processor 1 each hyper-period

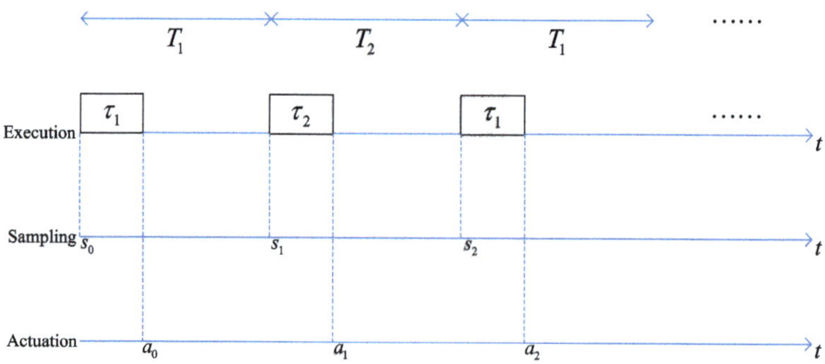


**Fig. 9.4** Timing diagram of control task 1 representing sampling, calculation, and actuation

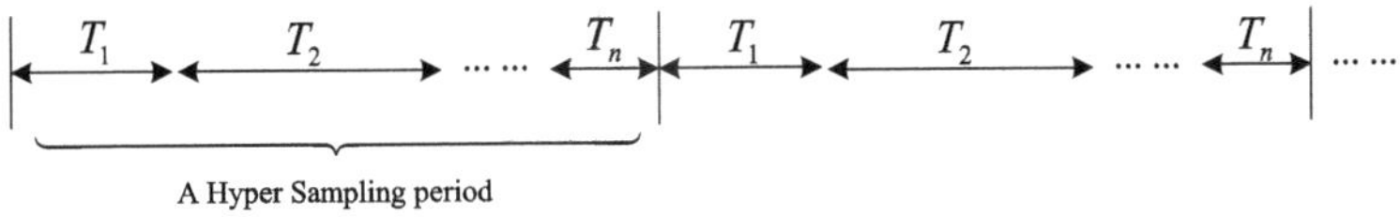


**Fig. 9.5** A hyper-sampling period of a controlled task consisting of $n$ sub-sampling periods

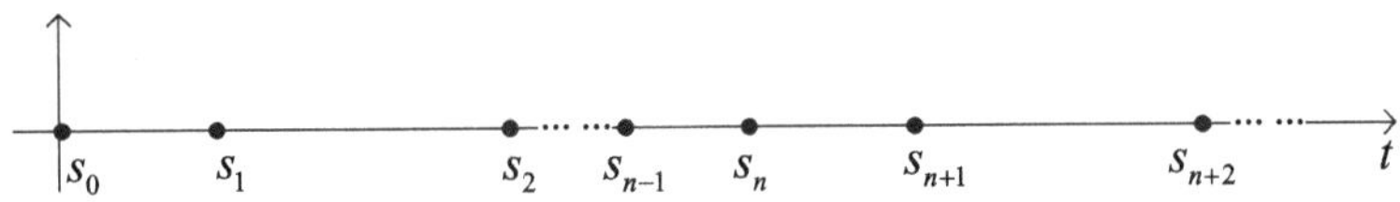

**Fig. 9.6** Sampling instants under the hyper-sampling mode

If we refer to the plant controlled by task $\upsilon^1$, its state feedback model includes necessarily the plant model, task calculation model and task communication model and is given by (9.3).

Define $z(k) = [x'(s_k)\ u'(a_{k-1})]'$. Then it follows that:

$$z(k+2) = \Psi(\tau_1, \tau_2, T_1, T_2)z(k) \tag{9.14}$$

where

$$\Psi(\tau_1, \tau_2, T_1, T_2) = \begin{cases} \Phi(\tau_1, T_1)\Phi(\tau_2, T_2), & k \text{ is odd}, \\ \Phi(\tau_2, T_2)\Phi(\tau_1, T_1), & k \text{ is even}, \end{cases}$$

where $\Phi(\tau_i, T_i), i \in 1, 2$ are defined in (9.10).

From above, it is easy to generalize the plant state feedback model in the case of a *hyper-sampling period* as given in Fig. 9.5.

Without any loss of generality, we suppose that a hyper-sampling period is composed of $n \in \mathbb{N}_+$ sub-sampling periods $T_i \in \mathbb{R}_+$, $i = 1, \ldots, n$. A general hyper-sampling period is depicted in Fig. 9.5. In particular, when $n = 1$, the hyper-sampling mode reduces to the well-known standard single-sampling mode. Thus, we may treat the standard single-sampling mode as a special case of the hyper-sampling mode.

*Remark 9.1* A larger $n$ allows a more flexible design and hence in general the corresponding hyper-sampling period leads to better stability and dynamic performance. Meanwhile, the analysis and optimization problems will become more involved as $n$ increases.

Under the hyper-sampling mode, the sampling instants will take place as shown in Fig. 9.6. It follows that $s_1 - s_0 = T_1$ ($s_0 = 0$), $s_2 - s_1 = T_2, \ldots, s_n - s_{n-1} = T_n$, $s_{n+1} - s_n = T_1$, $s_{n+2} - s_{n+1} = T_2, \ldots$. Once the hyper-sampling period is set (i.e., the $n$ sub-samplings $T_1, \ldots, T_k$ are determined), the sampling instants will occur periodically.

It is easy to see that a general rule of the sampling periods under the hyper-sampling mode can be described as follows:

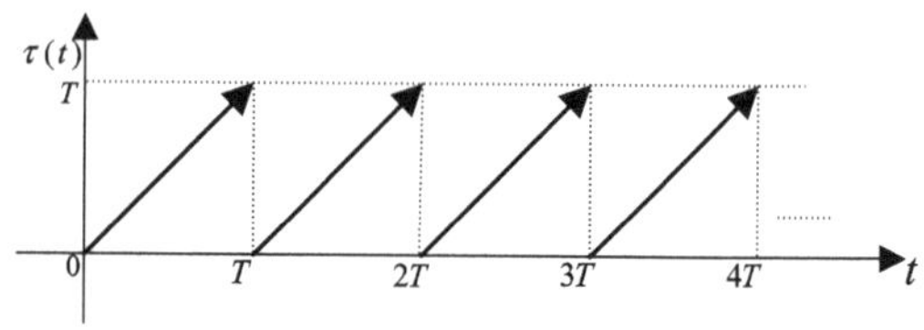


**Fig. 9.7** The artificial time delay $\tau(t)$

$$T(k) = s_k - s_{k-1} = \begin{cases} T_{k \bmod n}, & k \bmod n \neq 0,\ k \in \mathbb{N}_+, \\ T_n, & k \bmod n = 0,\ k \in \mathbb{N}_+. \end{cases} \tag{9.15}$$

Once a hyper-sampling sequence is designed (via an appropriate setting of the sub-sampling periods $T_1, \ldots, T_n$), the sampling instants can be triggered periodically by means of (9.15).

*Remark 9.2* In order to avoid confusion or misleading, we remember the use of $s_k$ to denote the sampling instants in order to take into account the *DCES-induced* delays and the use of $t_k$ to denote the sampling instants when the induced delays are neglected.

## 9.3 Macroscopic Time-Delay Model of *DCESs*

As we have already mentioned in the previous paragraphs, different delay modeling levels may be applied to *DCESs* in relation to its *Hardware/Software* architecture. In the precedent section, a delay modeling is applied in the case of static periodic scheduling. The input/output latency may be considered as an input time delay which represents a macroscopic time-delay model. In this case, the sampled-data state feedback model can be considered as a system with an *input delay* allowing to study a *DCES* in some appropriate time-delay system setting. For the simple case of control task single-sampling period, the sampled data controlled task model is given by associating the plant model (9.1) with the sampled-data control model (9.8). Such a simple *DCES* can by transformed into the following time-delay system

$$\dot{x}(t) = Ax(t) + BKx\big(t - \tau(t)\big), \quad kT \leq t < (k+1)T, \tag{9.16}$$

where $\tau(t) = t - kT$ for $kT \leq t < (k+1)T$ is the (artificial) input delay, with the sawtooth structure as shown in Fig. 9.7.

Thus, instead of the original model, we may study (9.16) by adopting the existing methods in the area of time-delay systems. Such an approach is called the *input-delay approach* which has some advantages when dealing with model uncertainties and/or external disturbance. However, to the best the authors knowledge, the input-delay approach appears to be more conservative than the discrete-time approach.

In Sect. 9.2, we show that the induced delay model of a *DCES* can be represented by the augmented model (9.9). In Chap. 10, the stability conditions will be studied based on some appropriate augmented model which includes the effect of induced delays.

## 9.4 Control Input Missings

As introduced previously, the sampled-data control input (9.7) is generated by the following steps: (1) at a sampling instant $t_k$, the state $x(t_k)$ is sampled by the sensor, (2) a data packet containing the information of $x(t_k)$ is sent to the controller, (3) the controller computes a new control value $Kx(t_k)$, (4) a data packet containing the information of control signal $Kx(t_k)$ is sent to the actuator, and (5) the actuator updates the control input to the controlled plant with the value of $Kx(t_k)$.

However, any failure among the steps mentioned above will make the control law (9.7) invalid. Such a case is known as the control input missing, as defined in *Zhang and Yu* [265]. The factors causing a control input missing include data-dropout, failures of sensor and/or controller and/or actuator nodes, scheduling pre-emption, and communication network congestion.

Since in the presence of a control input missing, the control input (9.7) can not be implemented, we have to define a control-input-missing compensator such that the *DCES* has an effective control signal when such a control input missing takes place. In the literature, two types of compensation strategies are mainly used: the hold-control and the zero-control ones. In Chap. 13, we will propose a new control-input-missing compensation strategy, a switched hold-zero compensation strategy. Such a new compensation strategy is a combination of the hold-control and the zero-control strategies and, in our opinion, it works better than both two strategies working independently. A detailed analysis in this sense will be given in Chap. 13.

## 9.5 Some Specific Problems of *DCESs* and Related Approaches

In this part of the book, four problems will be treated and discussed in detail in the corresponding chapters. We will give very brief presentation of the subjects treated and the study objectives.

- *Asymptotic stability of* DCESs *under the hyper-sampling mode*: To the best of the authors' knowledge, the stability of a *DCES* under the hyper-sampling mode was first studied in *Li et al.* [151]. In Chap. 10, we will introduce the related results. More precisely, we will give the ranges of the sub-sampling periods or the network-induced delays guaranteeing the asymptotic stability of the corresponding system. The approach we are proposing makes use of an appropriate *parameter-sweeping method*.
- *Optimal design of the hyper-sampling period*: To the best of the authors' knowledge, the analytical relation between the system's dynamic performance and the hyper-sampling period has not been reported in the open literature. If such a relation is known, we may use it to design a hyper-sampling period such that the optimal performance index can be achieved. This issue will be addressed in Chap. 11.
- *Improving system stability and dynamic properties by using a switched sampled-data control*: For a *DCES* under the standard single-sampling mode, we consider how to improve the system dynamics and/or performances. Through studying the

intersample dynamics of a *DCES* by taking into account the interpretation of the corresponding system as a time-delay system, we propose an easily-implemented switched sampled-data control approach. We will consider the related issues in Chap. 12.

- *Compensation strategy in the case of control input missings*: A new compensation strategy will be proposed and studied in Chap. 13. Our objective is to obtain a greater *admissible control input missing rate (ACIMR)* index for the *DCES* in order to enhance its stability robustness with respect to control input missings.

In the sequel, the mathematical models and tools used in *Part* III of the book are briefly discussed:

- *Discrete-time approach*: The *DCESs* considered in Chaps. 10 and 13 are non-nominal systems. More precisely, the systems considered in Chap. 10 are subject to time-varying hyper-sampling periods or *DCES-induced delays*, and the systems in Chap. 13 are subject to time-varying sampling periods and uncertain control input missings. We adopt the discrete-time approach and the discretization technique in deriving the Lyapunov-based stability conditions. We will see in Chaps. 10 and 13 that the discrete-time approach may lead to less conservative conditions than the ones derived by using some specific approaches in continuous-time framework.
- *Switched control approach*: When a *DCES* is sampled under the hyper-sampling mode or when it is subject to control input missings, we can treat the *DCES* as a *switched control system* consisting in several sub-systems. It is known that a switched system is asymptotically stable if all its sub-systems have a common Lyapunov function, which decreases with respect to time. However, when the considered system is nominal, we study the *DCES* by adopting the discrete-time model which involves no conservatism. For instance, in Chap. 10, we directly address the nominal case through analyzing the transition matrix of the corresponding *DCES*; in Chaps. 12 and 13, we obtain the analytic function of the performance index with respect to the hyper-sampling sequence or the switching parameter.
- *Analytical and numerical results*: If the considered *DCES* is nominal (i.e., the system's matrices, the DCES-induced delays, and the hyper-sampling sequence are all constant, and no control inputs missings happen), we can derive some appropriate necessary and sufficient stability conditions (see, for instance, the results proposed in Sect. 10.3.1 of Chap. 10). In addition, for a nominal *DCES*, the analytic relation between the performance index and the system parameters are derived in Chaps. 12 and 13. If the *DCES* under consideration is not nominal (e.g., the case with time-varying sampling periods or induced delays, and the case with control input missings), only numerical results can be derived. For a non-nominal *DCES*, we will study the stability by using the Lyapunov-based method in the time-domain framework. The resulting conditions are generally sufficient though not necessary and are expressed in terms of linear matrix inequalities (*LMIs*).

## 9.6 Notes and Comments

There exists an abundant literature devoted to the stability analysis as well as on the performance optimization of sample-data systems especially in the case of constant sampling periods. However, as explained in the introduction of this chapter, a *DCES* may be subject to sampling time variation due to limited computation and communication resources. This fact is underlined also by *Zhang and Branicky* [264], by *Bushnell et al.* [245], and by *Mounier* [179]. The sampling time variation has a destabilizing effect if it is not properly taken into account as clearly discussed by *Wittenmark* [252], *Zhang and Branicky* [264], and *Li et al.* [151]. Next, the characterization of all the delay values allowing the closed-loop stability of some SISO networked control systems as well as a further finer analysis of the existing links among the controller gain, the induced delay and the sampling period was proposed by [173]. The corresponding results have been derived by using an eigenvalue-based approach (see, also [58, 174] for further details). As it will be detailed in Chap. 10, the sampling time fluctuation increases the difficulty in analyzing the corresponding systems and the standard eigenvalue analysis is no more valid to conclude on the stability. To overcome these difficulties, the Lyapunov-based approaches have to be taken into account. The rationale behind is to consider the system dynamics induced by a piecewise continuous or an irregular sawtooth type delay as given in *Richard et al.* [202] and in *Mounier* [180]. Among them, one may cite the ones inspired by two distinct interpretations of the state notion specific to time-delays systems: Lyapunov–Razumikhin and Lyapunov–Krasovskii methods [104, 186]. Alternative approaches treating this problem are those based on small gain approaches. This class of methods consider the system delay variation as a perturbation with respect to some continuous control law $u(t) = Kx(s_k)$ and rewrite the continuous system as an appropriate interconnected one. For more detail we can refer to *Mirkin* [175] and to *Fujioka* [90].

Next, another class of methods devoted to the stability of sampled-data systems under variable delays is based on the so-called convex-embedding approaches, see, for instance, the contributions of *Hetel et al.* [120], *Fujioka* [89], *Cloosterman* [64], *Gielen et al.* [95]. In the case of a sampled-data *LTI* system with time-varying sampling intervals or delays, the stability conditions may be given by an infinite number of parameter dependent *LMI* conditions. They depend quadratically on the transition matrix which is continuous on the variable induced delay. The objective of this class of methods is the complexity reduction by transforming the infinite number of *LMIs* to a finite ones through an appropriate computation of some polytopic transition matrix over-approximation. Several over-approximation methods can be found in the literature such as those based on gridding and norm bounding given in *Donkers et al.* [76], *Fujioka* [89], *Skaf and Boyd* [214]. Other approximation methods are based on *Taylor* series expansion given in *Hetel et al.* [120, 121], on real *Jordan* form decomposition given in *Olaru et al.* [189], *Goebel et al.* [96], *Van de Wouw et al.* [240], *Cloosterman* [64], as well as those based on the *Cayley–Hamilton* theorem given in *Gielen et al.* [95], *Goebel* [96]. In *Heemels et al.* [112], a short comparison of these different approaches with numerical examples can be found. The simple intuitive approach of this class of methods is faced with complexity problems which may be, in some cases, numerically prohibitive.

# Chapter 10
# Stability of *DCESs* Under the Hyper-Sampling Mode

## 10.1 Introduction

The main objective of this chapter is to understand the way the periodic scheduling or the *hyper-sampling periods* and *DCES-induced delays*, interpreted as parameters of the system, affect the stability of the controlled plant. Several scenarios are considered and discussed, leading to three stability problems detailed in the Sect. 10.2. *First*, a delay-sweeping method is given in the case of constant parameters (*hyper-sampling periods* and *DCES-induced delays*). In this case, as expected, the necessary and sufficient stability conditions are given by the *Schur* stability of the transition matrix. Next, a special attention is paid to the cases when at least one of the uncertain parameters is assumed to be time-varying. In such a situation, we will first derive an appropriate sufficient stability condition expressed in terms of the existence of an appropriate Lyapunov matrix. Next, we will focus on two particular problems: the stability of a system including a constant *hyper-sampling period*, but time-varying uncertain *DCES-induced delays* and the stability of a system free of DCES-induced delays but including two time-varying uncertain *hyper-sampling periods*. In the case of time-varying uncertain DCES-induced delays, sufficient conditions expressed in terms of the feasibility of some appropriate linear matrix inequalities (*LMIs*) are presented. We will first check if there exists a Lyapunov matrix for some chosen parameter regions by an appropriate verification of the feasibility of the corresponding *LMI* conditions. If such a Lyapunov matrix exists, we can further compute and refine the whole stability region for this Lyapunov matrix by using an appropriate parameter-sweeping method. Finally, the last problem concerns the case without *DCES-induced delay* but subject to time-varying uncertain *hyper-sampling periods*. In this sense, in order to simplify the problem, a lemma is first derived allowing to connect the single-sampling and *hyper-sampling* cases. More precisely, the underlying idea is to appropriately generalize the single-sampling case in order to treat the general case. Finally, similar to the previous case, an appropriate parameter-sweeping method is employed to detect the whole stability region.

It is worth mentioning that the stability assessment of a *DCES* subject to periodic scheduling or *hyper-sampling periods* is generally much more complicated than the

A. Çela et al., *Optimal Design of Distributed Control and Embedded Systems*,
Communications and Control Engineering, DOI 10.1007/978-3-319-02729-6_10,

case of DCES including a single-sampling period (classical equidistant case). The rationale behind is that a *hyper-sampling period* with two sub-sampling periods $T_1$ and $T_2$ induces dynamics complexity due to the switching between the corresponding subsystem models: the first subsystem corresponding to the sub-sampling period $T_1$ and the second one to the sub-sampling period $T_2$, respectively. An interesting phenomenon observed concerns the relation between the stability of the individual subsystems and the stability of the overall system. This is not necessarily an equivalence relation. More precisely, the stability of each subsystem does not necessarily imply the stability of the overall system and vice-versa. Therefore, our approach mainly concerns the study of the overall system instead of finding a common stability condition for all subsystems.

Different examples are given to illustrate this point of view as well as the derived results. As it will be shown, it is advantageous to treat the overall system instead of each subsystem separately. The derived stability regions include sub-regions where some subsystems are unstable. It is also shown that the proposed stability regions are very close to the real ones. In addition, to further illustrate our approach, we address an example involving two inverted pendulums where each of them has two *sub-sampling periods*. A common *DCES-induced delay* bound guaranteeing the asymptotic stability for each inverted pendulum is explicitly derived.

## 10.2 Problem Formulation

As stated in the introduction, we address the stability of a *DCES* under periodic scheduling or *hyper-sampling* mode. We consider the generic *DCES* model (9.3) with the associated discrete-time model given by (9.14). Assume that the *hyper-sampling* period of this system includes two sub-sampling periods $T_1$ and $T_2$. Although this case appears quite simplistic, it depicts a lot of interesting properties that need a better understanding in order to further exploit them in various applications. It is worth mentioning that the underlying ideas proposed here for handling such a simple case may be appropriately extended to a more general situation (for instance, the case of a *hyper-sampling* period including more than two sub-sampling periods) without any difficulty.

Throughout this chapter, for numerical illustration, we choose the following matrices for the (second-order) controlled plant given by (9.1):

$$A = \begin{pmatrix} 0 & 1 \\ 0 & -0.1 \end{pmatrix}, \qquad B = \begin{pmatrix} 0 \\ 0.1 \end{pmatrix}, \qquad K = (-3.75 \quad -11.5). \tag{10.1}$$

According to the properties (time-invariant or time-variant) of the parameters (*hyper-sampling periods* or *DCES-induced delays*), we consider the following three problems:

**Problem 10.1** (Delay-parameter space: stability intervals) Assume that the *sub-sampling periods* $T_1$ and $T_2$ are known constant and the *DCES-induced delay* $\tau$ ($\tau_1 = \tau_2 = \tau$) is constant, but uncertain. *Find* the stability interval $S_\tau$ for $\tau$.

**Problem 10.2** (Delay-parameter space: uncertain and time-varying delays) Assume that the *sub-sampling periods* $T_1$ and $T_2$ are known constant while the *DCES-induced delays* $\tau_1$ and $\tau_2$ are uncertain and time-varying. *Find* the stability region $S_{\tau_1,\tau_2}$ in the $\tau_1$–$\tau_2$ parameter plane.

**Problem 10.3** (Hyper-sampling parameter space: time-varying hyper-sampling periods) Assume that the *sub-sampling periods* $T_1$ and $T_2$ are uncertain and time-varying and the system is free of *DCES-induced delays*. *Find* the stability region $S_{T_1,T_2}$ in the $T_1$–$T_2$ parameter plane.

*Remark 10.1* (Insights in computing stability regions in the delay-parameter space) Inspired by the existing works in computing the stability regions of continuous time-delay systems in the *delay-parameter space*, the first problem corresponds to the so-called $\tau$-decomposition method introduced by *Lee and Hsu* [147], where the delay is interpreted as a free parameter and the other parameters of the system are assumed to be fixed and known (see also [174] for further details on the existing approaches). In the case of a system involving two delays, an extension of the $\tau$-decomposition method can be found in [105], where the analysis was performed by using a geometric argument combined with some appropriate frequency-sweeping tests. In the case of DCESs, the frequency-sweeping method was appropriately adapted to the considered class (delay-sweeping method). A further refinement will be proposed in order to take into account the fact that the delays are assumed to be time-varying. Finally, the third problem is similar to the second one, but with the difference that the parameter-space to be considered is defined by the hyper-sampling periods.

As discussed in the previous section, the system (9.14) can be viewed as a switched system consisting in two subsystems, corresponding to the sampled-data control system with sampling periods $T_1$ and $T_2$, respectively.

*Example 10.1* The stability region depends on the hyper-sampling periods and the *DCES-induced delays*. In the following, we give some indications and results on this dependence:

(a) *Problem 10.1*: Let $T_1 = 0.8$ and $T_2 = 1.8$. Subsystem 1 is stable if $\tau \in [0, 0.763)$; subsystem 2 is stable if $\tau \in (0.045, 0.769)$ (these two results can be obtained by the method for single-sampling analysis, see, for instance, *Li et al.* [148]); the overall system is stable for $\tau \in [0, 0.292)$. One can see that if $\tau \in [0, 0.045]$, the subsystem 2 is *unstable* but the overall system is *stable*.
(b) *Problem 10.2*: Let $T_1 = 0.8$ and $T_2 = 1.8$. If $\tau_2 \in [0, 0.045]$, then the subsystem 2 is *unstable*, see Fig. 10.1. However, we can obtain the *stability region* shown in Fig. 10.3 for the overall system. The system initial state is chosen as $(1\ 1)'$.
(c) *Problem 10.3*: Each subsystem is stable if $T_1, T_2 \in (0, 1.72]$. Interestingly, we find that for some parameters outside the corresponding rectangle where either $T_1$ or $T_2$ is *greater* than 1.72, the system is still *stable*. For example, we make a simulation for $T_1 \in [1.3, 1.4]$ and $T_2 \in [2.1, 2.2]$, as later shown in Figs. 10.5 and 10.6.

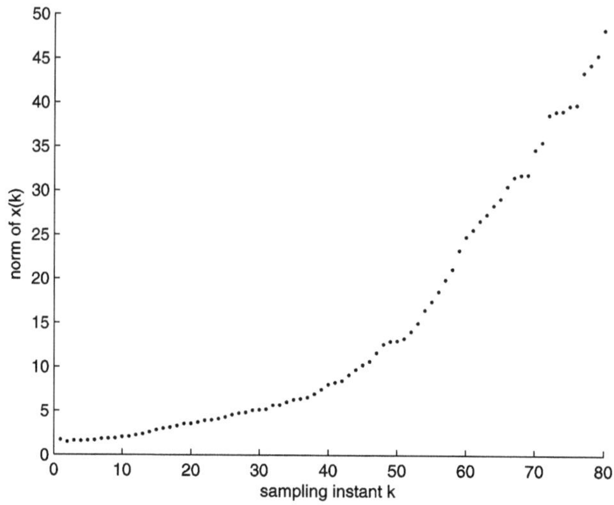


**Fig. 10.1** Simulation of subsystem 2 with the *DCES-induced delay* $\tau_2 \in [0, 0.045]$ (unstable case)

## 10.3 Insights in Computing Stability Regions

Let us study in more details the above stability problems separately.

### *10.3.1 Constant Delay Case*

Static scheduling of communication messages and calculation tasks for a given *DCES* may guarantee constant *DCES-induced delays* (see, e.g., Controller Area Network by Lian [152]). In addition, if the *DCES-induced delay* is not constant, one can make it constant by adding an appropriate buffer (see, for instance, *Luck and Ray* [162]). When the parameters (sampling periods and *DCES-induced delays*) are time-invariant, an *eigenvalue-based method* can be applied to obtain non-conservative stability bounds. For Problem 10.1, we have

$$z(k+2) = \Psi(\tau, \tau, T_1, T_2) z(k), \tag{10.2}$$

where

$$\Psi(\tau, \tau, T_1, T_2) = \begin{cases} \Phi(\tau, T_1)\Phi(\tau, T_2), & k \text{ is odd}, \\ \Phi(\tau, T_2)\Phi(\tau, T_1), & k \text{ is even}, \end{cases}$$

$$\Phi(\tau, T_i) = \begin{pmatrix} e^{AT_i} + \int_\tau^{T_i} e^{A\theta}\, d\theta\, BK & \int_0^\tau e^{A\theta}\, d\theta\, B \\ K & 0 \end{pmatrix}, \quad i = 1, 2.$$

We first introduce the following property which will be helpful in computing the stability region $S_\tau$:

*Property 10.1* For square matrices $Q_1, Q_2, \ldots, Q_\ell$, the matrices $Q_1 Q_2 \cdots Q_\ell$, $Q_2 \cdots Q_\ell Q_1, \ldots, Q_\ell Q_1 \cdots Q_{\ell-1}$ have the same eigenvalues.

*Proof* It is well known that: $\sigma(Q_1 Q_2) = \sigma(Q_2 Q_1)$ (see, e.g., *Poznyak* [194]), where $\sigma(\cdot)$ denotes the spectrum of the argument. Next, $\sigma(Q_1 Q_2 Q_3) = \sigma(Q_1(Q_2 Q_3)) = \sigma((Q_1 Q_2) Q_3) = \sigma(Q_2 Q_3 Q_1) = \sigma(Q_3 Q_1 Q_2)$. In this manner, the result for more matrices follows straightforwardly. □

In the sequel, we demonstrate the following theorem which gives the stability conditions for the Problem 10.1 and helps in finding the stability region $S_\tau$.

**Theorem 10.1** *System* (10.2) *is asymptotically stable for the constant* DCES-induced delay *satisfying* $\tau \in S_\tau$ *if and only if* $\Phi(\tau, T_1)\Phi(\tau, T_2)$ *(or equivalently* $\Phi(\tau, T_2)\Phi(\tau, T_1)$*) is Schur for* $\tau \in S_\tau$.

*Proof* The necessary and sufficient stability condition for system (10.2) is that the matrix $\Psi(\tau, \tau, T_1, T_2)$ is Schur. Further, $\Phi(\tau, T_1)\Phi(\tau, T_2)$ is Schur for $\tau \in S_\tau$ if and only if $\Phi(\tau, T_2)\Phi(\tau, T_1)$ is Schur for $\tau \in S_\tau$ because these two terms always have the same eigenvalues according to the Property 10.1. Thus, the proof is complete. □

*Remark 10.2* In order to find the stability interval, we can check the *Schurness* of either $\Phi(\tau, T_1)\Phi(\tau, T_2)$ or $\Phi(\tau, T_2)\Phi(\tau, T_1)$ according to Theorem 10.1.

*Remark 10.3* Theorem 10.1 can be extended to systems with multiple sub-sampling periods by using the Property 10.1: $n$ sub-sampling periods lead to $n$ forms of transition matrices. We only need to check the *Schurness* of any one of them simplifying thus the stability analysis.

The method used to find the stability region $S_\tau$ is the *delay-sweeping* one. It consists in sweeping the values of $\tau$ and testing the spectral radius of the matrix $\Psi(\tau, \tau, T_1, T_2)$. The interval with the spectral radius less than 1 (i.e., $\Psi(\tau, \tau, T_1, T_2)$ is a Schur matrix) defines the corresponding stability interval. It is worth to mention that this sweeping method can cover also the cases with single sampling period. When the single sampling period $T$ is 0.8, 1.5, and 1.8, the stability interval for delay $\tau$ is $[0, 0.763)$, $[0, 0.761)$, and $(0.045, 0.769)$, respectively.

Consider now an illustrative example for the *Problem* 10.1.

*Example 10.2* Let $T_1 = 0.8$ and $T_2 = 1.8$. By employing the above delay-sweeping method, we can obtain the relationship between the spectral radius and the delay, as depicted in Fig. 10.2. It is seen that the system is stable for $\tau \in [0, 0.292)$.

In the sequel, we focus on the systems with time-varying uncertain parameters. Unlike the case with constant parameters, the eigenvalue-based method is no longer valid. A *DCES* with time-variant transition matrix is not necessarily stable, even if

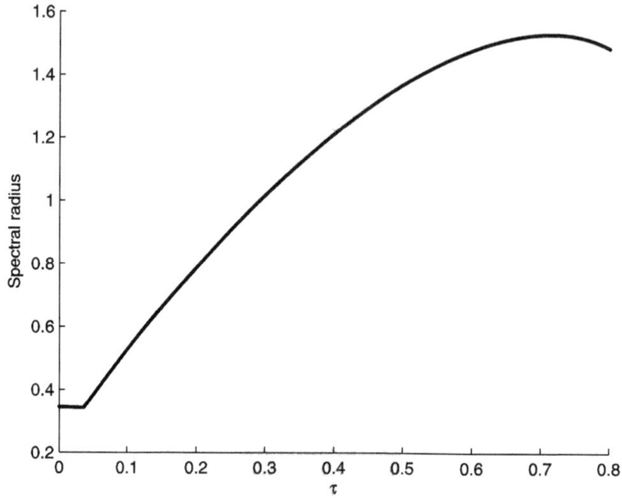


**Fig. 10.2** Spectral radius vs delay for Example 10.2

the transition matrix is Schur at every sampling instant. Counterexamples can be found in *Zhang and Branicky* [263] as well as in Cloosterman et al. [65].

### *10.3.2 Time-Varying Delay Case*

As explained in Chap. 9 the *DCES-induced delays* are caused by many factors which are directly related to their calculation and communication model, overload conditions as well as to resources allocation optimization strategy. The analysis of the simple communication and calculation model given in *Ben Gaid et al.* [31] leads to the conclusion that the *DCES-induced delays* are more likely to be uncertain and *time-varying* than constant. This fact completely justifies the study of *Problem* 10.2 and we give in the following the related stability condition.

**Theorem 10.2** *For given $T_1$ and $T_2$, the system* (9.3) *is asymptotically stable for $(\tau_1, \tau_2) \in S_{\tau_1,\tau_2}$, if there exists a positive-definite matrix $P$, such that $\Omega(\tau_1, \tau_2, T_1, T_2, P) < 0$ for $(\tau_1, \tau_2) \in S_{\tau_1,\tau_2}$, where $\Omega(\tau_1, \tau_2, T_1, T_2, P)$ can be chosen as*:

$$\Omega(\tau_1, \tau_2, T_1, T_2, P) \triangleq \left(\Phi(\tau_2, T_2)\Phi(\tau_1, T_1)\right)' P\Phi(\tau_2, T_2)\Phi(\tau_1, T_1) - P, \quad (10.3)$$

*or*

$$\Omega(\tau_1, \tau_2, T_1, T_2, P) \triangleq \left(\Phi(\tau_1, T_1)\Phi(\tau_2, T_2)\right)' P\Phi(\tau_1, T_1)\Phi(\tau_2, T_2) - P, \quad (10.4)$$

*with*

$$\Phi(\tau_i, T_i) = \begin{pmatrix} e^{AT_i} + \int_{\tau_i}^{T_i} e^{A\theta}\, d\theta\, BK & \int_0^{\tau_i} e^{A\theta}\, d\theta\, B \\ K & 0 \end{pmatrix}, \quad i = 1, 2.$$

*Proof* The result is obtained by constructing an appropriate Lyapunov function candidate $V(k) : k \mapsto z'(k)Pz(k)$. The asymptotic stability can be ensured if for any even $k$[1] and $z(0) \neq 0$

$$V(k+2) - V(k) < 0.$$

If the corresponding sampling periods are $T(1) \triangleq s_1 - s_0 = T_1$, $T(2) \triangleq s_2 - s_1 = T_2$, $T(3) \triangleq s_3 - s_2 = T_1, \ldots$, for an even $k$, it follows that

$$V(k+2) = z'(k)\big(\Phi(\tau_2, T_2)\Phi(\tau_1, T_1)\big)' P\Phi(\tau_2, T_2)\Phi(\tau_1, T_1)z(k).$$

Thus, under this sampling sequence,

$$V(k+2) - V(k) = z'(k)\Omega(\tau_1, \tau_2, T_1, T_2, P)z(k), \tag{10.5}$$

where the corresponding $\Omega(\tau_1, \tau_2, T_1, T_2, P)$ is given by (10.3).

If the corresponding sampling periods are $T(1) \triangleq s_1 - s_0 = T_2$, $T(2) \triangleq s_2 - s_1 = T_1$, $T(3) \triangleq s_3 - s_2 = T_2, \ldots$, for an even $k$, it follows that

$$V(k+2) = z'(k)\big(\Phi(\tau_1, T_1)\Phi(\tau_2, T_2)\big)' P\Phi(\tau_1, T_1)\Phi(\tau_2, T_2)z(k).$$

Therefore, under this sampling sequence, we have the relation (10.5) where the corresponding $\Omega(\tau_1, \tau_2, T_1, T_2, P)$ is given by (10.4).

No matter which sampling sequence is adopted, the system is asymptotically stable if

$$\Omega(\tau_1, \tau_2, T_1, T_2, P) < 0,$$

in the light of (10.5). □

*Remark 10.4* In the sequel, we will derive the stability condition in terms of LMIs. The conservatism of the stability conditions derived depends on the chosen form of $\Omega(\tau_1, \tau_2, T_1, T_2, P)$. Thus, unlike Problem 10.1, different forms of $\Omega(\tau_1, \tau_2, T_1, T_2, P)$ may lead to different stability regions. In the sequel of this chapter, we choose the expression of $\Omega(\tau_1, \tau_2, T_1, T_2, P)$ given by (10.3).

In general, it is not easy to directly check if there exists a Lyapunov matrix $P$ for a given region in the $\tau_1$–$\tau_2$ plane. In the sequel, some computation-oriented conditions are presented. The basic idea is similar to the one proposed by *Li et*

[1] We have multiple lines to derive the stability condition. We may also consider "The asymptotic stability can be ensured if for any odd $k$ and $z(0) \neq 0$ such that $V(k+2) - V(k) < 0$".

*al.* [148], where the robust stability (with respect to time-varying uncertain sampling period and delay) for single-sampling system is studied.

**Theorem 10.3** *Define a rectangular region* $S_{\tau_1,\tau_2}: \tau_1 = \tau_{10} + \Delta\tau_1, \tau_2 = \tau_{20} + \Delta\tau_2$ ($\tau_{10}$ *and* $\tau_{20}$ *are known constants;* $\Delta\tau_1$ *and* $\Delta\tau_2$ *are uncertain scalars whose lower and upper bounds are available*).

*Then the system* (9.3) *is asymptotically stable for* $(\tau_1, \tau_2) \in S_{\tau_1,\tau_2}$, *if there exist a positive-definite matrix* $P$ *and positive scalars* $\varepsilon_r$ $(r = 1, 2, 3)$ *satisfying the following linear matrix inequality*

$$\begin{pmatrix} L_{11} & * & * & * & * \\ P\Phi(\tau_{20}, T_2)\Phi(\tau_{10}, T_1) & -P & * & * & * \\ 0 & \begin{pmatrix} e^{A\tau_{10}} & 0 \\ 0 & 0 \end{pmatrix}' \Phi'(\tau_{20}, T_2)P & -\frac{\varepsilon_1}{\beta_1^2} I & * & * \\ 0 & \begin{pmatrix} e^{A\tau_{20}} & 0 \\ 0 & 0 \end{pmatrix}' P & 0 & -\frac{\varepsilon_2}{\beta_2^2} I & * \\ 0 & \begin{pmatrix} e^{A\tau_{20}} & 0 \\ 0 & 0 \end{pmatrix}' P & 0 & 0 & -\frac{\varepsilon_3}{\beta_3^2} I \end{pmatrix} < 0 \tag{10.6}$$

*with*

$$\begin{aligned} L_{11} &= -P + (\varepsilon_1 + \varepsilon_3)\begin{pmatrix} K'B'BK & * \\ -B'BK & 0 \end{pmatrix} \\ &\quad + \varepsilon_2 \Phi'(\tau_{10}, T_1)\begin{pmatrix} K'B'BK & * \\ -B'BK & 0 \end{pmatrix}\Phi(\tau_{10}, T_1), \\ &\left\| J(\Delta\tau_1) \right\| \le \beta_1, \\ &\left\| J(\Delta\tau_2) \right\| \le \beta_2, \\ &\beta_3 = \beta_1\beta_2 \left\| \begin{pmatrix} -BK & B \\ 0 & 0 \end{pmatrix}\begin{pmatrix} e^{A\tau_{10}} & 0 \\ 0 & 0 \end{pmatrix} \right\|, \\ &J(\cdot) \triangleq \int_0^{\cdot} e^{A\theta}\, d\theta. \end{aligned}$$

*Proof* The transition matrix $\Phi(\tau_1, T_1)$ can be expressed as:

$$\Phi(\tau_1, T_1) = \Phi(\tau_{10}, T_1) + \Delta\Phi(\tau_1),$$

with

$$\Delta\Phi(\tau_1) = \begin{pmatrix} e^{A\tau_{10}} & 0 \\ 0 & 0 \end{pmatrix}\begin{pmatrix} J(\Delta\tau_1) & 0 \\ 0 & 0 \end{pmatrix}\begin{pmatrix} -BK & B \\ 0 & 0 \end{pmatrix}.$$

By similarity,

$$\Phi(\tau_2, T_2) = \Phi(\tau_{20}, T_2) + \Delta\Phi(\tau_2),$$

with

$$\Delta\Phi(\tau_2) = \begin{pmatrix} e^{A\tau_{20}} & 0 \\ 0 & 0 \end{pmatrix} \begin{pmatrix} J(\Delta\tau_2) & 0 \\ 0 & 0 \end{pmatrix} \begin{pmatrix} -BK & B \\ 0 & 0 \end{pmatrix}.$$

The stability of the system can be ensured if $z'(k+2)Pz(k+2) - z'(k)Pz(k) < 0$ for any $k$. Noting that $z(k+2) = \Phi(\tau_2, T_2)\Phi(\tau_1, T_1)z(k)$, we have a sufficient stability condition that $(\Phi(\tau_2, T_2)\Phi(\tau_1, T_1))'P\Phi(\tau_2, T_2)\Phi(\tau_1, T_1) - P < 0$, which, by the *Schur complement* properties (see *Boyd et al.* [40]), can be equivalently transformed into the following matrix inequality:

$$\begin{pmatrix} -P & * \\ P\Phi(\tau_2, T_2)\Phi(\tau_1, T_1) & -P \end{pmatrix} < 0,$$

where

$$P\Phi(\tau_2, T_2)\Phi(\tau_1, T_1) = P\Phi(\tau_{20}, T_2)\Phi(\tau_{10}, T_1) + P\Phi(\tau_{20}, T_2)\Delta\Phi(\tau_1)$$
$$+ P\Delta\Phi(\tau_2)\Phi(\tau_{10}, T_1) + P\Delta\Phi(\tau_2)\Delta\Phi(\tau_1),$$

$$P\Delta\Phi(\tau_2)\Delta\Phi(\tau_1) = P\begin{pmatrix} e^{A\tau_{20}} & 0 \\ 0 & 0 \end{pmatrix} \Delta_{21} \begin{pmatrix} -BK & B \\ 0 & 0 \end{pmatrix},$$

$$\Delta_{21} = \begin{pmatrix} J(\Delta\tau_2) & 0 \\ 0 & 0 \end{pmatrix} \begin{pmatrix} -BK & B \\ 0 & 0 \end{pmatrix} \begin{pmatrix} e^{A\tau_{10}} & 0 \\ 0 & 0 \end{pmatrix} \begin{pmatrix} J(\Delta\tau_1) & 0 \\ 0 & 0 \end{pmatrix}.$$

The norm of $\Delta_{21}$ is upper bounded by:

$$\|\Delta_{21}\| \leq \left\| \begin{pmatrix} J(\Delta\tau_2) & 0 \\ 0 & 0 \end{pmatrix} \right\| \left\| \begin{pmatrix} -BK & B \\ 0 & 0 \end{pmatrix} \begin{pmatrix} e^{A\tau_{10}} & 0 \\ 0 & 0 \end{pmatrix} \right\| \left\| \begin{pmatrix} J(\Delta\tau_1) & 0 \\ 0 & 0 \end{pmatrix} \right\|$$
$$\leq \beta_1\beta_2 \left\| \begin{pmatrix} -BK & B \\ 0 & 0 \end{pmatrix} \begin{pmatrix} e^{A\tau_{10}} & 0 \\ 0 & 0 \end{pmatrix} \right\|.$$

Note that

$$\begin{pmatrix} -P & * \\ P\Phi(\tau_2, T_2)\Phi(\tau_1, T_1) & -P \end{pmatrix}$$
$$= \begin{pmatrix} -P & * \\ P\Phi(\tau_{20}, T_2)\Phi(\tau_{10}, T_1) & -P \end{pmatrix} + \sum_{q=1}^{3} (D_q F_q E_q + E_q' F_q' D_q'),$$

where

$$D_1' = D_3' = \left( \begin{pmatrix} -BK & B \\ 0 & 0 \end{pmatrix} \quad 0 \right),$$
$$E_1 = \left( 0 \quad \begin{pmatrix} e^{A\tau_{10}} & 0 \\ 0 & 0 \end{pmatrix}' \Phi'(\tau_{20}, T_2)P \right),$$

$$F_1 = \begin{pmatrix} J(\Delta\tau_1) & 0 \\ 0 & 0 \end{pmatrix}',$$

$$D_2' = \left( \left(\begin{smallmatrix} -BK & B \\ 0 & 0 \end{smallmatrix}\right) \Phi(\tau_{10}, T_1) \quad 0 \right),$$

$$E_2 = \left( 0 \quad \left(\begin{smallmatrix} e^{A\tau_{20}} & 0 \\ 0 & 0 \end{smallmatrix}\right)' P \right),$$

$$F_2 = \begin{pmatrix} J(\Delta\tau_2) & 0 \\ 0 & 0 \end{pmatrix}',$$

$$E_3 = \left( 0 \quad \left(\begin{smallmatrix} e^{A\tau_{20}} & 0 \\ 0 & 0 \end{smallmatrix}\right)' P \right),$$

$$F_3 = \begin{pmatrix} \Delta_{21} & 0 \\ 0 & 0 \end{pmatrix}'.$$

For any positive scalars $\varepsilon_1$, $\varepsilon_2$, and $\varepsilon_3$, the following inequalities always hold

$$D_1 F_1 E_1 + E_1' F_1' D_1' \leq \varepsilon_1 D_1 D_1' + \frac{\beta_1^2}{\varepsilon_1} E_1' E_1,$$

$$D_2 F_2 E_2 + E_2' F_2' D_2' \leq \varepsilon_2 D_2 D_2' + \frac{\beta_2^2}{\varepsilon_2} E_2' E_2,$$

$$D_3 F_3 E_3 + E_3' F_3' D_3' \leq \varepsilon_3 D_3 D_3' + \frac{\beta_3^2}{\varepsilon_3} E_3' E_3.$$

The condition (10.6) is thus obtained by using an appropriate *Schur complement* argument. □

By the above theorem, we can check if there exists a Lyapunov matrix for some parameter region $S_{\tau_1,\tau_2} : \tau_1 = \tau_{10} + \Delta\tau_1, \tau_2 = \tau_{20} + \Delta\tau_2$. If it does, $S_{\tau_1,\tau_2}$ represents a stability region. If a larger stability region is desired, we can first choose some small independent parameter regions and then check if there exists a common Lyapunov matrix for these (parameter) regions. If a common Lyapunov matrix exists, the combination of these small parameter regions defines a stability parameter region. We are now ready to propose the following theorem which formalizes this idea:

**Theorem 10.4** *Define $N$ rectangular regions $S_{\tau_{j1},\tau_{j2}} : \tau_{j1} = \tau_{j10} + \Delta\tau_{j1}, \tau_{j2} = \tau_{j20} + \Delta\tau_{j2}$, $j = 1, \ldots, N$ ($\tau_{j10}$ and $\tau_{j20}$ are known constants; $\Delta\tau_{j1}$ and $\Delta\tau_{j2}$ are uncertain scalars whose lower and upper bounds are available).*

*Then the system (9.3) is asymptotically stable for $(\tau_1, \tau_2) \in \bigcup_{j=1}^{N} S_{\tau_{j1},\tau_{j2}}$, if there exist a positive-definite matrix $P$ and positive scalars $\varepsilon_{jm}$ ($m = 1, 2, 3, j = 1, \ldots, N$) satisfying the following linear matrix inequalities*

$$\begin{pmatrix} L_{j11} & * & * & * & * \\ P\Phi(\tau_{j20},T_2)\Phi(\tau_{j10},T_1) & -P & * & * & * \\ 0 & \begin{pmatrix} e^{A\tau_{j10}} & 0 \\ 0 & 0 \end{pmatrix}' \Phi'(\tau_{j20},T_2)P & -\frac{\varepsilon_{j1}}{\beta_{j1}^2}I & * & * \\ 0 & \begin{pmatrix} e^{A\tau_{j20}} & 0 \\ 0 & 0 \end{pmatrix}' P & 0 & -\frac{\varepsilon_{j2}}{\beta_{j2}^2}I & * \\ 0 & \begin{pmatrix} e^{A\tau_{j20}} & 0 \\ 0 & 0 \end{pmatrix}^T P & 0 & 0 & -\frac{\varepsilon_{j3}}{\beta_{j3}^2}I \end{pmatrix}$$

$$< 0, \quad j = 1, \ldots, N, \tag{10.7}$$

*with*

$$L_{j11} = -P + (\varepsilon_{j1} + \varepsilon_{j3}) \begin{pmatrix} K'B'BK & * \\ -B'BK & 0 \end{pmatrix}$$

$$+ \varepsilon_{j2}\Phi'(\tau_{j10},T_1) \begin{pmatrix} K'B'BK & * \\ -B'BK & 0 \end{pmatrix} \Phi(\tau_{j10},T_1),$$

$$\|J(\Delta\tau_{j1})\| \le \beta_{j1},$$

$$\|J(\Delta\tau_{j2})\| \le \beta_{j2},$$

$$\beta_{j3} = \beta_{j1}\beta_{j2} \left\| \begin{pmatrix} -BK & B \\ 0 & 0 \end{pmatrix} \begin{pmatrix} e^{A\tau_{j10}} & 0 \\ 0 & 0 \end{pmatrix} \right\|.$$

The result is an extension of Theorem 10.3, and the proof is omitted.

The procedure to find a stability region for Problem 10.2 can be summarized as follows:

**Algorithm 10.1** (Computing stability regions)

- *Step 1*: Choose some regions $S_{\tau_{j1},\tau_{j2}}$ ($j = 1, \ldots, N$), and find a $P$ such that (10.7) holds for $(\tau_1, \tau_2) \in \bigcup_{j=1}^N S_{\tau_{j1},\tau_{j2}}$ by using Theorem 10.4.
- *Step 2*: By using the Lyapunov matrix $P$ obtained from Step 1, detect the whole stability region (corresponding to this Lyapunov matrix $P$) by using the parameter-sweeping method.

*Remark 10.5* Theorems 10.3 and 10.4 allow finding an appropriate Lyapunov matrix (if any) for one or some corresponding parameter region(s). It is clear that increasing the value of $N$, the derived stability region is closer and closer to the real one. This method can also be used to address Problem 10.3 or similar ones with multiple uncertain parameters. However, it is worth mentioning that its complexity increases with the number of parameters to be considered.

*Remark 10.6* One method to compute the upper bounds of $\|J(\Delta\tau_1)\|$ and $\|J(\Delta\tau_2)\|$ is represented by the *analytical approximation*, as studied by *Suh* [222].

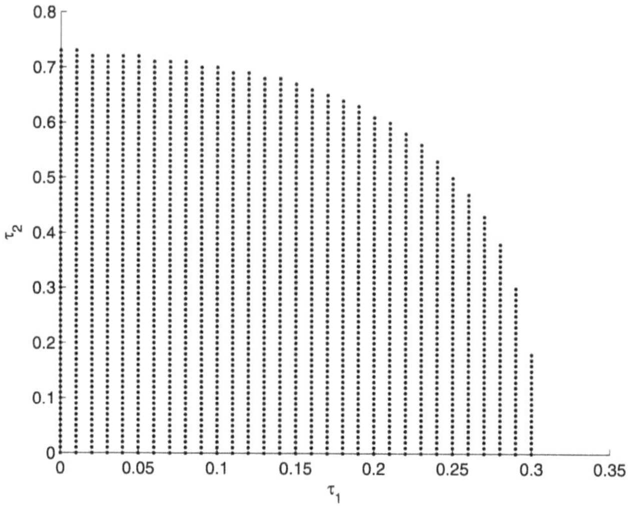


**Fig. 10.3** Stability region on the $\tau_1$–$\tau_2$ plane for Example 10.3

An alternative method is the *parameter-sweeping approach*. It is worth mentioning that such bounds are actually not necessarily important when using the Algorithm 10.1 above. Indeed, we can choose some bounds $\beta_{j1}, \beta_{j2}$ $(j = 1, \dots, N)$, without knowing the accurate range of $\Delta_{\tau_{j1}}, \Delta_{\tau_{j2}}$. After finding a Lyapunov matrix, the whole stability region corresponding to this Lyapunov matrix will be found by explicitly applying the parameter-sweeping approach. Therefore, this bound is not strictly required when computing the stability region.

*Example 10.3* Assume that the system (9.3) with matrices (10.1) has two constant sub-sampling periods ($T_1 = 0.8$, $T_2 = 1.8$) and two time-varying *DCES-induced delays* $\tau_1$ and $\tau_2$.

Choose $N = 3$, $\tau_{110} = 0.01$, $\tau_{120} = 0.01$, $\tau_{210} = 0.05$, $\tau_{220} = 0.07$, $\tau_{310} = 0.28$, $\tau_{320} = 0.28$, $\beta_{j1} = \beta_{j2} = 0.001$ $(j = 1, 2, 3)$. In this case, the use of Theorem 10.4 leads to the Lyapunov matrix

$$P = 10^7 \times \begin{pmatrix} 1.3536 & -0.1006 & 0.0315 \\ -0.1006 & 0.1945 & -0.0068 \\ 0.0315 & -0.0068 & 0.0017 \end{pmatrix}$$

and the corresponding stability region, derived via the parameter-sweeping method, as depicted in Fig. 10.3. To show that this stability region is reliable and very close to the real one, we choose two small regions near the boundary of the computed stability region, one inside and the other outside. If the system is stable for the inner parameter region meanwhile unstable for the outer parameter region, the derived stability region can be considered sufficiently good. We choose an inner region $\tau_1, \tau_2 \in [0.28, 0.29]$ and an outer region $\tau_1, \tau_2 \in [0.29, 0.30]$, both closely near the stability boundary. The simulations for two cases are shown in Fig. 10.4.

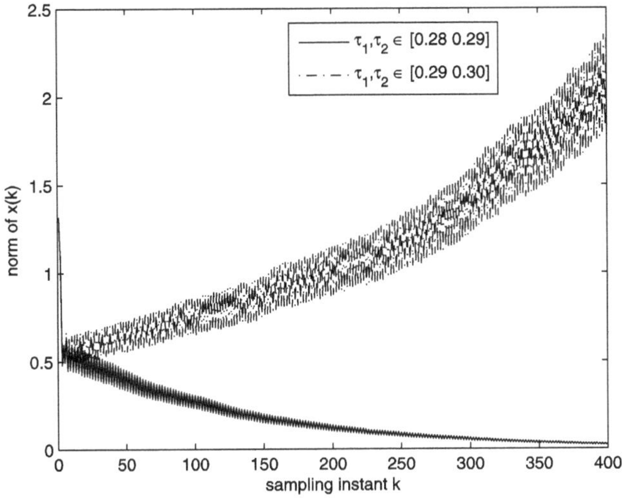


**Fig. 10.4** Simulations when $\tau_1, \tau_2 \in [0.28, 0.29]$ and $\tau_1, \tau_2 \in [0.29, 0.30]$, showing that the system is stable with the inner parameter region while unstable with the outer parameter region

## *10.3.3 Time-Varying Hyper-Sampling Periods*

In a *DCES*, the task executions may be preempted by some other tasks with higher priorities. In this situation, the sampling periods are in general uncertain. For a delay-free system with single uncertain sampling period $T$, some methods are proposed to find the stability conditions. A *Lyapunov–Krasovskii* approach was proposed by *Fridman et al.* [88]; *Mirkin* [175] studied the problem by using the small gain theorem; next, *Zhang and Branicky* [263] and *Fujioka* [89] focused on finding a common Lyapunov function in a discrete-time framework. To the best of our knowledge, the result of *Fujioka* [89] appears to be the least conservative one. As stated before, Problem 10.3 is more complicated than the single-sampling case. For Example 10.1, although the system is stable if both $T_1$ and $T_2$ belong to (01.72] (see, e.g., *Zhang and Branicky* [263] and *Fujioka* [89]), this result is rather conservative, as indicated earlier.

The stability region $S_{T_1,T_2}$ is crucial not only for theoretical analysis but also for practical applications. Indeed, a larger stability region provides us more flexibility in the *DCES* design. In addition, if we apply larger sampling periods (for which the stability is guaranteed by theoretical analysis), more system resources can be reserved for other tasks without upgrading its communication bandwidth and calculation power. Consequently, a *DCES* can handle more tasks, fact that, in turn, reduces the costs. Thus, we are always interested to have a stability region $S_{T_1,T_2}$ as large as possible. In the sequel, we give a stability condition as well as an appropriate algorithm for the Problem 10.3, which are in line with the objectives stated above.

**Theorem 10.5** *A delay-free system described by* (9.1) *and* (9.7) *with sub-sampling periods $T_1$ and $T_2$ is asymptotically stable for $(T_1, T_2) \in S_{T_1,T_2}$, if there exists a positive-definite matrix $P$, such that $\Upsilon(T_1, T_2, P) < 0$ for $(T_1, T_2) \in S_{T_1,T_2}$, where $\Upsilon(T_1, T_2, P)$ can be chosen as*

$$\Upsilon(T_1, T_2, P) \triangleq \big(\Phi(T_2)\Phi(T_1)\big)' P\Phi(T_2)\Phi(T_1) - P, \tag{10.8}$$

*or*

$$\Upsilon(T_1, T_2, P) \triangleq \big(\Phi(T_1)\Phi(T_2)\big)' P\Phi(T_1)\Phi(T_2) - P. \tag{10.9}$$

This result is similar to the one proposed in Theorem 10.2 for handling the Problem 10.2.

*Remark 10.7* As in the case of the Problem 10.2, different forms of $\Upsilon(T_1, T_2, P)$ may lead to different stability regions for Problem 10.3. In this chapter, we choose the former form (10.8).

*Remark 10.8* It is worth mentioning that in solving Problem 10.3, it is not so simple to find directly a Lyapunov matrix $P$ and the associated stability region, because it depends on two parameters $(T_1, T_2)$, and both are assumed to be time-varying.

We now give a helpful lemma, by which a Lyapunov matrix and a stability region can be guaranteed:

**Lemma 10.1** *If there is a $P$ satisfying $\Phi'(T)P\Phi(T) - P < 0$ for $T \in [\underline{T}, \overline{T}]$, then $\Upsilon(T_1, T_2, P) < 0$ for $T_1, T_2 \in [\underline{T}, \overline{T}]$.*

*Proof* If $\Phi'(T_2)P\Phi(T_2) - P < 0$ for $T_2 \in [\underline{T}, \overline{T}]$, $\Phi'(T_1)\Phi'(T_2)P\Phi(T_2)\Phi(T_1) - \Phi'(T_1)P\Phi(T_1) < 0$ for any non-negative $T_1$ and $T_2 \in [\underline{T}, \overline{T}]$. If $T_1 \in [\underline{T}, \overline{T}]$, it follows that $\Phi'(T_1)P\Phi(T_1) < P$. We now have that $(\Phi(T_2)\Phi(T_1))'P\Phi(T_2)\Phi(T_1) - P < (\Phi(T_2)\Phi(T_1))'P\Phi(T_2)\Phi(T_1) - \Phi'(T_1)P\Phi(T_1) < 0$. The proof is complete. □

*Remark 10.9* Lemma 10.1 can be extended to the case with more sub-sampling periods. More precisely, in the case of $l$ *sub-sampling period*, it holds that

$$\big(\Phi(T_l)\cdots\Phi(T_1)\big)' P\Phi(T_l)\cdots\Phi(T_1) - P < 0$$

for $T_1, \dots, T_l \in [\underline{T}, \overline{T}]$ if $\Phi'(T)P\Phi(T) - P < 0$ for $T \in [\underline{T}, \overline{T}]$.

*Remark 10.10* If we find a matrix $P$ satisfying $\Phi'(T)P\Phi(T) - P < 0$ for $T \in [\underline{T}, \overline{T}]$, then, in handling the Problem 10.3 and according to Lemma 10.1, we find a first estimation of the stability region in the parameter-space $(T_1, T_2)$: the rectangle defined by $T_1, T_2 \in [\underline{T}, \overline{T}]$. A Lyapunov matrix $P$ associated with the stability interval $[\underline{T}, \overline{T}]$ for a single-sampling system can be found in *Fujioka* [89] and *Zhang and Branicky* [263].

*Remark 10.11* In the proof of Lemma 10.1, we replace $\Phi'(T_1)P\Phi(T_1)$ with $P$, fact that unfortunately introduces some further degree of conservatism. Hence, for some parameters outside the region $T_1, T_2 \in [\underline{T}, \overline{T}]$, the inequality $\Upsilon(T_1, T_2, P) < 0$ may still hold. Thus, in order to detect the whole stability region corresponding to the Lyapunov matrix $P$ described by the inequality, $\Upsilon(T_1, T_2, P) < 0$, the parameter-sweeping method appears as an appropriate approach.

A *parameter-sweeping* method is given in order to detect the whole stability region.

**Algorithm 10.2**

- Step 1: Find a matrix $P$ satisfying the matrix inequality $\Phi'(T)P\Phi(T) - P < 0$ for $T \in [\underline{T}, \overline{T}]$ with the interval $[\underline{T}, \overline{T}]$ as large as possible, by using the existing result for the single-sampling systems, e.g., *Fujioka* [89].
- Step 2: Sweep parameters $T_1$ and $T_2$ to check if $\Upsilon(T_1, T_2, P) < 0$, and mark the region where $\Upsilon(T_1, T_2, P) < 0$ in the corresponding $T_1$–$T_2$ plane.

The obtained region represents an appropriate stability region. In general, a larger interval $[\underline{T}, \overline{T}]$ (obtained via Step 1) leads to a larger stability region, since the square stability region $T_1, T_2 \in [\underline{T}, \overline{T}]$ is included according to Lemma 10.1.

*Example 10.4* In the context of the Problem 10.3, the delay-free system has two time-varying uncertain sub-sampling periods $T_1$ and $T_2$. According to *Fujioka* [89], we choose

$$P = \begin{pmatrix} 4.0300 & 5.0900 \\ 5.0900 & 13.4900 \end{pmatrix}.$$

Algorithm 10.2 leads to the stability region depicted in Fig. 10.5. With the choice of the rectangle region $T_1 \in [1.3, 1.4]$ and $T_2 \in [2.1, 2.2]$ as the inner one and the rectangle region $T_1 \in [1.4, 1.5]$ and $T_2 \in [2.2, 2.3]$ as the outer one, the corresponding simulations covering both cases are shown in Fig. 10.6.

*Remark 10.12* An appropriate combination of some of the cases mentioned in Example 10.1 leads to the following interesting observation: the *overall system* can be *stable* even if *some subsystems* are *unstable*.

In our opinion, the method applied to solve the Problem 10.3 appears to be satisfactory (as shown in the simulation) with an essential advantage in terms of *simplicity* because we can directly use the existing result for single-sampling system according to Lemma 10.1. An alternative solution for handling the Problem 10.3 is to use the method given for solving the Problem 10.2, which is derived by using appropriate robust stability arguments. Such conditions (in terms of LMIs developed by using the discretization technique) developed for solving the Problem 10.3 are similar to the ones stated in Theorems 10.3, 10.4, and hence are not specially given.

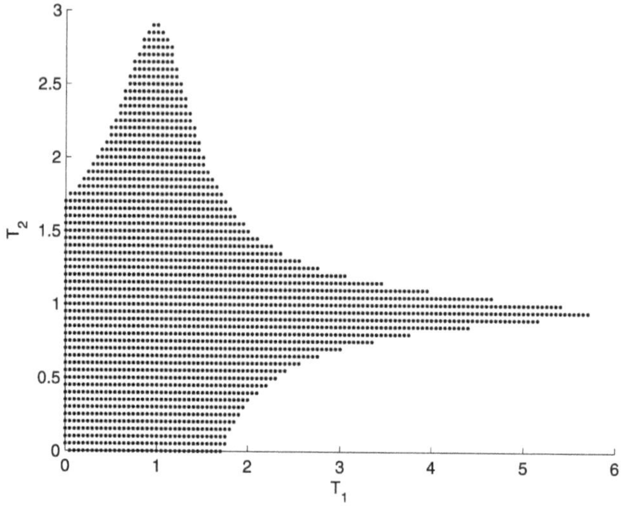


**Fig. 10.5** Stability region on the $T_1$–$T_2$ plane for Example 10.4

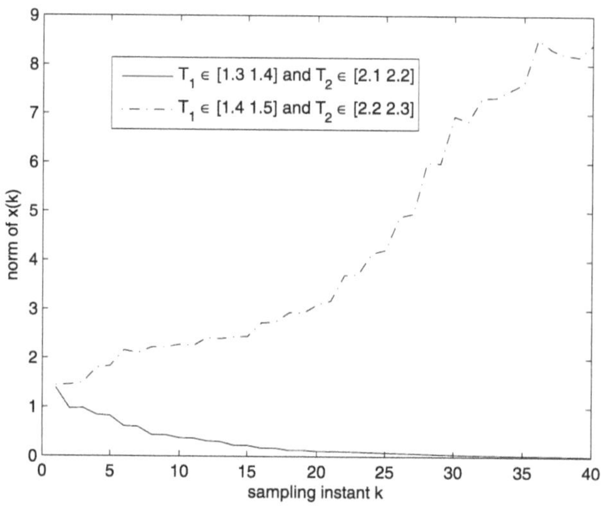


**Fig. 10.6** Simulations for the regions ($T_1 \in [1.3, 1.4]$, $T_2 \in [2.1, 2.2]$) and ($T_1 \in [1.4, 1.5]$, $T_2 \in [2.2, 2.3]$), showing that the system is stable with the inner parameter region while unstable with the outer parameter region

## 10.4 An Illustrative Application

We now consider an inverted pendulum example to illustrate the methods proposed in this chapter. An inverted pendulum model borrowed from *Gao and Chen* [91] is considered. We study the case when two inverted pendulum systems (see Fig. 10.7)

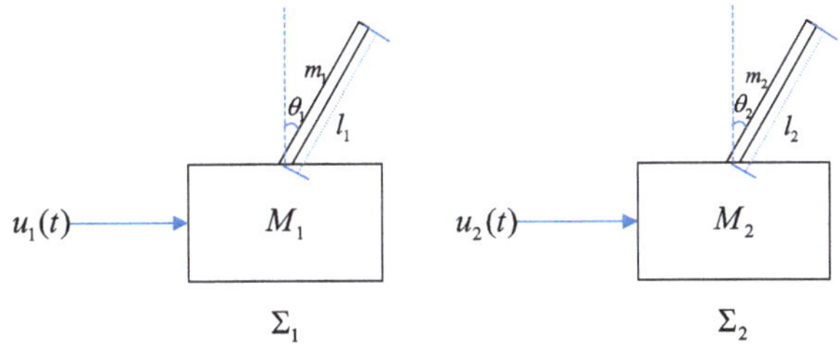


**Fig. 10.7** Two inverted pendulums ($\Sigma_1$ and $\Sigma_2$)

are controlled in the *DCES* framework, and each inverted pendulum has two sub-sampling periods.

By linearizing the inverted pendulum model at the equilibrium point, we obtain the corresponding systems ($\Sigma_1$ and $\Sigma_2$) described as follows:

$$\Sigma_1 : \dot{z}_1(t) = \begin{pmatrix} 0 & 1 \\ \frac{3(M_1+m_1)g}{l_1(4M_1+m_1)} & 0 \end{pmatrix} z_1(t) + \begin{pmatrix} 0 \\ \frac{-3}{l_1(4M_1+m_1)} \end{pmatrix} u_1(t), \tag{10.10}$$

$$\Sigma_2 : \dot{z}_2(t) = \begin{pmatrix} 0 & 1 \\ \frac{3(M_2+m_2)g}{l_2(4M_2+m_2)} & 0 \end{pmatrix} z_2(t) + \begin{pmatrix} 0 \\ \frac{-3}{l_2(4M_2+m_2)} \end{pmatrix} u_2(t). \tag{10.11}$$

For the system $\Sigma_1$, the parameters are selected as $M_1 = 8.0$ kg, $m_1 = 2.0$ kg, $l_1 = 0.5$ m, and $g = 9.8$ m/s$^2$. Suppose that the feedback gain is $K_1 = (102.91\ 80.7916)$ and two sub-sampling periods are $T_{11} = 0.1$ s and $T_{12} = 0.15$ s. Assume now that, for the system $\Sigma_2$, the parameters are selected as $M_2 = 8.0$ kg, $m_2 = 4.0$ kg, $l_2 = 1.0$ m, and $g = 9.8$ m/s$^2$. Suppose also that the feedback gain is $K_2 = (127.2\ 21.6)$ and two sub-sampling periods are respectively $T_{21} = 0.12$ s and $T_{22} = 0.17$ s.

The *DCES-induced delays* $\tau_{11}$, $\tau_{12}$, $\tau_{21}$ and $\tau_{22}$ (corresponding to $T_{11}$, $T_{12}$, $T_{21}$, and $T_{21}$, respectively) are time-varying and uncertain. Our objective is to find a common upper bound $\overline{\tau}$ such that systems $\Sigma_1$ and $\Sigma_2$ are stable if $0 \leq \tau_{11}, \tau_{12}, \tau_{21}, \tau_{22} < \overline{\tau}$.

With the choice of $N = 3$, $\tau_{1110} = 0.0001$, $\tau_{1120} = 0.0001$, $\tau_{2110} = 0.01$, $\tau_{2120} = 0.01$, $\tau_{3110} = 0.05$, $\tau_{3120} = 0.05$, $\beta_{j1} = \beta_{j2} = 0.001$ $(j = 1, 2, 3)$, Theorem 10.4 leads to the following Lyapunov matrix:

$$P_1 = 10^7 \times \begin{pmatrix} 2.4918 & -0.2397 & -0.0091 \\ -0.2397 & 0.4103 & 0.0038 \\ -0.0091 & 0.0038 & 0.0001 \end{pmatrix},$$

and the corresponding stability region is shown in Fig. 10.8 for the system $\Sigma_1$. Inside the stability region, we can find a square region $0 \leq \tau_{11}, \tau_{12} \leq 0.05$, hence the system $\Sigma_1$ is stable if $0 \leq \tau_{11}, \tau_{12} \leq 0.05$.

Now, with the choice of $N = 3$, $\tau_{1210} = 0.0001$, $\tau_{1220} = 0.0001$, $\tau_{2210} = 0.01$, $\tau_{2220} = 0.01$, $\tau_{3210} = 0.04$, $\tau_{3220} = 0.04$, $\beta_{j1} = \beta_{j2} = 0.001$ $(j = 1, 2, 3)$, Theo-

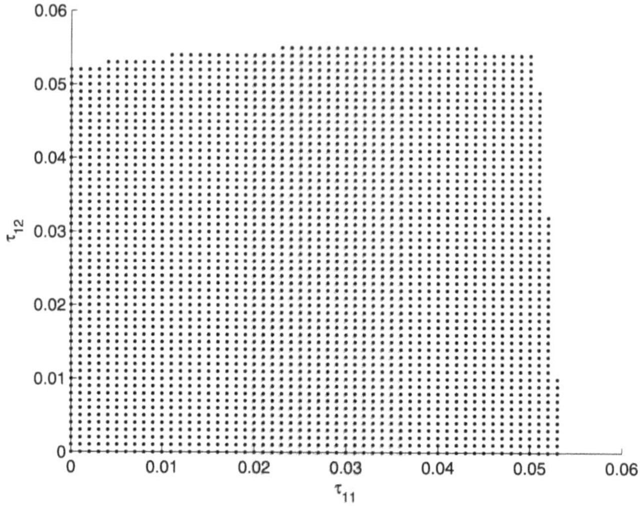


**Fig. 10.8** Stability region on the $\tau_{11}$–$\tau_{12}$ plane for the inverted pendulum $\Sigma_1$

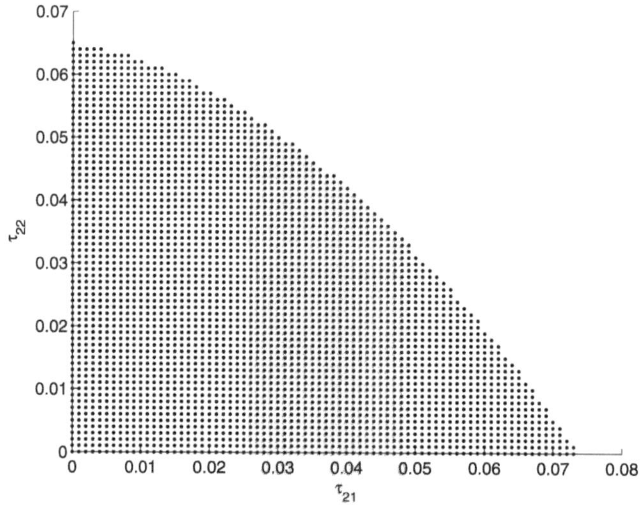


**Fig. 10.9** Stability region on the $\tau_{21}$–$\tau_{22}$ plane for the inverted pendulum $\Sigma_2$

rem 10.4 leads to the following Lyapunov matrix:

$$P_2 = 10^8 \times \begin{pmatrix} 2.9642 & 0.1012 & -0.0192 \\ 0.1012 & 0.1969 & -0.0001 \\ -0.0192 & -0.0001 & 0.0001 \end{pmatrix},$$

and its associated stability region for the system $\Sigma_2$ is shown in Fig. 10.9. Inside the stability region, we can find a square region $0 \le \tau_{21}, \tau_{22} \le 0.04$, hence the system

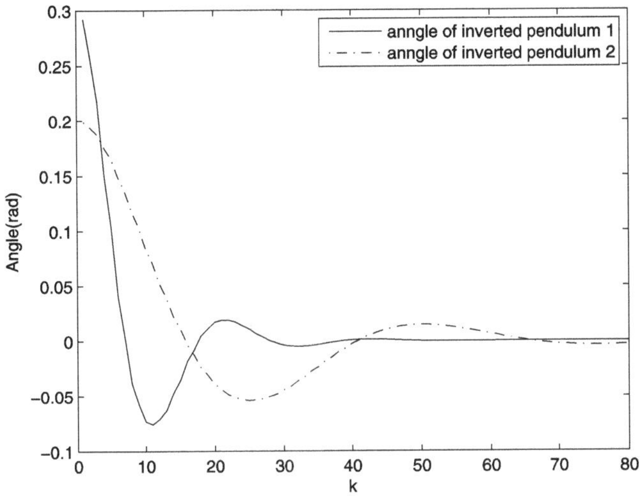


**Fig. 10.10** Simulations of two inverted pendulums when $0.039 \leq \tau_{11}, \tau_{12}, \tau_{21}, \tau_{22} \leq 0.040$, showing the asymptotic stability of two inverted pendulums

$\Sigma_2$ is stable if $0 \leq \tau_{21}, \tau_{22} \leq 0.04$. Thus, we know that system $\Sigma_1$ and $\Sigma_2$ are stable if $0 \leq \tau_{11}, \tau_{12}, \tau_{21}, \tau_{22} \leq 0.04$.

Assume now that the initial conditions for the inverted pendulums (described by $\Sigma_1$ and $\Sigma_2$) are $(0.3\ 0)'$ and $(0.2\ 0)'$, respectively. The simulations when $0.039 \leq \tau_{11}, \tau_{12}, \tau_{21}, \tau_{22} \leq 0.040$ are shown in Fig. 10.10.

Through the above results, we see that the inverted pendulum systems can be stabilized if the DCES-induced delays can be appropriately upper bounded. The objective of this example belongs to Problem 10.2, as we assume that the sub-sampling periods are given and that the DCES-delays are time-varying and uncertain. We may also verify the methods proposed for Problems 10.1 and 10.3 in this widely used experimental platform.

## 10.5 Notes and Comments

In this chapter, we addressed some stability problems specific to a *DCES* with the *hyper-sampling periods* and *DCES-induced delays*. Constant and time-varying *DCES-induced delays* and/or *sub-sampling periods* have been considered. It is well known that the main stability analysis difficulties are related to the uncertain variation of the sampling periods and the *DCES-induced delays*, and there exists an abundant literature devoted to such problems. Among the existing contributions, one may cite the papers *Hespanha et al.* [119], *Hristu-Varsakelis and Levine* [124], *Zampieri* [261], and *Zhang et al.* [264].

As a simple case of the hyper-sampling mode, the single-sampling mode has been largely used and treated in the literature and an important number of results have been reported on the corresponding stability issues. To mention a few, Lyapunov-based methods have been considered by *Gao et al.* [92], *Jian et al.* [133], *Seuret and Richard* [209], and *Yue et al.* [260]; next, *Estrada and Antsaklis* [81] and *Montestruque and Antsaklis* [178] have used the model-based method to stabilize the system; next, *Mirkin* [175] adopted the input-output method to develop the stability condition. Compared with the single-sampling mode, only a few studies have been reported in the control literature for handling the hyper-sampling mode (see, for instance, optimal scheduling and control approaches considered by *Ben Gaid et al.* [28] and *Cervin and Alriksson* [54]). To the best of the authors' knowledge, the influence induced by these parameters (*DCES-induced delays* and *hyper-sampling periods*) on the corresponding stability was not sufficiently addressed and no explicit stability conditions have been reported on how to find the ranges of *DCES-induced delays* and *hyper-sampling periods* guaranteeing the stability of the corresponding *DCES* under the hyper-sampling mode. Especially, when taking the effects induced by the use of networks into consideration, the *DCES-induced delays* and/or *hyper-sampling periods* may be *uncertain*, *time-varying*, and consequently, the analysis becomes significantly more complicated. In our opinion, such stability results are of paramount importance with which scientists and engineers will be faced more and more in the future. One of the main reason is related to the production cost. If a controlled task can stabilize a given plant with greater sampling periods (guaranteed by the theoretical analysis), less system resources are necessary. This fact implies that more tasks can be processed and share the network at the same time without upgrading the *Hardware/Software* architecture of *DCES*.

The approach proposed here is similar to the one proposed by *Li et al.* [148], where the problem of robust stability for single-sampling *DCES* with time-varying uncertain sampling periods and *DCES-induced delays* is investigated. The basic idea of *Li et al.* [148] is to model a *DCES* with time-varying uncertain sampling period and *DCES-induced delay* as a discrete-time model with norm-bounded uncertainties and then employ an appropriate robust control method to derive the corresponding stability condition. The modeling method used here represents an extension of the one proposed by *Suh* [222]. An alternative approach is to model a *DCES* as a discrete-time model with polytopic uncertainties, see, for example, *Hetel et al.* [120] and *Gielen et al.* [94].

As seen above, the common feature of the methods we are proposing is the *parameter-sweeping method*. Such parameter-sweeping methods have been widely used in the stability analysis of continuous time-delay systems, see, for example, *Chen and Latchman* [61], *Fazelinia et al.* [83], *Gu et al.* [105], and *Olgac and Sipahi* [190], and they are particularly adapted to our problems.

Finally, it is important to mention that under a switched-system angle, we can treat the system with *sub-sampling period* $T_1$ ($T_2$) as subsystem 1 (2). As discussed in this chapter, the system is stable even if one of the subsystem is not necessarily stable. Thus, estimating only a common stability bound for both subsystems appears to be conservative. Although such switched-type approaches are appealing for handling the stability of *DCESs* (see, e.g., *Sun et al.* [224] and the references therein), to the best of the authors' knowledge, the existing results on switched systems do not lead to less conservative results. Finally, the simulation results obtained for different types of problems confirm the theoretical results and give some credit to the previous statement.

# Chapter 11
# Optimization of the Hyper-Sampling Sequence for *DCESs*

## 11.1 Introduction

In Chap. 10, we have addressed the stability conditions of *DCESs* under the *hyper-sampling mode or periodic scheduling*, and we have proposed appropriate ranges of the *sub-sampling periods* guaranteeing the system stability. From the study of the stability regions and the stability properties of the subsystems composing the *DCES*, interesting stability characteristics have been derived. In particular, one concerns the close relation between stability and *hyper-sampling mode*. We have observed also that, in some special cases, stability domain can be enlarged by an appropriate choice of *sub-sampling periods*. Naturally, these observations, absolutely necessary and important to handle the overload state of node processors, can constitute a base to calculate a *hyper-sampling sequence* consisting in an ordered number of control task executions in the hyper-sampling period. Even if the stability is the most important property for a controlled system, we always require dynamic performance enhancement. The interplay between system dynamic performances and the hyper-sampling sequence is the main object of the study proposed in this chapter.

Such an interplay has also been discussed in Chaps. 4, 5, and 6, but the formalism and the model used are different. The objective is to address once more the close relation between the state of the system and the *hyper-sampling sequence*. A different point of view is adopted here in order to shed light on the relation between system's dynamic characteristics and node processor computation constraints given by an average value of the system sampling frequency. More precisely, we will derive an analytical relation between the hyper-sampling sequence and the system performance index in order to optimally design the hyper-sampling sequence.

The problem formulation is given in Sect. 11.2, where we choose a continuous-time cost function. Next, we first address the standard single-sampling case in Sect. 11.3. Even if this case seems relatively simple and well understood, we address this problem by using a different angle. Intuitively, increasing the sampling period, which in turn saves system resources, leads to some decrease of the dynamic performance. It is worth mentioning that this claim is not necessarily valid. In Sect. 11.3, we will see that, for some systems, increasing the sampling period may lead to

A. Çela et al., *Optimal Design of Distributed Control and Embedded Systems*, Communications and Control Engineering, DOI 10.1007/978-3-319-02729-6_11,

some improvement concerning the system performance. This fact is very useful in practical applications implying that with less system resources we can achieve better performance. In Sect. 11.4, we extend the results proposed in Sect. 11.3 to the hyper-sampling sequence design. We will obtain the analytical function of the performance index with respect to the hyper-sampling periods, based on which we can optimally design the hyper-sampling sequence. As expected, we will see that, for the same average value of calculation resources consumption, a hyper-sampling mode can improve the dynamic performance over the standard single-sampling mode.

In order to verify the proposed approach, we built an appropriate experimental platform. In this experimental platform, we control the angular velocity of a direct current (*DC*) motor. The hyper-sampling period for the *DC* motor has two sub-sampling periods and an average sampling frequency (*ASF*) is given. The objective is to optimally set the hyper-sampling periods under the fixed *ASF*. Detailed description of the applications as well as the results obtained are given in Sect. 11.5. Finally, some notes and comments complete this chapter.

## 11.2 Problem Formulation

Consider a simple *DCES* with the controlled plant (9.1)

$$\dot{x}(t) = Ax(t) + Bu(t),$$

and the sampled-data controller (9.7)

$$u(t) = Kx(t_k), \quad t_k \le t < t_{k+1},\ k \in \mathbb{N}.$$

The closed-loop model of such a simple *DCES* is as follows

$$\dot{x}(t) = Ax(t) + BKx(t_k), \quad t_k \le t < t_{k+1}. \tag{11.1}$$

As mentioned in the Introduction, we focus our study on the hyper-sampling mode execution of the related control task, for which the corresponding stability issue has been studied in Chap. 10. The mathematical expression of the hyper-sampling mode is given by (9.15) in Chap. 9:

$$T(k) = t_k - t_{k-1} = \begin{cases} T_{k \bmod n}, & k \bmod n \ne 0, k \in \mathbb{N}_+, \\ T_n, & k \bmod n = 0,\ k \in \mathbb{N}_+. \end{cases}$$

Since we ignore the sub-sampling periods variation as well as the *DCES-induced delays*, we denote the sampling instants by $t_k$ rather than $s_k$. For brevity, we adopt the following notations: the hyper-sampling period with $n$ sub-sampling periods $T_1, \ldots, T_n$ can be expressed as an $n$-tuple, $S_n = \{T_1, \ldots, T_n\}$; $T_\Sigma \triangleq \sum_{i=1}^{n} T_i$ represents the length of the hyper-sampling period.

As stated earlier, we design the hyper-sampling period under a given *average sampling frequency* (*ASF*) constraint, as defined below, such that the system performance is optimized.

The concept of the *ASF* is easy to understand. Suppose a hyper-sampling sequence is allowed to have $n$ sub-sampling periods $T_1, \dots, T_n$. Then, the *ASF* is defined as follows

$$ASF = \frac{n}{T_\Sigma}. \tag{11.2}$$

That is, for every $T_\Sigma$ seconds, the system is sampled $n$ times.

We may also introduce the concept of *average sampling period* (*ASP*) as

$$ASP = \frac{1}{ASF}, \tag{11.3}$$

which implies that the system is sampled once every *ASP* seconds on average.

To compare the system dynamic performance with respect to the choice of sub-sampling periods, we adopt the performance index $\eta$ given by:

$$\eta \triangleq \sup\left\{\frac{J}{x'(0)x(0)} : \forall x(0) \neq 0\right\}, \tag{11.4}$$

$$J = \int_0^\infty x'(t)Px(t)\,dt, \tag{11.5}$$

where $P$ is a positive semi-definite matrix and $x(0)$ is the initial state of the system.

*Remark 11.1* It is known that a system under the sampled-data control can be rewritten as a discrete-time system and hence one can also use a discrete-time cost function $J$ of the form $\sum x'(kT)Px(kT)$. We choose the continuous-time cost function (11.5) instead of a discrete-time one since the former considers the intersample behavior while the latter only considers the system state at the sampling instants. It was pointed out by Åström and Wittenmark [14] that the system's continuous-time state may have oscillations that are not seen at the sampling points. These are called hidden oscillations or intersample ripple.

Now, we can formulate our problem: given the number of sub-sampling periods $n$ and the *average sampling frequency ASF*, design the hyper-sampling period $S_n$ subject to the condition $T_\Sigma = \frac{n}{ASF}$ such that the corresponding performance index $\eta$, defined in (11.4), is optimized.

We can see that the above optimization problem is not only of theoretical but also of practical importance. In practice, with the same consumption of the system resources (i.e., the same *ASF*), we always search for optimal system performance.

In the sequel, we give the following lemma, which will be used in the remaining part of this book:

**Lemma 11.1** *Given matrices $M$ and $N$ where $M$ is Schur and $N$ is positive semi-definite, $\sum_{k=0}^{\infty}(M^k)'NM^k$ is the unique solution of the Lyapunov equation*

$$M'XM - X = -N,$$

*where $X$ is the variable to be determined.*

The proof of this lemma can be found in Anderson and Moore [4], pp. 64–65.

## 11.3 Design of the Standard Single-Sampling Period

In this section, we study the optimization of the standard single-sampling period with respect to system performance. With this setup, the *DCES* is described by (9.1) and (9.8). More precisely, the control signal expression is given by:

$$u(t) = Kx(kT), \quad kT \leq t < (k+1)T, \ k \in \mathbb{N}.$$

As the cost function $J$ defined in (11.5) is an integration over the unbounded interval $[0, \infty)$, we first need to find the closed form of its expression. Between two consecutive sampling instants, the system state is given by:

$$x(t) = \Phi(t - t_k)x(t_k), \quad t_k \leq t < t_{k+1},$$

where the one-argument operator $\Phi(\cdot)$ is defined in (9.12)

$$\Phi(\beta) = e^{A\beta} + \int_0^\beta e^{A\theta}\, d\theta\, BK.$$

We recall that throughout this section we consider the single-sampling mode and consequently the sampling period is denoted by $T$. For a given $T$, $\Phi(T)$ represents the corresponding transition matrix.

It is well known that a system under the single-sampling mode is asymptotically stable if and only if $\Phi(T)$ is Schur. Thus, $T$ has to be assigned in the interval where $\Phi(T)$ is Schur. Otherwise, the performance index $\eta$ goes to $\infty$.

A closed form of the cost function $J$ is given by the following:

**Lemma 11.2** *Assume that the single-sampling mode is adopted. If $\Phi(T)$ is Schur, the cost function $J$ defined in* (11.5) *for the system described by* (9.1) *and* (9.8) *can be expressed as*

$$J = x'(0)F(T)x(0), \tag{11.6}$$

*where $F(T)$ is the unique solution of the Lyapunov equation*

$$\Phi'(T)X\Phi(T) - X = -\Omega(T), \tag{11.7}$$

*where*

$$\Omega(T) = \int_0^T \Phi'(s)P\Phi(s)\, ds.$$

*Proof* The cost function $J$ can be rewritten as

$$J = \sum_{k=0}^{\infty}\left(\int_{kT}^{(k+1)T} x'(t)Px(t)\, dt\right).$$

Furthermore,

$$J = \sum_{k=0}^{\infty}\left(\int_0^T \left(\Phi(s)\Phi^k(T)x(0)\right)' P\left(\Phi(s)\Phi^k(T)x(0)\right) ds\right).$$

Therefore, we have (11.6) with

$$F(T) = \sum_{k=0}^{\infty} \left(\Phi^k(T)\right)' \Omega(T) \Phi^k(T).$$

It is easy to see that $F(T)$ and $\Omega(T)$ are symmetric and positive semi-definite. If $\Phi(T)$ is Schur, $F(T)$ is the unique solution of the Lyapunov equation (11.7), according to Lemma 11.1. The proof is now complete. □

*Remark 11.2* The cost function $J$ can not be directly handled in the form (11.5) since it is an integration over an unbounded interval. By Lemma 11.2, we equivalently transform it into the expression (11.6), which amounts to compute the solution matrix $F(T)$ of the Lyapunov equation (11.7), based on effective algorithms and software, for example, MATLAB.

We are now in a position to present the main result of this section.

**Theorem 11.1** *Assume that the single-sampling mode is adopted. The performance index* $\eta$*, defined in* (11.4)*, for the system described by* (9.1) *and* (9.8) *is*

$$\eta = \lambda_{\max}\big(F(T)\big),$$

*if* $\Phi(T)$ *is Schur.*

*Proof* By the definition (11.4), it is equivalent to find the minimum $\eta$ such that

$$J - \eta x'(0)x(0) \le 0, \quad \forall x(0) \ne 0.$$

Furthermore, in the light of Lemma 11.2, it is equivalent to

$$x'(0)\big(F(T) - \eta I\big)x(0) \le 0, \quad \forall x(0) \ne 0,$$

if $\Phi(T)$ is Schur. Therefore, $\eta$ can be precisely expressed as

$$\eta = \inf\big\{\eta : F(T) - \eta I \le 0\big\}.$$

It is true that $\eta = \lambda_{\max}(F(T))$, and thus, the proof is completed. □

*Remark 11.3* For a given sampling period $T$, we may use the symbolic manipulation toolbox in MATLAB to obtain the expression of $\eta$ as a function of $T$.

The sampling period assigned to a control task is closely related to the system (computation and communication) resources consumed by this task. A smaller (larger) sampling period implies more (less) system resources used. Intuitively, the performance index should be in direct proportion to the sampling period (i.e., if we demand smaller performance index $\eta$, smaller sampling period $T$ should be chosen

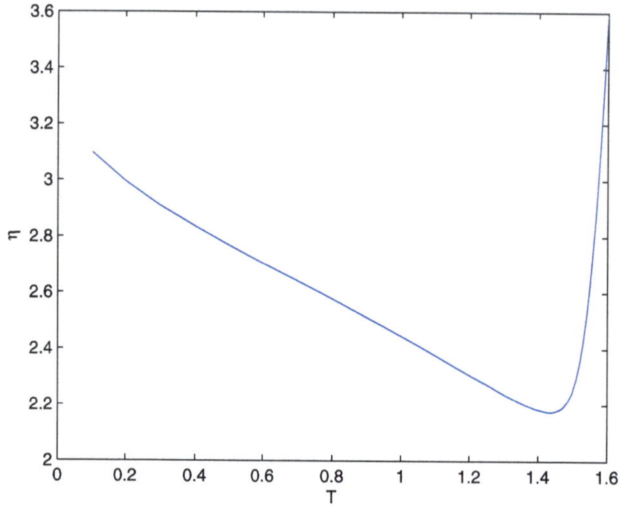


**Fig. 11.1** The performance index $\eta$ vs sampling period $T$ for *Example 11.1*

and, consequently, the required system resources should increase). In the following illustrative example, we will observe that this intuition does not always hold.

*Example 11.1* Consider the *DCES* described by (9.1) and (9.8) under the single-sampling mode, with matrices $A$, $B$, and $K$ given in (10.1)

$$A = \begin{pmatrix} 0 & 1 \\ 0 & -0.1 \end{pmatrix}, \qquad B = \begin{pmatrix} 0 \\ 0.1 \end{pmatrix}, \qquad K = (-3.75 \quad -11.5).$$

We choose the cost function $J$ (11.5) with $P = I$. It can be analytically calculated that the system is asymptotically stable if and only if $T \in (0, 1.72]$, see *Li et al.* [151]. Since a real sampling period can not be zero, assume the starting point of sampling period at $T = 0.1$. The use of Theorem 11.1 leads to the relation between sampling period and the performance index as depicted in Fig. 11.1.

It is interesting to see that for $T \leq 1.44$, the performance index $\eta$ monotonously decreases (increasing the sampling period may improve the system performance). This means that we can obtain better performance and consume less system resources at the same time.

To illustrate the results above, we compare the intersample behavior when sampled with $T = 0.1$ and $T = 0.6$. Intuitively, the system's dynamic behavior when sampled with $T = 0.1$ should be better than when sampled with $T = 0.6$, as the former sampling frequency is 5 times higher than the latter. From the simulation shown in Fig. 11.2, we see that this intuition is not necessarily true. The fact is that the system dynamic behavior when sampled with $T = 0.6$ is "better" than when sampled with $T = 0.1$. In this simulation, the initial state was assumed as: $x(0) = (1\ \ 1)'$. By Lemma 11.2, we have that $J = 2.2115$ when $T = 0.6$ while $J = 2.5392$ when $T = 0.1$.

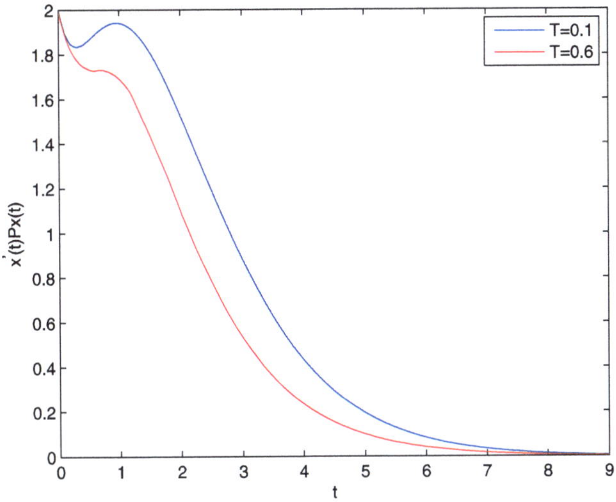


**Fig. 11.2** $x'(t)Px(t)$ vs $t$ for *Example 11.1*: a larger sampling period may lead to better dynamics

## 11.4 Design of the Hyper-Sampling Sequence

In this section, we study the optimization of the hyper-sampling sequence $S_n$. We first study the stability condition of the *DCESs* under the hyper-sampling mode. Given a hyper-sampling sequence $S_n$, the state at sampling instant $t_k$ is given by the following expression

$$x(t_k) = \Phi\big(T(k)\big)\cdots\Phi\big(T(2)\big)\Phi\big(T(1)\big)x(t_0),$$

where $t_0 = 0$ and $T(1), \ldots, T(k)$ are generated according to (9.15).

For $z \in N$, it follows that:

$$x(t_{(z+1)n}) = \widetilde{\Phi}(S_n)x(t_{zn}),$$

where

$$\widetilde{\Phi}(S_n) = \Phi(T_n)\cdots\Phi(T_2)\Phi(T_1).$$

The stability condition of a *DCES* under the hyper-sampling mode is given by the following theorem.

**Theorem 11.2** *Under a hyper-sampling period $S_n$, the system described by* (9.1) *and* (9.7) *is asymptotically stable if and only if $\widetilde{\Phi}(s_n)$ is Schur.*

*Proof* It follows that

$$x(t) = R(\theta)\widetilde{\Phi}^p(S_n)x(0),$$

with

$$
\begin{aligned}
p &= \left\lfloor \frac{t}{T_\Sigma} \right\rfloor, \\
\theta &= t - pT_\Sigma, \quad \theta \in [0, T_\Sigma), \\
R(\theta) &= \begin{cases} R_0(\theta), & 0 \le \theta < T_1, \\ R_1(\theta), & T_1 \le \theta < T_1 + T_2, \\ R_2(\theta), & T_1 + T_2 \le \theta < T_1 + T_2 + T_3, \\ \vdots & \\ R_{n-1}(\theta), & T_\Sigma - T_n \le \theta < T_\Sigma, \end{cases} \\
R_0(\theta) &= \Phi(\theta), \\
R_1(\theta) &= \Phi(\theta - T_1)\Phi(T_1), \\
R_2(\theta) &= \Phi(\theta - T_1 - T_2)\Phi(T_2)\Phi(T_1), \\
&\vdots \\
R_{n-1}(\theta) &= \Phi(\theta - T_\Sigma + T_n)\Phi(T_{n-1}) \cdots \Phi(T_1).
\end{aligned}
$$

As $t \to +\infty$, $p \to +\infty$, while $R(\theta)x(0)$ is always bounded for any $x(0) \neq 0$. One can see that $\lim_{t\to\infty} x(t) = 0$ if and only if $\widetilde{\Phi}(S_n)$ is Schur. □

Similar to Lemma 11.2 for the single-sampling mode, we have the following lemma for the hyper-sampling mode.

**Lemma 11.3** *Assume that the hyper-sampling sequence is composed of n sub-sampling periods $T_1, \ldots, T_n$. If $\widetilde{\Phi}(S_n)$ is Schur, the cost function $J$ defined in* (11.5) *for the system described by* (9.1) *and* (9.7) *can be expressed as*

$$
J = x'(0)F(S_n)x(0), \tag{11.8}
$$

*where $F(S_n)$ is the unique solution of the Lyapunov equation*

$$
\widetilde{\Phi}'(S_n)X\widetilde{\Phi}(S_n) - X = -\Omega(S_n), \tag{11.9}
$$

*where*

$$
\Omega(S_n) = \sum_{j=1}^{n} W'(j-1) \int_0^{T_j} \Phi'(s)P\Phi(s)\,ds\, W(j-1),
$$

$$
W(j) = \begin{cases} \Phi(T_j) \cdots \Phi(T_2)\Phi(T_1), & j = 1, \ldots, n, \\ I, & j = 0. \end{cases}
$$

*Proof* The cost function $J$ can be rewritten as

$$
J = \sum_{k=0}^{\infty} \left( \int_{kT_\Sigma}^{(k+1)T_\Sigma} x'(t)Px(t)\,dt \right).
$$

Furthermore,

$$\begin{aligned} J = \sum_{k=0}^{\infty} \bigg( & \int_0^{T_1} \big(\Phi(s)W^k(n)x(0)\big)' P \big(\Phi(s)W^k(n)x(0)\big)\, ds \\ & + \int_0^{T_2} \big(\Phi(s)W(1)W^k(n)x(0)\big)' P \big(\Phi(s)W(1)W^k(n)x(0)\big)\, ds + \cdots \\ & + \int_0^{T_n} \big(\Phi(s)W(n-1)W^k(n)x(0)\big)' P \big(\Phi(s)W(n-1)W^k(n)x(0)\big)\, ds \bigg). \end{aligned}$$

Equivalently,

$$J = \sum_{k=0}^{\infty} \left( \sum_{j=1}^{n} \int_0^{T_j} \big(\Phi(s)W(j-1)W^k(n)x(0)\big)' P \big(\Phi(s)W(j-1)W^k(n)x(0)\big)\, ds \right).$$

Therefore, we have (11.8) with

$$F(S_n) = \sum_{k=0}^{\infty} \big(W^k(n)\big)' \Omega(S_n) W^k(n).$$

It is easy to see that $F(S_n)$ and $\Omega(S_n)$ are symmetric and positive semi-definite. If $\widetilde{\Phi}(S_n)$ (note that $\widetilde{\Phi}(S_n) = W(n)$) is Schur, $F(S_n)$ is the unique solution of the Lyapunov equation (11.9), by Lemma 11.1. The proof is now complete. □

We are now in a position to present the main result of this chapter.

**Theorem 11.3** *Assume that the hyper-sampling sequence is composed of $n$ sub-sampling periods $T_1, \ldots, T_n$. The performance index $\eta$, defined in* (11.4) *for the system described by* (9.1) *and* (9.7) *is*

$$\eta = \lambda_{\max}\big(F(S_n)\big),$$

*if $\widetilde{\Phi}(S_n)$ is Schur.*

The proof is in the same line of that for Theorem 11.1, and hence is omitted here. We now give a numerical example to illustrate the proposed design method:

*Example 11.2* Consider the *DCES* described by (9.1) and (9.7) under the hyper-sampling mode, with matrices $A$, $B$, and $K$ given in (10.1). The matrix $P$ in (11.5) is given by

$$P = \begin{pmatrix} 4.03 & 5.09 \\ 5.09 & 13.49 \end{pmatrix}.$$

Suppose the *average sampling frequency* is $\frac{1}{1.5}$. We will design the hyper-sampling period or sequence when $n = 2$ and $n = 3$, respectively.

For the case $n = 2$, two sub-sampling periods, $T_1$ and $T_2$, satisfy $T_1 + T_2 = 3$. One can see that, in this case, in fact only one parameter has to be determined. If

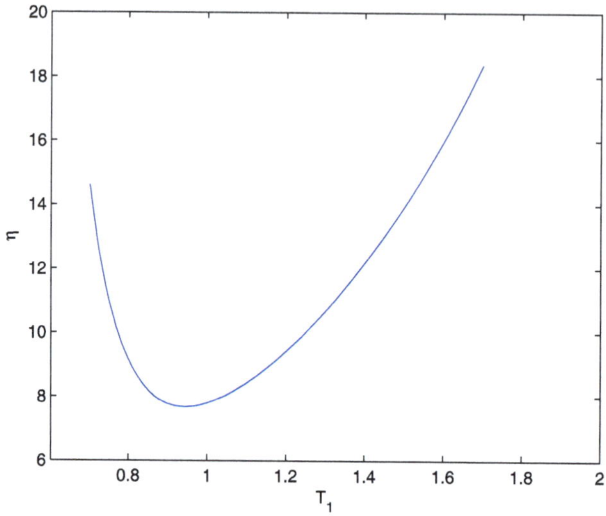


**Fig. 11.3** $\eta$ vs $T_1$ for Example 11.2

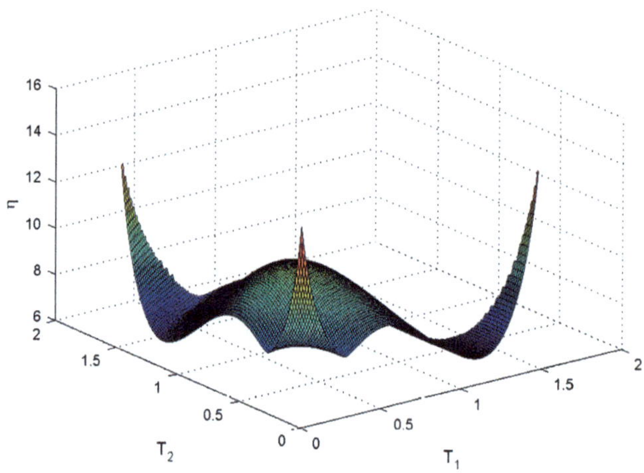


**Fig. 11.4** $\eta$ vs $T_1$, $T_2$ for Example 11.2

we consider $T_1$ such a parameter, then $T_2 = 3 - T_1$. By Theorem 11.3, the relation between $\eta$ and $S_n$ is given by a 2-D plot, as shown in Fig. 11.3.

Now, in the case $n = 3$, three sub-sampling periods, $T_1$, $T_2$, and $T_3$, satisfy $T_1 + T_2 + T_3 = 4.5$, and there are two parameters to be determined. If we consider $T_1$ and $T_2$ such parameters, then $T_3 = 4.5 - T_1 - T_2$. By Theorem 11.3, the relation between $\eta$ and $S_n$ is given by a 3-D plot, as shown in Fig. 11.4.

To illustrate the result, we observe the system dynamics, under the single-sampling mode (namely $n = 1$), the hyper-sampling mode with $n = 2$ and the hyper-

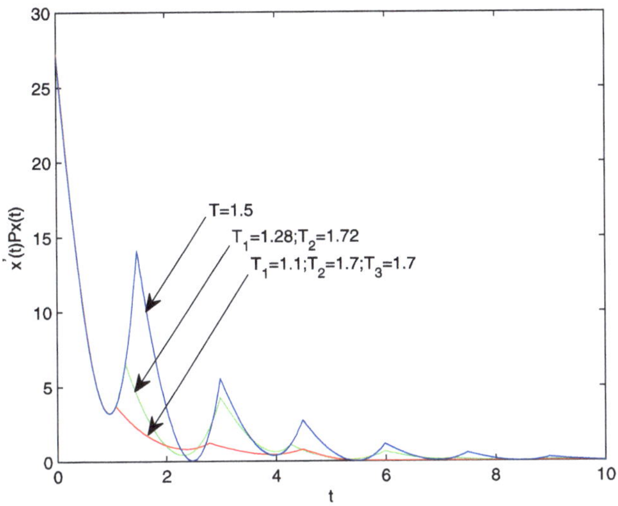


**Fig. 11.5** Dynamic simulation for Example 11.2

sampling mode with $n = 3$, respectively. The simulation is given in Fig. 11.5, where we choose the initial state as $x(0) = (1\ 1)'$. It is clearly seen that as $n$ increases, the system dynamics can be improved.

In this example, it is easy to see that the cases $T_1 = T_2 = 1.5$ when $n = 2$ and $T_1 = T_2 = T_3 = 1.5$ when $n = 3$ correspond to the standard single-sampling mode. One can see that under the same *average sampling frequency* the hyper-sampling mode can enhance the system dynamics.

In Example 11.1, we observed that the performance index $\eta$ is not directly proportional to the sampling period. Here, we observe how the optimal performance index, denoted by $\eta^*$, varies with respect to the average sampling period (*ASP*).

Consider the hyper-sampling mode with $n = 2$. The relation between $\eta^*$ and *ASP* is shown in Fig. 11.6. Through the relation, we see that, the optimal performance index is neither directly proportional to the ASP under the hyper-sampling mode.

*Remark 11.4* The approach proposed in this chapter covers the stability analysis. If $\widetilde{\Phi}(S_n)$ is not Schur (i.e., the system is not asymptotically stable), no proper solution of $\lambda_{\max}(F(S_n))$ will be obtained. Thus, the plots in Figs. 11.3 and 11.4 also indicate the stability ranges of the sub-sampling periods.

## 11.5 An Experimental Platform

In the sequel, we will illustrate the approach proposed by an experiment. The experimental platform, shown in Fig. 11.7, consists in four primary components: (1) a direct current (DC) brushed motor (from Maxon, model A-116105), driven by an analog power amplifier, (2) a data acquisition (DAQ) card (from National Instrument,

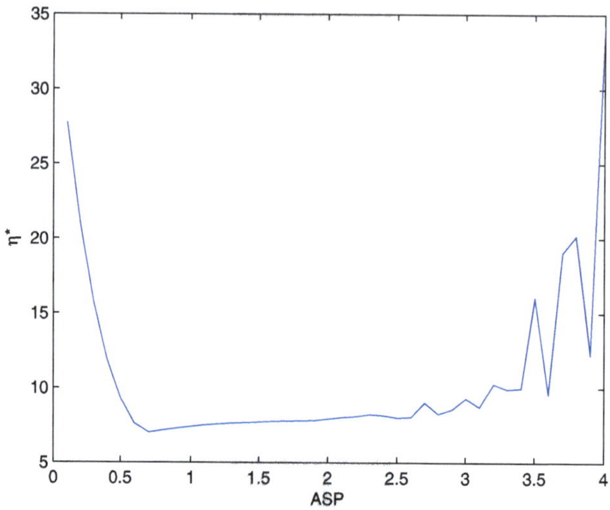


**Fig. 11.6** $\eta^*$ vs *ASP* for Example 11.2

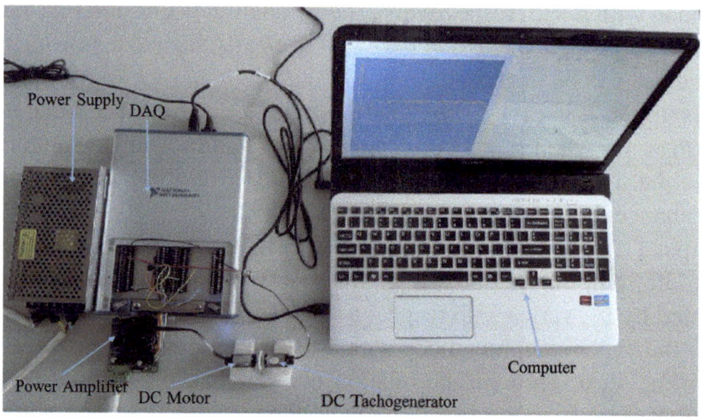


**Fig. 11.7** An experimental platform

model USB X-6343), (3) a DC tachogenerator, and (4) a laptop. Our objective is to control the angular velocity of the DC motor with respect to a given reference.

The communication is implemented by the DAQ card: (1) the output voltage of the DC tachogenerator, serving as the measurement of the angular velocity of the DC motor, is connected to the DAQ card, (2) the control input to the DC motor is generated by the DAQ card and the power amplifier, (3) the DAQ card can communicate bidirectionally with the laptop, on which the Matlab/Simulink-based control algorithms run. In this way, the closed-loop *DCES* architecture is established and the control parameters (the feedback gain and the hyper-sampling periods) can be set by the control algorithms.

The dynamics of a DC motor can be approximately described as the following linear[1] differential equation (see Ramirez et al. [198]):

$$\dot{\omega}(t) = A\omega(t) + Bu(t), \tag{11.10}$$

where $\omega(t)$ and $u(t)$ denote respectively, the angular velocity and the control input voltage and the parameters $A$ and $B$ are estimated as $A = -0.66$ and $B = 1000$.

We choose the reference angular velocity (denoted by $\omega_r$) as $\omega_r = 2000$ r/min. Let $e(t) = \omega(t) - \omega_r$. Then, we have that

$$\dot{e}(t) = Ae(t) + B\widetilde{u}(t),$$

where

$$\widetilde{u}(t) = u(t) + \frac{A}{B}\omega_r.$$

We adopt the state-feedback sampled-data control

$$\widetilde{u}(t) = Ke(t_k), \quad t_k \leq t < t_{k+1},$$

and we set $K = -0.0033$. Let us define the performance index $\eta$ as in (11.4) with the cost function $J = \int_0^\infty e^2(t)\,dt$.

We first use the single-sampling mode with the sampling period $T = 0.3$ s. By Theorem 11.1, the corresponding optimal performance index is $\eta = 8.78$.

We next use the hyper-sampling mode with $n = 2$. We let $T_1 + T_2 = 0.6$ s (where $T_1$ and $T_2$ are the two sub-sampling periods) in order to make the average sampling frequency equal to that of the single-sampling mode.

Finally, by using Theorem 11.3, we obtain the relation between the hyper-sampling period and the performance index, see, for instance, Fig. 11.8. We find the optimal hyper-sampling period with $T_1 = 0.28$ s, $T_2 = 0.32$ s and the corresponding $\eta = 8.70$. The simulations for two cases with $\omega(0) = 0$ are shown in Fig. 11.9.

## 11.6 Notes and Comments

In this chapter, we proposed a systematic method to design the hyper-sampling sequence or period (including the standard single-sampling period as a special case) for a *DCES*. We obtained the analytic relation between the performance index and the hyper-sampling period. Thus, we may obtain the optimal hyper-sampling period for a given *average sampling frequency* (*ASF*). That is, with the same system resources utilization, the *DCES* dynamic performance are optimized.

For the single-sampling mode, an interesting phenomenon was observed: the performance index is not necessarily in direct proportion to the sampling frequency. A similar phenomenon was also noticed in the hyper-sampling mode: the performance index may be not in direct proportion to the *average sampling frequency*

[1] If a strict linear model is demanded, we may adopt a friction compensation strategy, see, e.g., Canudas et al. [51].

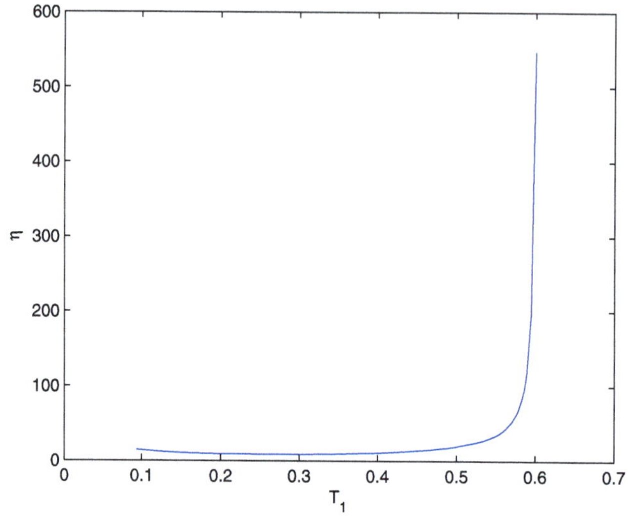


**Fig. 11.8** $\eta$ vs $T_1$ for Sect. 11.5

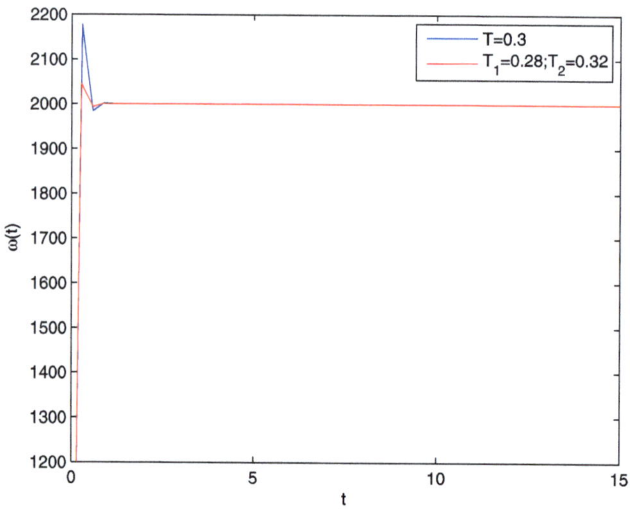


**Fig. 11.9** $\omega(t)$ vs $t$: The hyper-sampling mode leads to better dynamics

(*ASP*). Consequently, in some situations, we may spend less system resources and obtain better system dynamics at the same time. Such an observation is important for both theoretical studies and practical applications.

Our idea to solve the optimal problem was as follows: first, deriving the analytical relation between the performance index and the design parameter (i.e., the hyper-sampling period in this chapter) and next, applying the parameter-sweeping technique based on the derived analytical function to find the optimal parameter (i.e., the optimal hyper-sampling period in this case). Such a framework used in this chapter may be also adopted to other nominal cases, see, for example, the *DCES* with switched sampled-data control (to be investigated in the next chapter). However, if the system under consideration is non-nominal, the analysis becomes more complicated, and only some sufficient though not necessary conditions can be obtained, see, for example, the study for non-nominal sampled-data control systems by Skaf and Boyd [214].

In this chapter, we have deliberately made the choice to focus on the effect of the hyper-sampling period on the *DCES'* dynamics. We have treated the sub-sampling periods as the design parameters while the feedback gain matrix was fixed. One may predict a further performance improvement if the hyper-sampling period and the feedback gain matrix are designed simultaneously. Such a *co-design* problem, more complicated and challenging, is not considered in this book.

# Chapter 12
# A Switched Sampled-Data Control Strategy for *DCESs*

In Chaps. 10 and 11, we addressed the stability and dynamic performance of distributed control and embedded systems (*DCESs*) under the periodic scheduling or *hyper-sampling mode*. We have considered the stability of *DCESs* under the *hyper-sampling mode* as well as the scheduling design in order to enhance the performance. The stability analysis of this class of systems is motivated by their increasing practical importance and the necessity to handle faulty and overload situations. The results obtained prove the usefulness of their stability analysis by reducing the conservatism and so increasing their stability domain as well as pointing out some contradiction with the generally accepted intuition. It is worth noting that reducing the system communication and calculation resources does not necessarily mean the reduction of the dynamic performance.

In this chapter, we will follow the same line by further focusing our study "inside" the sampling period. The rationale behind it is the system stability together with the system performance enhancement. We will propose an easily-implemented switched sampled-data (*SD*) control strategy which may enhance the stability as well as dynamic properties. Regarding the stability issue, we will show that the use of the switched *SD* control strategy allows to enlarge the stability bound on the sampling period. As a consequence, a controlled task can be stabilized with less system resources by using the switched *SD* controller. We will also study the dynamic performance of a *DCES* with the switched *SD* control. An optimization method is proposed to optimally set the switching parameter of the switched *SD* controller. Thus, given a sampling period, we may obtain the optimal performance index through an appropriate setting of the switching parameter. We will show that the switched *SD* control may lead to a much better performance index than the standard single-sampling control. In other words, with the same utilization of computation and communication resources, the dynamic performance of a controlled task may be considerably enhanced by using the switched *SD* control.

A. Çela et al., *Optimal Design of Distributed Control and Embedded Systems*, Communications and Control Engineering, DOI 10.1007/978-3-319-02729-6_12,

## 12.1 Introduction

A *DCES* with the controlled plant (9.1)

$$\dot{x}(t) = Ax(t) + Bu(t),$$

and the standard sampled-data controller (9.8)

$$u(t) = Kx(kT), \quad kT \le t < (k+1)T,\ k \in \mathbb{N},$$

represents both the continuous-time and discrete-time dynamics. The controlled plant (9.1) is a continuous-time system, while its control input $u(t)$ is updated only at discrete sampling instants. Between two sampling instants $kT$ and $(k+1)T$ (where $T$ is the constant sampling period), the control input $u(t)$ is fixed as $Kx(kT)$.

Since a *DCES* described by (9.1) and (9.8) can be modeled by a discrete-time system (9.13)

$$x\big((k+1)T\big) = \Phi(T)x(kT),$$

its dynamics at the sampling instants can be easily studied. In Sect. 12.2 of this chapter, we will address the "intersample" dynamics (i.e., the dynamic behavior within a sampling interval) of the *DCESs*, through modeling it as a continuous-time system with an input delay.

Let us first give the following motivating example in order to observe the intersample dynamics of the system under consideration, without giving a detailed mathematical expression.

*Example 12.1* Consider system (9.1) with matrices given in (10.1)

$$A = \begin{pmatrix} 0 & 1 \\ 0 & -0.1 \end{pmatrix}, \qquad B = \begin{pmatrix} 0 \\ 0.1 \end{pmatrix}, \qquad K = (-3.75 \quad -11.5).$$

Let us adopt the standard sampled-data controller (9.8)[1] and assume a sampling period $T = 1.7$. Let us choose also the following Lyapunov function $V(t) = x'(t)Px(t)$ where

$$P = \begin{pmatrix} 4.03 & 5.09 \\ 5.09 & 13.49 \end{pmatrix}.$$

The evolution of $V(t)$ under the standard sampled-data control and for the initial state $x(0) = (1\ 1)'$ is shown in Fig. 12.1. It can be easily seen that at the sampling instants, marked with $*$ in the figure, $V(t)$ monotonously decreases. However, $V(t)$ does not monotonously decrease between sampling instants. More precisely, we

[1] In this chapter, we will propose a new sampled-data controller, a switched sampled-data (SD) controller. Throughout this chapter, in order to distinguish it from the switched SD controller, we call the sampled-data controller (9.8) the *standard* sampled-data controller.

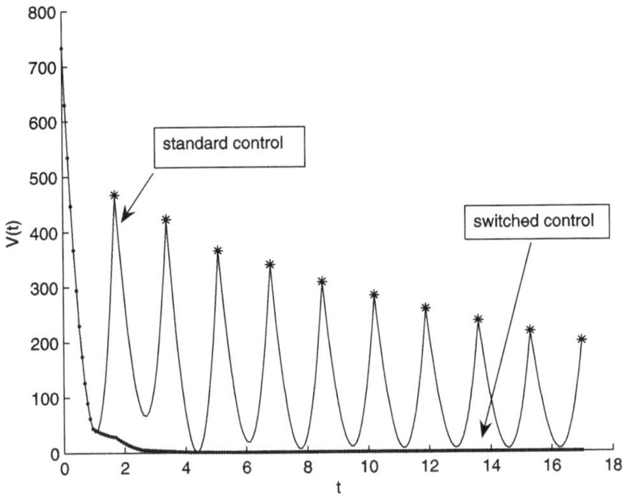


**Fig. 12.1** $V(t)$ vs $t$ for Example 12.1

observe that the sampling interval for the system considered in this example is composed of two parts depicting distinct behaviors: in the first part of it $V(t)$ decreases whereas in the second part it increases.

Based on such an observation, we may divide each sampling interval into two parts: the so-called *"convergence part"* (the first interval where $V(t)$ decreases) and the *"divergence part"* (the second interval where $V(t)$ increases), respectively. In the divergence part within a sampling interval, the state feedback control signal, in fact, loses its stabilizing effect. This motivates us to ask the following question: *Could the dynamics be improved if we "turn off" the control input $u(t)$ (i.e., let the system be open-loop) in the latter part within a sampling interval?*

Let us see what happens with the system performance if we switch from control signal produced by the feedback control law, $u(t) = Kx(kT)$, to the zero control signal at instant $KT + a$ inside each sampling interval. For the chosen switching time parameter $a = 0.94$, the evolution of $V(t)$ under the switched control is also given (by bold curve) in Fig. 12.1. It is interesting to see that the switched control law provides better performance than the standard one.

From the above motivating example, the following questions may naturally arise:

1. *How to explain the fact that the standard sampled-data control loses its stabilizing effect in the second part of the sampling interval?*
2. *How to identify the first part and the second part of the sampling interval? In other words, how to set the switching time parameter $a$?*

In the remaining part of this chapter, we will study the above questions in detail and we will give a systematic approach to design the proposed switched sampled-data (*SD*) controller. Concerning the first question, we will address it from a delay-system perspective in Sect. 12.2. Next in Sect. 12.3, we will propose a new control

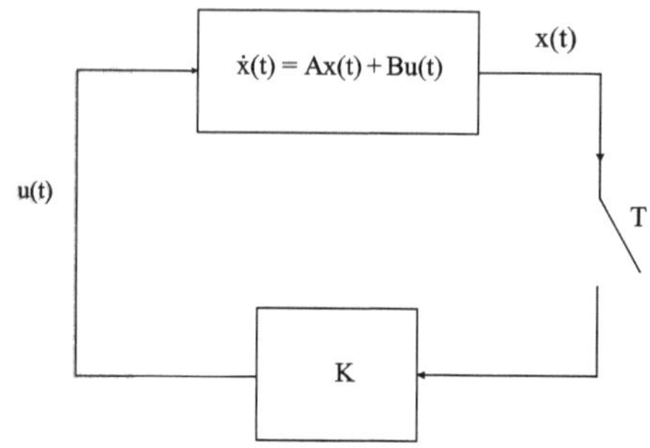


**Fig. 12.2** The standard sampled-data control

strategy: a switched sampled-data (*SD*) control law. In Sect. 12.4, we will analyze the stability under the switched *SD* control strategy. Next, in Sect. 12.5, we will optimize the dynamic performance through setting the switching time parameter of the switched *SD* controller. In Sect. 12.6, an illustrative application will be presented. Finally, some notes and comments complete this chapter.

## 12.2 Intersample Dynamics: A Delay-System Perspective

The scheme of the standard sampled-data control is given in Fig. 12.2. As we already know, in such a configuration, the state of the controlled plant $x(t)$ may be sampled only at discrete sampling instants. In this chapter, we consider the standard single-sampling case with a constant sampling-period $T$ and thus the sampling instants are denoted by $kT, k \in \mathbb{N}$. Further, we adopt the state-feedback control (9.8) with feedback gain $K$.

As proposed by *Fridman et al.* [88], the closed-loop form of a sampled-data system can be rewritten as the input-delayed system (9.16)

$$\dot{x}(t) = Ax(t) + BKx\big(t - \tau(t)\big), \quad kT \leq t < (k+1)T,$$

with delay $\tau(t) = t - kT, kT \leq t < (k+1)T$. Such a delay (also called the artificial delay in Part III of the book) is illustrated in Fig. 9.7.

Let us focus now on the intersample dynamics by observing $\dot{x}(kT + \theta)$, $0 \leq \theta < T$. When $\theta$ is small, $\dot{x}(kT + \theta)$ may be approximated by $(A + BK)x(kT)$ as the artificial delay $\tau(t)$ is small, see Fig. 9.7. Since $A + BK$ is designed to be Hurwitz, the system state converges for small $\theta$ (as the closed-loop system is close to the continuous-time system $\dot{x}(t) = (A + BK)x(t)$). As $\theta$ increases, the artificial delay $\tau(t)$ increases accordingly (see Fig. 9.7) and its effect can no longer be neglected. Thus, for a large $\theta$, the system dynamics diverge from the continuous-time system dynamics given by $\dot{x}(t) = (A + BK)x(t)$ and hence we may observe performance deterioration.

By the above analysis from a delay-system angle, an explicit answer to the question concerning the "strange" behavior of the Lyapunov function $V(t)$ for the DCES considered in Example 12.1[2] can be given.

## 12.3 A Switched Sampled-Data (*SD*) Control Approach

According to the arguments in Sect. 12.2, it appears potentially possible that, at the second (divergent) part of the sampling interval, the open-loop dynamics $\dot{x}(t) = Ax(t)$ may give *"less divergent"* state evolution than the feedback system dynamics. If such an argument may hold, from an appropriate instant $kT + a$, we can set the control input to 0 (which is easy to implement in practice) and the resulting system dynamics may be enhanced. This motivates us to propose the following switched sampled-data (*SD*) controller:

$$u(t) = \begin{cases} Kx(kT), & kT \le t < kT + a, \\ 0, & kT + a \le t < (k+1)T, \end{cases} \tag{12.1}$$

where $a$ is the switching time parameter (*STP*), satisfying $0 \le a \le T$. It is important to see that the switched *SD* controller (12.1) has two modes:

- for $kT \le t < kT + a$, its mathematical expression, given by $Kx(kT)$, is same as the "standard" feedback controller,
- for $kT + a \le t < (k+1)T$, its value is set to 0.

Thus, the system (9.1) together with the controller (12.1) can be expressed by the following switched system:

$$\dot{x}(t) = \begin{cases} Ax(t) + BKx(kT), & kT \le t < kT + a, \\ Ax(t), & kT + a \le t < (k+1)T. \end{cases} \tag{12.2}$$

In order to clarify the structure of the switched *SD* control (12.1) we give it schematically in Fig. 12.3.

## 12.4 Stability Analysis

In this section, we will study the stability condition for a controlled plant (9.1) under the switched *SD* control (12.1). In this case, the corresponding closed-loop system is described by (12.2).

The discrete-time expression of the system is given by:

$$x\big((k+1)T\big) = \Phi_s(a, T)x(kT), \tag{12.3}$$

---

[2]First decreases and then increases inside a sampling interval.

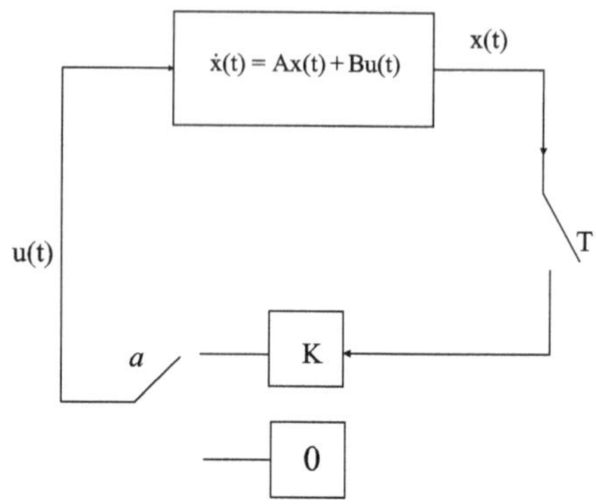


**Fig. 12.3** The switched sampled-data control

where the transition matrix function $\Phi_s(\alpha, \beta)$ is given by

$$\Phi_s(\alpha, \beta) = e^{A(\beta-\alpha)}\left(e^{A\alpha} + \int_0^\alpha e^{A\theta}\, d\theta\, BK\right).$$

For an initial state $x(0)$, the system has the following response:

$$x(t) = \begin{cases} \Phi_s(t-kT, t-kT)\Phi_s^k(a, T)x(0), & kT \le t < kT + a, \\ \Phi_s(a, t-kT)\Phi_s^k(a, T)x(0), & kT + a \le t < (k+1)T. \end{cases} \tag{12.4}$$

It is easy to see that when $a = T$, the switched *SD* control law (12.1) reduces to the "standard" one. Now, for the system (12.3), which is the discrete-time model of a DCES with the switched *SD* control (12.1), we can formulate the following stability condition:

**Theorem 12.1** *System* (9.1) *under the switched sampled-data controller* (12.1) *is asymptotically stable if and only if* $\Phi_s(a, T)$ *is Schur.*

By using Theorem 12.1 and the *parameter-sweeping* technique, we may find the stability ranges for the parameters $a$ and $T$. Let us illustrate this result through a numerical example.

*Example 12.2* Consider the system (9.1) with the following state-space model matrices:

$$A = \begin{pmatrix} -1 & -1 & 0 \\ 1 & -1 & 0 \\ 0 & 0 & 1 \end{pmatrix}, \qquad B = \begin{pmatrix} 0 \\ 0 \\ 1 \end{pmatrix}, \qquad K = \begin{pmatrix} -2 & -2 & -2 \end{pmatrix}.$$

If we use the standard sampled-data control law (9.8), the *stability interval* for the sampling period $T$ is: $T \in [0, 1.09)$. When applying the switched *SD* control (12.1), the stability region is shown in Fig. 12.4. The largest sampling period guaranteeing the stability is $T^* = 5.75$, associated to the *STP* $a^* = 0.69$. More precisely, if the parameter $a$ is chosen as $a = 0.69$, the corresponding *stability interval* is

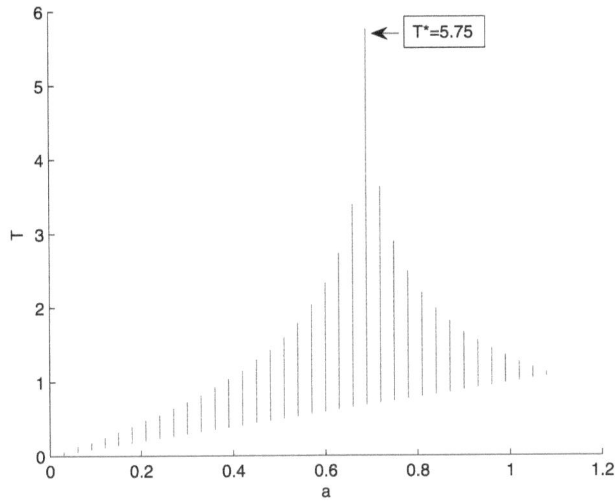


**Fig. 12.4** Stability region of Example 12.2

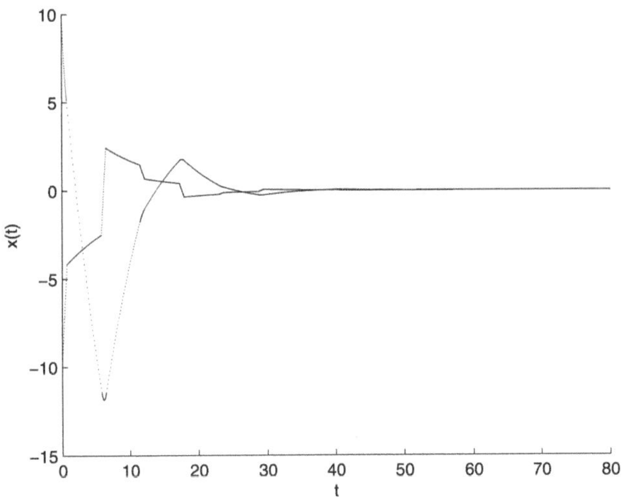


**Fig. 12.5** Simulation for Example 12.2

$T \in [0.69, 5.75]$. Now, with the choice $a = 0.69$ and $T = 5.75$, the system response with an initial condition $x(0) = (10 \; -10)'$ is shown in Fig. 12.5. From the results obtained we can easily see the system performance enhancement by the switched *SD* control strategy.

For some special class of systems, an interesting phenomenon is observed when the switched *SD* control (12.1) is used: this class of systems may be *stabilized* by an *infinitely large sampling period T*. That is, there may exist a switching parameter

$a$ such that $\Phi_s(a, T)$ is Schur for all $T \geq a$. It is not easy to fully investigate such a phenomenon and the relevant properties. However, a sufficient condition for characterizing this phenomenon and the related analysis can be found in *Li et al.* [149]. Here, we only give an example with an analytical illustration.

*Example 12.3* Consider the system (9.1) with the following matrices

$$A = \begin{pmatrix} -\lambda & 0 \\ 0 & 0 \end{pmatrix}, \qquad B = \begin{pmatrix} 0 \\ 1 \end{pmatrix}, \qquad K = \begin{pmatrix} -k_1 & -k_2 \end{pmatrix},$$

where $\lambda$ is a positive number. This system is asymptotically stable under the standard sampled-data control if $T < \frac{2}{k_2}$ and $k_2 > 0$. If the switched *SD* control (12.1) is applied, the corresponding transition matrix becomes

$$\Phi_s(a, T) = \begin{pmatrix} e^{-\lambda T} & 0 \\ -ak_1 & 1 - ak_2 \end{pmatrix}.$$

Due to the specific structure of the above transition matrix $\Phi_s(a, T)$, it is easy to see that the system is asymptotically stable if and only if

$$T > 0, \qquad 0 < a < \frac{2}{k_2}, \qquad k_2 > 0.$$

Therefore, the system can be stabilized with an infinitely large sampling period $T$ by introducing the switched *SD* control (12.1).

One may notice that the system considered in this example has an appropriate structure: the matrix $A$ has only one eigenvalue on the imaginary axis, while the other (remaining) one is in the left-half plane. Thus, this system is not divergent (marginally stable) even when it is in open-loop.

## 12.5 Optimization of the Dynamic Performance

In this section, we will study the dynamic performance of a *DCES* under the switched *SD* control (12.1). In particular, the problem is described as follows: For a controlled plant (9.1) with a given sampling period $T$, the switched *SD* control (12.1) is adopted, and the performance index $\eta$ as defined in (11.4)

$$\eta \triangleq \sup\left\{\frac{J}{x'(0)x(0)} : \forall x(0) \neq 0\right\},$$

is considered with the corresponding cost function $J$ defined in (11.5)

$$J = \int_0^\infty x'(t)Px(t)\,dt,$$

to measure the system dynamics. Our objective is to design the switching time parameter (*STP*) $a$ so as to achieve the minimum $\eta$.

Since $\eta$ is a function of $a$ and $T$, for the rest of this chapter, we will use the notation $\eta(a, T)$. In addition, the optimal *STP* is denoted by $a^*$ and the corresponding minimum $\eta$ is denoted by $\eta(a^*, T)$. Obviously, if $a^* \neq T$, the switched *SD* control can lead to better performance than the standard sampled-data controller.

Let us first give the following technical lemma, which allows to express the corresponding cost function $J$ in a closed form.

**Lemma 12.1** *If $\Phi_s(a, T)$ is Schur, the cost function $J$ defined in* (11.5) *for the* DCES *described by* (9.1) *and* (12.1) *can be rewritten as*

$$J = x'(0) F_s(a, T) x(0),$$

*where $F_s(a, T)$ is the unique solution of the Lyapunov equation*

$$\Phi_s'(a, T) X \Phi_s(a, T) - X = -\Omega_s(a, T), \tag{12.5}$$

*with*

$$\begin{aligned}
\Omega_s(a, T) = {} & \int_0^a \left(e^{A\theta} + \int_0^\theta e^{A\vartheta}\, d\vartheta\, BK\right)' P \left(e^{A\theta} + \int_0^\theta e^{A\vartheta}\, d\vartheta\, BK\right) d\theta \\
& + \left(e^{Aa} + \int_0^a e^{A\vartheta}\, d\vartheta\, BK\right)' \\
& \times \int_0^{T-a} \left(e^{A\theta}\right)' P e^{A\theta}\, d\theta \left(e^{Aa} + \int_0^a e^{A\vartheta}\, d\vartheta\, BK\right).
\end{aligned}$$

*Proof* The cost function $J$ can be rewritten as

$$J = \sum_{k=0}^{\infty} \int_{kT}^{(k+1)T} x'(t) P x(t)\, dt.$$

In the light of (12.4), we have:

$$\begin{aligned}
J = \sum_{k=0}^{\infty} \Bigg( & \int_0^a \left(\Phi_s(t, t) \Phi_s^k(a, T) x(0)\right)' P \Phi_s(t, t) \Phi_s^k(a, T) x(0)\, dt \\
& + \int_a^T \left(\Phi_s(a, t) \Phi_s^k(a, T) x(0)\right)' P \Phi_s(a, t) \Phi_s^k(a, T) x(0)\, dt \Bigg).
\end{aligned}$$

Thus,

$$J = x'(0) F_s(a, T) x(0),$$

$$F_s(a,T)=\sum_{k=0}^{\infty}\left(\Phi_s^k(a,T)\right)'\Omega_s(a,T)\Phi_s^k(a,T).$$

Let $\Omega_s(a,T)=\Omega_{s1}(a,T)+\Omega_{s2}(a,T)$ with

$$\Omega_{s1}(a,T)=\int_0^a\left(e^{A\theta}+\int_0^\theta e^{A\vartheta}\,d\vartheta\,BK\right)'P\left(e^{A\theta}+\int_0^\theta e^{A\vartheta}\,d\vartheta\,BK\right)d\theta,$$

$$\Omega_{s2}(a,T)=\left(e^{Aa}+\int_0^a e^{A\vartheta}\,d\vartheta\,BK\right)'$$
$$\times\int_0^{T-a}\left(e^{A\theta}\right)'Pe^{A\theta}\,d\theta\left(e^{Aa}+\int_0^a e^{A\vartheta}\,d\vartheta\,BK\right).$$

For any $\theta$, $(e^{A\theta}+\int_0^\theta e^{A\vartheta}\,d\vartheta\,BK)'P(e^{A\theta}+\int_0^\theta e^{A\vartheta}\,d\vartheta\,BK)$ is positive semi-definite since $P$ is positive semi-definite. Thus, the integration $\Omega_{s1}(a,T)$ must be also positive semi-definite. Similarly, it follows that $\Omega_{s2}(a,T)$ is necessarily positive semi-definite. Therefore, $\Omega_s(a,T)$ is positive semi-definite.

If $\Phi_s(a,T)$ is Schur, $F_s(a,T)$ is the unique solution of the Lyapunov equation (12.5) according to Lemma 11.1. The proof is completed. □

In the sequel, we give an important result which helps us to design the switching time parameter.

**Theorem 12.2** *For a sampling period $T$ and a switching time parameter $a$, the corresponding performance index $\eta(a,T)$ for the system described by* (9.1) *and* (12.1) *is*

$$\eta(a,T)=\lambda_{\max}\left(F_s(a,T)\right),$$

*if $\Phi_s(a,T)$ is Schur.*

The proof is similar to that of Theorem 11.1, and hence is omitted.

*Example 12.4* Consider the controlled plant (9.1), with matrices $A$, $B$, and $K$ given in (10.1). Assume now that the sampling period $T=1.7$. For each STP $a$, $\eta$ can be analytically calculated by Theorem 12.2. The variation of $\eta$ with respect to $a$ is plotted in Fig. 12.6.

It is explicitly found that the minimum $\eta$ is reached at $a^*=1.409$ with $\eta^*=1.871$. For $a=T=1.7$, $\eta=16.15$, which means that the performance index under the standard sampled-data control law is $\eta=16.15$. The switched *SD* control law (12.1) can improve significantly the system performance. To illustrate the result, let us see the simulation obtained with the initial state $x(0)=(0.5\ 2)'$ given in Fig. 12.7. The corresponding control signals are shown in Fig. 12.8. By Lemma 12.1, we have that $J=7.4828$ when $a=1.409$ while $J=65.2610$ when $a=T=1.7$.

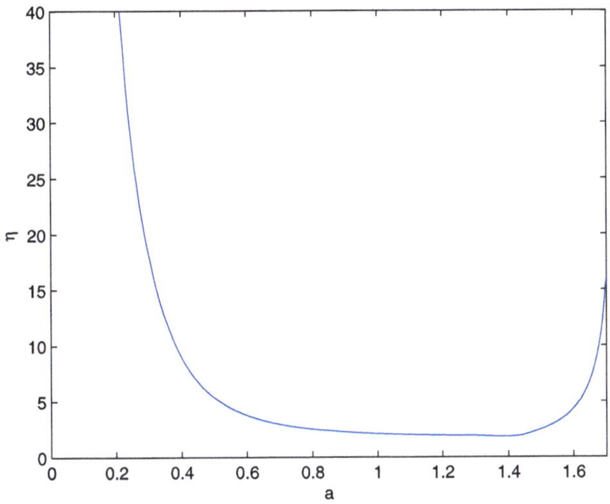


**Fig. 12.6** $\eta$ vs $a$ for Example 12.4

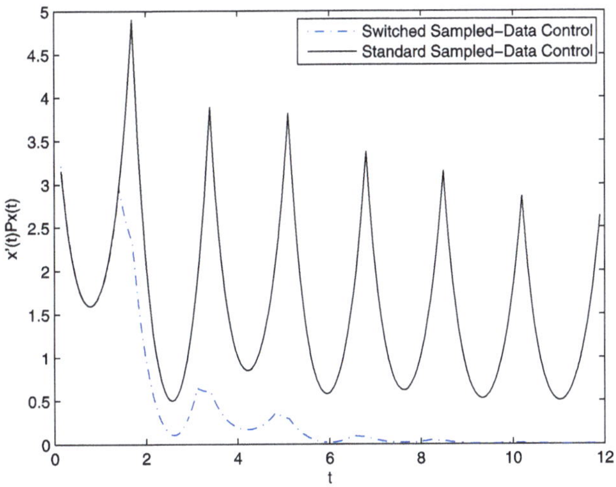


**Fig. 12.7** $x'(t)Px(t)$ vs $t$ for Example 12.4: the switched sampled-data control leads to better system dynamics over the standard sampled-data control

## 12.6 An Illustrative Application

In this section, we choose three *Miabots* (Miabot is a miniature mobile robot, see [1]) as the controlled plants. According to the parameter identification of the *Miabot*, when the speed is no more than 2 m/sec, the dynamics of the Miabots can be approximately expressed as

$$\ddot{x}_i(t) = -10\dot{x}_i(t) + u_i(t), \quad i = 1, 2, 3,$$

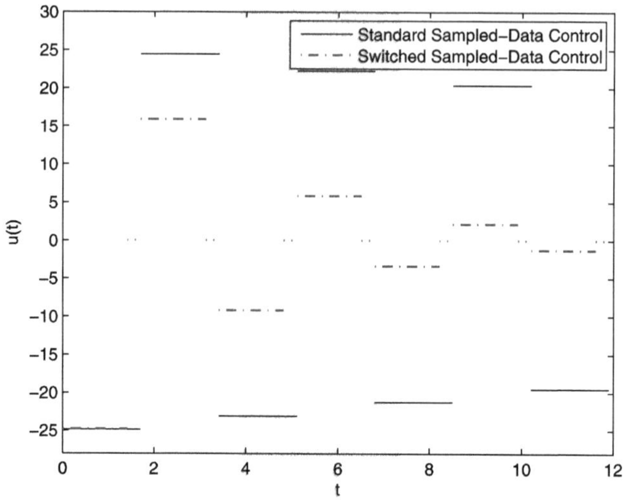


**Fig. 12.8** $u(t)$ vs $t$ for Example 12.4

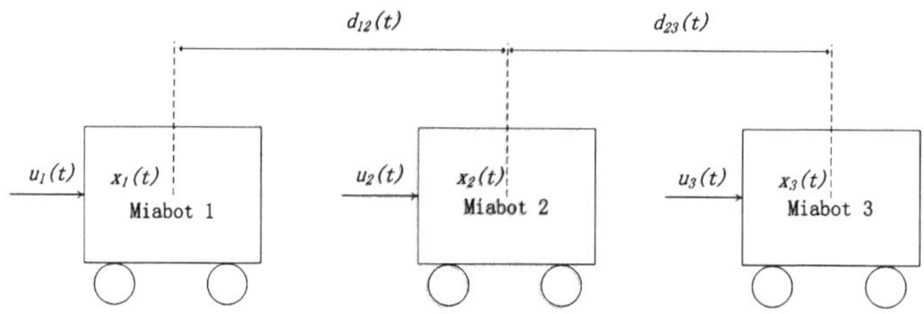


**Fig. 12.9** The distance control of 3 Miabots

where $x_i(t)$ and $u_i(t)$ denote the position and control signal of Miabot $i$, respectively.

Our objective, as shown in Fig. 12.9, is to control the distance between *Miabot 1* and *Miabot 2*, $d_{12}(t) = x_2(t) - x_1(t)$ and the distance between *Miabot 2* and *Miabot 3*, $d_{23}(t) = x_3(t) - x_2(t)$ to be $r_{12}$ and $r_{23}$ ($r_{12}$ and $r_{23}$ are given constants), respectively. In other words, our objective is to design the control laws of $u_1(t)$, $u_2(t)$, and $u_3(t)$ such that $\lim_{t\to\infty} e_{12}(t) = 0$ and $\lim_{t\to\infty} e_{23}(t) = 0$ ($e_{12}(t) = x_2(t) - x_1(t) - r_{12}$, $e_{23}(t) = x_3(t) - x_2(t) - r_{23}$).

Defining the state and the control variables by

$$Z(t) = \begin{pmatrix} e_{12}(t) \\ e_{23}(t) \\ \dot{e}_{12}(t) \\ \dot{e}_{23}(t) \end{pmatrix}, \qquad U(t) = \begin{pmatrix} u_1(t) \\ u_2(t) \\ u_3(t) \end{pmatrix},$$

we obtain the following state-space equation

$$\dot{Z}(t) = \begin{pmatrix} 0 & 0 & 1 & 0 \\ 0 & 0 & 0 & 1 \\ 0 & 0 & -10 & 0 \\ 0 & 0 & 0 & -10 \end{pmatrix} Z(t) + \begin{pmatrix} 0 & 0 & 0 \\ 0 & 0 & 0 \\ -1 & 1 & 0 \\ 0 & -1 & 1 \end{pmatrix} U(t).$$

In order to control the three *Miabots* we adopt, as expected, the following switched sampled-data control

$$U(t) = \begin{cases} KZ(kT), & kT \leq t < kT + a, \\ 0, & kT + a \leq t < (k+1)T, \end{cases}$$

where $T$ is the constant sampling period and the feedback gain $K$ is chosen as

$$K = \begin{pmatrix} 4.1656 & 0 & -0.6151 & 0 \\ -2.7181 & 1.0915 & 4.2536 & 1.3561 \\ 0 & -4.4518 & 0 & 4.9218 \end{pmatrix}.$$

This form of $K$ ensures that *Miabot 1* only needs to communicate with *Miabot 2* and that *Miabot 3* only needs to communicate with *Miabot 2*. From the practical point of view, each *Miabot* has a position sensor and a velocity sensor and hence each *Miabot* knows its own position and velocity. At each sampling instant $kT$, *Miabot 1* and *Miabot 3* send their position and velocity information to *Miabot 2* and meanwhile *Miabot 2* sends its position and velocity information to *Miabot 1* and *Miabot 3*. In this way, *Miabot 1* knows $e_{12}(kT)$ and $\dot{e}_{12}(kT)$; *Miabot 3* knows $e_{23}(kT)$ and $\dot{e}_{23}(kT)$; and *Miabot 2* knows $Z(kT)$ and therefore the control signal $U(kT)$ can be implemented.

With $T = 1.5$, the relationship between the *STP* $a$ and the performance index $\eta$ is given in Fig. 12.10. The optimal *STP* is $a^* = 1.005$ with $\eta^* = 1.9471$. When $a = T = 1.5$, $\eta = 7.7316$. One can see clearly the performance improvement of the switched *SD* control with respect to the standard sampled-data control.

To illustrate the result above, we let $x_1(0) = 0$, $x_2(0) = 6$, $x_3(0) = 9$, $\dot{x}_1(0) = \dot{x}_2(0) = \dot{x}_3(0) = 0$, and $r_{12} = r_{23} = 2$ and we operate a system simulation. In order to compare the results obtained for the standard and the switched *SD* control laws, the system state evolution or dynamics under two control laws are given in Fig. 12.11. The control signals of two controllers are shown in Figs. 12.12 and 12.13.

If we use the standard sampled-data control to achieve the performance index $\eta = 1.9471$ (the one when using the switched sampled-data control), we should set the sampling period as $T = 0.1186$.[3] It is seen that, if we require the same performance index, the switched sampled-data control occupies much less computation and communication resources than the standard sampled-data control. Notice that, the used computation and communication resources of a sampled-data system are proportional to the sampling frequency.

[3]This result can be computed by using the approach proposed in Chap. 10.

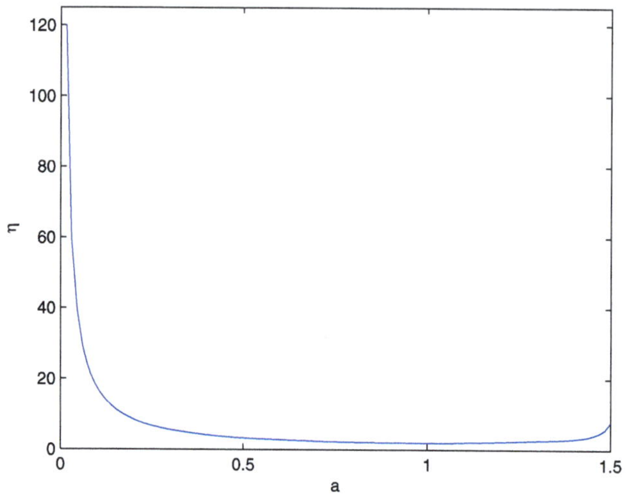


**Fig. 12.10** $\eta$ vs $a$ for Sect. 12.6

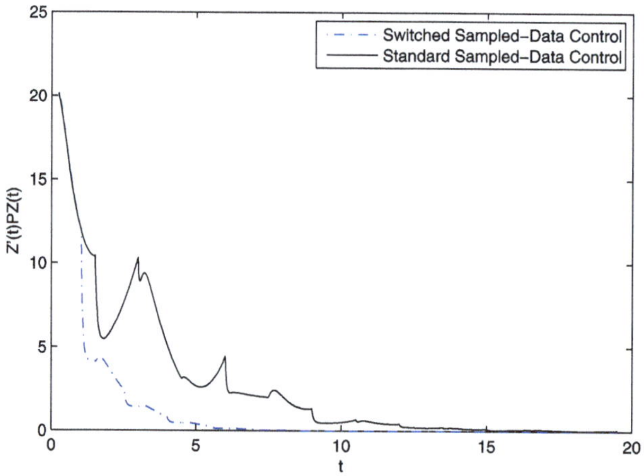


**Fig. 12.11** $Z'(t)PZ(t)$ vs $t$: the switched sampled-data control leads to better system dynamics over the standard sampled-data control

## 12.7 Notes and Comments

In this chapter, we proposed a switched sampled-data (*SD*) control strategy for a simple and particular configuration of *DCESs*, which can be easily implemented in practice. Through studying its intersample dynamics, we argue that for some time interval between sampling instants the open-loop dynamics can have better dynamic properties than the closed-loop ones, due to the effect of the induced delay. Thus, switching the control signal value to zero at a proper switching instant within each sampling interval may improve the system performance.

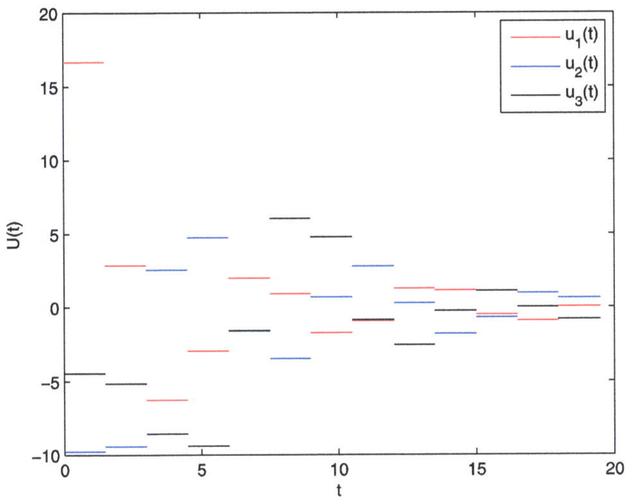


**Fig. 12.12** $U(t)$ of the standard sampled-data control

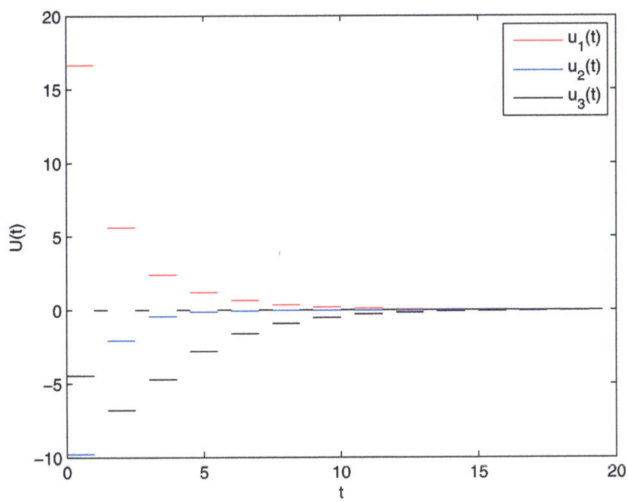


**Fig. 12.13** $U(t)$ of the switched sampled-data control

We first study the stability condition when it is controlled by the proposed switched *SD* controller. It is shown that the switched *SD* control may lead to greater upper bound on the sampling period guaranteeing the stability than the standard sampled-data controller.

In Example 12.3, we have observed an interesting phenomenon that a particular class of *DCES* can be stabilized with an infinitely large sampling period when using the switched *SD* control. It is worth mentioning that the system considered in Example 12.3 is very specific, since we may directly calculate the two eigenvalues with respect to the switching time parameter (*STP*) and the sampling period. For general systems, the analysis may become much more complicated. In Li et al. [149], some disjoint necessary and sufficient conditions are proposed for the analysis of such a phenomenon.

The stability conditions studied here concern only the nominal case, which means that the sampling period is constant. When the sampling period is time-varying and uncertain, no analytical results can be obtained in general. One may refer to *Li et al.* [149], where the discretization technique is used to deal with the time-varying sampling period. The resulting stability condition is however sufficient though not necessary.

Next, the optimization of the *STP* of the switched *SD* controller is addressed. We derived the analytical relation between the *STP* and the index of the system dynamic performance. Based on this analytical relation, we may set the *STP* optimally. From the numerical simulations, we see that the switched *SD* control strategy can lead to better dynamic properties than the standard sampled-data one.

From the optimization perspective, there is still much room to further improve the derived results. We let the *STP* be the only design parameter, while fixing the sampling period and the state-feedback gain matrix. It is natural to predict that both the stability and dynamic properties may be improved if we *co-design* the *STP*, sampling period and the state-feedback gain matrix at the same time. However, the related optimization problem will be much more complicated and will be faced with calculation complexity problems.

Various optimization problems have been investigated in the literature related to the problem treated here. The optimal state-feedback gain for the standard sampled-data control can be found in *Åström and Wittenmark* [14] and *Chen and Francis* [63]. The co-design of optimal real-time scheduling and optimal feedback gains for *DCESs* has been studied by *Ben Gaid et al.* [27]. In our opinion, integrating the optimal calculation of *STP* under this formalism can enhance the dynamic performance as well as the stability robustness.

Though a *DCES* controlled by the proposed switched *SD* controller is a switched system, we did not apply the theory of switched systems (see, e.g., *Liberzon* [154]) to its analysis and design. For the stability issue, we simply used the discrete-time model as proposed in Sect. 12.4. In this way, the stability can be investigated by studying the corresponding transition matrix, without conservatism. For the optimization issue, we obtain the analytical function in Sect. 12.5. The resulting approach involves no conservatism either.

# Chapter 13
# A Switched Hold-Zero Compensation Strategy for *DCESs* Subject to Control Input Missings

As studied in the previous chapters, a *DCES* may be subject to some undesired factors like *DCES-induced delays* and sampling period jitters. These factors, in general, deteriorate the dynamic performance, but it is well known that a controlled system has inherent stability robustness. Thus, a *DCES* with a proper sampled-data controller may tolerate induced delays and sampling jitters to a certain degree, without losing its stability.

However, some factors like data dropout, overload of calculation nodes and control tasks preemptions may cause standard sample-and-control execution failures at some sampling instants. Such a phenomenon of controller failure is called the *control input missing* which has recently attracted a special attention. With the generalization of *DCESs Hardware/Software control input missings* are likely to happen more frequently implying thus the necessity to use an appropriate adapted controller design methodology.

In the literature, there are two commonly used compensation strategies called the *zero-control strategy* and the *hold-control strategy*. If a control input missing occurs, the *zero-control strategy* sets the control input to zero as given in *Zhang and Yu* [265]. In the case of *hold-control strategy*, if a control input missing occurs, the previous control input is held as given in *Schenato* [207]. In *Zhang and Yu* [265], the stability of sampled-data control systems under the zero-control strategy is studied and the *admissible control input missing rate* (*ACIMR*)[1] is computed.

In this chapter, we will propose a novel compensation strategy: a *switched hold-zero* (*HZ*) *control* approach. The idea is to combine the benefits of hold-control law with the zero-control law. If a control input missing occurs, the actuator node of a *DCES* has two optional natural strategies: the hold-control and the zero-control ones. In this context, a natural question may arise: what are the appropriate control strategies guaranteeing larger *ACIMRs*? To the best of our knowledge, no answer has been given to this question. This motivates our study on the *control input missing* phenomenon and consequently, we propose to use both of them in an ordered and

[1]The definition of the ACIMR will be given later in this chapter.

A. Çela et al., *Optimal Design of Distributed Control and Embedded Systems*, Communications and Control Engineering, DOI 10.1007/978-3-319-02729-6_13,

precise temporal switched manner defined according to the structural properties of the *DCES* under study.

This chapter starts with the study of system dynamics under the zero-control and the hold-control strategies, respectively. The system dynamics under the two strategies present different characteristics. The variation rate of the Lyapunov function[2] under the zero control is *independent* of the distribution of the control input missings. While, under the hold control, such a quantity varies with respect to the distribution of the control input missings. From the delay-system point of view, the input delay is proportional to the number of consecutive control input missings. Thus, when a small number of consecutive control input missings happen, the hold control may lead to a smaller variation rate of the Lyapunov function than the zero control. As the number of consecutive control input missings increases, the variation rate of Lyapunov function of the hold control tends to increase.

The above observation inspires us to present the switched hold-zero control: first, we apply the hold control, and then, we switch to the zero control. The switching instant will be considered as a *design parameter* and it will be computed based on the system structural properties.

## 13.1 Modeling and Problem Formulation

### *13.1.1 Mathematical Expressions of Control Input Missings*

As in the precedent chapters of *Part* III, we represent a *DCES* by the controlled plant (9.1)

$$\dot{x}(t) = Ax(t) + Bu(t),$$

and the standard single-sampling sampled-data controller (9.7)

$$u(t) = Kx(t_k), \quad t_k \leq t < t_{k+1}.$$

Without any loss of generality, the sampling periods $T(k)$ $(T(k) = t_{k+1} - t_k)$ are time-variant due to sampling jitters. In the sequel, we will adopt the following model borrowed from *Skaf and Boyd* [214]:

$$T_{\text{nom}} - \Delta \leq T(k) \leq T_{\text{nom}} + \Delta, \tag{13.1}$$

where $T_{\text{nom}}$ and $\Delta$ are two positive constants with $T_{\text{nom}} > \Delta$.

We use a binary variable $\upsilon_k \in \{0, 1\}$ to describe whether or not a control input missing occurs at a sampling instant $t_k$: $\upsilon_k = 1$ if a control input missing occurs at $t_k$,

---

[2]For some chosen Lyapunov function $V : \mathbb{R}^n \mapsto \mathbb{R}$, $V(t_k) = x'(t_k)Px(t_k)$, where $t_k$ denote the sampling instants, if $V(t_{k+1}) \leq \rho V(t_k)$, $\rho > 0$, the scalar $\rho$ is called a *variation rate* of the Lyapunov function.

and $\upsilon_k = 0$ otherwise. A sampling (instant) at $t_k$ is called an *ineffective sampling* (instant) if $\upsilon_k = 1$. While, a sampling (instant) at $t_k$ is called an *effective sampling* (instant) if $\upsilon_k = 0$.

With the notations above, the zero-control strategy is formally given described by

$$u(t_k) = 0, \quad \text{if } \upsilon_k = 1, \tag{13.2}$$

which means that if a control input missing occurs, the control signal is simply set to zero.

Whereas, the hold-control strategy is given by

$$u(t_k) = u(t_{k-1}), \quad \text{if } \upsilon_k = 1. \tag{13.3}$$

Namely, when a control input missing occurs, the current control signal is held.

In order to measure the occurrence frequency of control input missings, we define the *control input missing rate* (*CIMR*) as:

$$CIMR = \lim_{k\to\infty} \frac{M(k)}{k+1}, \tag{13.4}$$

where $M(k) = \sum_{j=0}^{k} \upsilon_j$ represents the number of control input missing over the total $k+1$ sampling instants. Clearly, $CIMR \in [0, 1]$. The definition of CIMR (13.4) refers to the definition of data dropout rate used in *Guan et al.* [106].

We are interested in maximizing the value of *CIMR* tolerated by the *DCES* under study without losing its stability. As a measure of this tolerance we introduce the concept of *admissible control input missing rate* (*ACIMR*), which implies that a *DCES* is asymptotically stable if *CIMR* is upper-bounded by *ACIMR* (i.e., $CIMR < ACIMR$).

In fact, to further analyze the dynamic behavior of a *DCES* subject to control input missings, we need to consider also the distribution of control input missings. Later in this chapter, we will see that such an information affects considerably the dynamic behavior. Under the same value of *CIMR*, a given *DCES* may exhibit different *ACIMR* due to different distribution of control input missings.

Assume $t_k$ is an ineffective sampling instant. In order to denote how many ineffective samplings happen between the last effective sampling instant and $t_k$, we introduce the following notation:

$$\theta_k = \begin{cases} 0, & \upsilon_k = 0, \\ k - \varsigma_k, & \upsilon_k = 1, \end{cases} \tag{13.5}$$

where $\varsigma_k = \max\{j < k : \upsilon_j = 0\}$ denotes the last effective sampling instant before $t_k$. For an ineffective sampling instant $t_k$, $\theta_k$ means that $t_k$ is the $\theta_k$th ineffective sampling after $\varsigma_k$.

Introduce now the following quantities $M_1(k), M_2(k), \ldots,$ as

$$M_1(k) = \sum_{j=0}^{k} \vartheta_j(1), \qquad M_2(k) = \sum_{j=0}^{k} \vartheta_j(2), \ldots,$$

$$\vartheta_j(i) = \begin{cases} 1, & \theta_j = i, \\ 0, & \theta_j \neq i. \end{cases}$$

Here, $M_1(k), M_2(k), \ldots,$ denote, respectively, the number of *one control input missing after an effective control*, the number of *two consecutive control input missings after an effective control*, ..., in the total $k+1$ sampling instants. It is not difficult to see that $M_1(k) \geq M_2(k) \geq \cdots$.

Let us define $R_1 = \lim_{k\to\infty} \frac{M_1(k)}{M(k)}$, $R_2 = \lim_{k\to\infty} \frac{M_2(k)}{M(k)}, \ldots,$ denoting respectively, the ratio of $M_1(k), M_2(k), \ldots,$ as the *distribution indices*. It holds that $R_1 + R_2 + \cdots = 1$. It is clear that the distribution indices make sense when $M(k) \neq 0$, which will be supposed in the sequel. It is important to point out that such indices play an important role due to the fact that the system dynamics depend not only on *CIMR* but also on the related distribution indices.

### 13.1.2 A Switched Hold-Zero (HZ) Compensation Strategy

When controlled by the zero control (13.2) from an ineffective sampling instant $t_k$, the *DCES*, in fact, simply represents the open-loop system. Before a new effective sampling instant, the state $x(t)$ satisfies that $x(t) = e^{A(t-t_k)}x(t_k)$ (independent on the "old" control input). In addition, since the matrix $A$ is assumed not to be Hurwitz, the zero control (13.2) has *no stabilizing effect*. If from an ineffective sampling instant $t_k$, the *DCES* is controlled by the hold control (13.3), the system can be viewed as a time-delay system as detailed in the discussion given in Chap. 9. More precisely, before a new effective sampling instant, $\dot{x}(t) = Ax(t) + BKx(t-\tau(t))$ with the (artificial) input delay $\tau(t) = t - t_k$.

When the input delay $\tau(t)$ is sufficiently small, the hold control (13.3) necessarily has a stabilizing effect (as the system dynamics are close to those of the closed-loop system $\dot{x}(t) = (A+BK)x(t)$) and hence it works better than the zero control (13.2). On the other hand, if the input delay $\tau(t)$ increases, from a delay-system perspective, the hold control (13.3) generally tends to lose its stabilizing effect. Moreover, as the input delay $\tau(t)$ becomes sufficiently large, it is likely that the dynamics under the hold control (13.3) are worse than the open-loop counterpart (i.e., when controlled by the zero control (13.2)).

Based on the above remarks, we propose the following switched hold-zero (*HZ*) control as a new compensation strategy:

$$u(t_k) = \begin{cases} u(t_{k-1}), & \text{if } \upsilon_k = 1, \text{ and } \theta_k \leq \sigma, \\ 0, & \text{if } \upsilon_k = 1, \text{ and } \theta_k > \sigma, \end{cases} \tag{13.6}$$

or equivalently,

$$u(t_k) = \begin{cases} u(t_{\varsigma_k}), & \text{if } \upsilon_k = 1, \text{ and } \theta_k \leq \sigma, \\ 0, & \text{if } \upsilon_k = 1, \text{ and } \theta_k > \sigma, \end{cases}$$

where $\sigma \in \mathbb{N}_+$ is the switching parameter to be designed.

We will explain in more details how the switched *HZ* controller (13.6) works, taking the case $\sigma = 2$ as an example. Assume that $t_{k-1}$ is an effective sampling instant and a control input missing occurs at $t_k$. For $t_k$, $\theta_k = 1$. As $\theta_k < \sigma$, the control input to the controlled plant keeps the last effective control value, i.e., $u(t_k) = u(t_{\varsigma_k}) = u(t_{k-1})$. If $t_{k+1}$ is also an ineffective sampling instant (in this case, $t_{k+1}$ is the second consecutive ineffective sampling), $\theta_{k+1} = 2$. As $\theta_{k+1} = \sigma$, the control input still keeps the last effective control value, i.e., $u(t_{k+1}) = u(t_{\varsigma_{k+1}}) = u(t_{k-1})$. Further, if $t_{k+2}$ is also an ineffective sampling instant (i.e., $t_{k+2}$ is the third consecutive ineffective sampling), $\theta_{k+1} = 3$. As $\theta_{k+2} > \sigma$, the control input is set to zero until an effective sampling instant.

We can summarize our idea of the switched *HZ* control (13.6) strategy: After an ineffective sampling instant, we try to first use the hold-control strategy, as this strategy may still stabilize the system. Next, if consecutive ineffective sampling instants follow, the hold-control strategy may gradually lose its stabilizing effect or even renders the system more divergent than the zero-control strategy. Thus, starting with an appropriate time, we shall apply the zero-control strategy instead of the hold-control strategy. One may see that the idea of the switched *HZ* control (13.6) is similar to that of the switched control given in Chap. 12.

*Remark 13.1* The switched *HZ* controller is simple to implement: basically it is a hold-control one whose output may be turned off at the switching instant.

One of our objectives in the next section is the design of the switching parameter $\sigma$ in order to make the *ACIMR* as large as possible.

*Remark 13.2* In practice, a *DCES* is composed of multiple controlled tasks sharing the limited calculation and communication resources. Notice that, in some situations, control input missings may be positively generated by the processor to handle the overload situation. This positive control input missing may lead to a hyper-sampling periods scheduling, see, for example, *Li et al.* [151]. Therefore, a larger *ACIMR* allows more positive control input missings and consequently a less resource consumption. This point will be illustrated through some numerical examples.

## 13.2 Problem Analysis and Solution

This section is composed of three subsections. In the first one, we will compute the *ACIMR* under the zero-control strategy. In the second subsection, we will discuss the dynamics under the hold-control strategy. Finally, in the third subsection, we will give a method to optimally set the switching parameter $\sigma$ of the switched *HZ* control strategy. Independently of the adopted control strategy, *ACIMR* always exists. However, its *real ACIMR* is generally not available. As discussed in Chap. 9, we may only obtain sufficient though not necessary stability conditions for *DCESs* subject to control input missings. It is known that a sufficient though not necessary condition involves conservatism. Thus, we can not analytically calculate the *ACIMR*. Our objective is to render the numerical results as close as possible to the corresponding *real ACIMR* bounds.

### *13.2.1* **ACIMR** *Under the Zero-Control Strategy*

A method to calculate the *ACIMR* under the zero-control strategy was recently presented in *Zhang and Yu* [265], in a continuous-time framework whereas in this subsection we will present a method in the discrete-time framework. Let us recall here the one-argument transition matrix defined in (9.12)

$$\Phi(\beta) = e^{A\beta} + \int_0^{\beta} e^{A\theta}\, d\theta\, BK.$$

It will be seen that the method proposed in the sequel provides results which are less conservative than those proposed in the literature. Let us begin our analysis with the following result:

**Proposition 13.1** *If there exist a positive-definite matrix $P$, positive scalars $a_s$ and $a_u$ satisfying the following conditions*:

$$\Phi'(T)P\Phi(T) < a_s P, \quad T \in [T_{\text{nom}} - \Delta, T_{\text{nom}} + \Delta], \tag{13.7}$$

$$\left(e^{AT}\right)' P e^{AT} < a_u P, \quad T \in [T_{\text{nom}} - \Delta, T_{\text{nom}} + \Delta], \tag{13.8}$$

*then the system* (9.1), *under the zero-control strategy, has the following ACIMR*

$$ACIMR = \frac{\overline{D}}{\overline{D}+1}, \tag{13.9}$$

*where*

$$\overline{D} = -\log_{a_u}(a_s).$$

*Proof* Define a Lyapunov function

$$V(t_k) = x'(t_k)Px(t_k). \tag{13.10}$$

In the case of $\upsilon_k = 0$, $V(t_{k+1}) < a_s V(t_k)$, if (13.7) holds. Similarly, in the case of $\upsilon_k = 1$, $V(t_{k+1}) < a_u V(t_k)$, if (13.8) holds. It follows that

$$V(t_k) < a_s^{k-M(k-1)} a_u^{M(k-1)} V(t_0),$$

and

$$V(t_k) < \left(a_u^D a_s\right)^{k-M(k-1)} V(t_0),$$

where

$$D = \frac{M(k-1)}{k - M(k-1)}.$$

Thus, the system is asymptotically stable if $a_u^D a_s \leq 1$. The proof is completed. □

Proposition 13.1 defines a set of conditions where the parameter $T$ lies in an interval $[T_{\text{nom}} - \Delta, T_{\text{nom}} + \Delta]$. It is worth mentioning that such conditions are infinite-dimensional, as we need to check (13.7) and (13.8) for each value of $T$ lying in the interval $[T_{\text{nom}} - \Delta, T_{\text{nom}} + \Delta]$. Since a real interval does not represent a countable set (of points), Proposition 13.1 can not be directly used. To overcome such a problem, we will employ the discretization technique, which has been used in Chap. 10, such that the obtained conditions are finite-dimensional. As a result, we can establish the conditions in terms of linear matrix inequalities, which are numerically tractable.

Divide the interval $[T_{\text{nom}} - \Delta, T_{\text{nom}} + \Delta]$ into $N$ sub-intervals $[T_{\text{nom},j} - \frac{\Delta}{N}, T_{\text{nom},j} + \frac{\Delta}{N}]$ with $T_{\text{nom},j} = T_{\text{nom}} - \Delta + \frac{(2j-1)\Delta}{N}$, $j = 1, \ldots, N$.

Further, for $T \in [T_{\text{nom},j} - \frac{\Delta}{N}, T_{\text{nom},j} + \frac{\Delta}{N}]$, it follows (similar to Lemma 2 in Suh [222]) that

$$\begin{aligned}
\Phi(T) &= \Phi(T_{\text{nom},j}) + \Delta(\upsilon)(A \quad B)F(T_{\text{nom},j})(I \quad K)', \\
e^{AT} &= e^{AT_{\text{nom},j}} + \Delta(\upsilon)Ae^{AT_{\text{nom},j}},
\end{aligned}$$

with

$$\begin{aligned}
F(T_{\text{nom},j}) &= \exp\left(\left(\begin{smallmatrix} A & B \\ 0 & 0 \end{smallmatrix}\right)T_{\text{nom},j}\right), \\
\Delta(\upsilon) &= \int_0^{\upsilon} e^{A\theta}\, d\theta, \quad \upsilon \in \left[-\frac{\Delta}{N}, \frac{\Delta}{N}\right].
\end{aligned}$$

For the discretization above, it is easy to see that there exists a positive real number $\beta$, such that $\|\Delta(\upsilon)\| \leq \beta$.

Let us give the following theorem which states the condition to obtain the *ACIMR* in the case of zero-control strategy.

**Theorem 13.1** *If there exist a positive-definite matrix $P$, positive scalars $\varepsilon_{1,j}, \varepsilon_{2,j}$ $(j = 1, \ldots, N)$, $a_s$ and $a_u$ satisfying the following matrix inequalities*

$$\begin{pmatrix}
-a_s P + \varepsilon_{1,j}(A\Phi(T_{\text{nom},j}) + BK)'(A\Phi(T_{\text{nom},j}) + BK) & * & * \\
P\Phi(T_{\text{nom},j}) & -P & * \\
0 & P & -\frac{\varepsilon_{1,j}}{\beta^2} I
\end{pmatrix} < 0,$$

$$j = 1, \ldots, N, \tag{13.11}$$

$$\begin{pmatrix} -a_u P + \varepsilon_{2,j}(Ae^{AT_{\mathrm{nom},j}})'(Ae^{AT_{\mathrm{nom},j}}) & * & * \\ Pe^{AT_{\mathrm{nom},j}} & -P & * \\ 0 & P & -\frac{\varepsilon_{2,j}}{\beta^2} I \end{pmatrix} < 0,$$

$$j = 1, \ldots, N, \tag{13.12}$$

*system* (9.1) *under the zero-control strategy has the ACIMR* (13.9).

*Proof* We need to prove that (13.7) and (13.8) can be ensured by (13.11) and (13.12), respectively. Thus, it is easy to see that (13.7) holds if and only if the following conditions hold simultaneously

$$\begin{cases} \Phi'(T)P\Phi(T) < a_s P, \quad T \in \left[T_{\mathrm{nom},1} - \dfrac{\Delta}{N}, T_{\mathrm{nom},1} + \dfrac{\Delta}{N}\right], \\ \vdots \\ \Phi'(T)P\Phi(T) < a_s P, \quad T \in \left[T_{\mathrm{nom},N} - \dfrac{\Delta}{N}, T_{\mathrm{nom},N} + \dfrac{\Delta}{N}\right]. \end{cases}$$

The inequality condition for each sub-interval

$$\Phi'(T)P\Phi(T) < a_s P, \quad T \in \left[T_{\mathrm{nom},j} - \frac{\Delta}{N}, T_{\mathrm{nom},j} + \frac{\Delta}{N}\right],$$

still has an infinite-dimensional character. Let us demonstrate how it can be transformed into the form of (13.11). By Schur Complements (*Boyd et al.* [40]), the condition $\Phi'(T)P\Phi(T) < a_s P$ is equivalent to

$$\begin{pmatrix} -a_s P & * \\ P\Phi(T) & -P \end{pmatrix} < 0. \tag{13.13}$$

For $T \in [T_{\mathrm{nom},j} - \frac{\Delta}{N}, T_{\mathrm{nom},j} + \frac{\Delta}{N}]$, the left-hand side of (13.13) is

$$\begin{pmatrix} -a_s P & * \\ P\Phi(T_{\mathrm{nom},j}) & -P \end{pmatrix} + \begin{pmatrix} 0 & * \\ P\Delta(\upsilon)(A\ B)F(T_{\mathrm{nom},j})\binom{I}{K} & 0 \end{pmatrix}.$$

Next, for any positive scalar $\varepsilon_{1,j}$, it is true that

$$\begin{aligned} &\begin{pmatrix} 0 & * \\ P\Delta(\upsilon)(A\ B)F(T_{\mathrm{nom},j})\binom{I}{K} & 0 \end{pmatrix} \\ &\quad \le \varepsilon_{1,j}\left((A\ B)F(T_{\mathrm{nom},j})\tbinom{I}{K} \quad 0\right)'\left((A\ B)F(T_{\mathrm{nom},j})\tbinom{I}{K} \quad 0\right) \\ &\qquad + \frac{\beta^2}{\varepsilon_{1,j}}\begin{pmatrix} 0 \\ P \end{pmatrix}(0 \quad P). \end{aligned}$$

Notice that

$$\begin{aligned}&\left((A\ B)F(T_{\text{nom},j})\binom{I}{K}\ \ 0\right)'\left((A\ B)F(T_{\text{nom},j})\binom{I}{K}\ \ 0\right)\\&=\begin{pmatrix}(A\Phi(T_{\text{nom},j})+BK)'(A\Phi(T_{\text{nom},j})+BK) & 0\\ 0 & 0\end{pmatrix}.\end{aligned}$$

By applying the Schur complement property, for $T \in [T_{\text{nom},j} - \frac{\Delta}{N}, T_{\text{nom},j} + \frac{\Delta}{N}]$, (13.13) holds if there exists a positive scalar $\varepsilon_{1,j}$ such that (13.11) holds.

Similarly, for $T \in [T_{\text{nom},j} - \frac{\Delta}{N}, T_{\text{nom},j} + \frac{\Delta}{N}]$, (13.8) holds if there exist positive scalars $\varepsilon_{2,j}$ such that (13.12) holds, and thus, the proof is now completed. □

In the case of a constant sampling period, i.e., $T(K) \equiv T_{\text{nom}}$, we have the following corollary:

**Corollary 13.1** *Suppose the constant sampling period of system* (9.1) *is* $T_{\text{nom}}$. *If there exist a positive-definite matrix* $P$, *positive scalars* $a_s$ *and* $a_u$ *satisfying the following matrix inequalities*

$$\begin{pmatrix}-a_s P & *\\ P\Phi(T_{\text{nom}}) & -P\end{pmatrix} < 0, \tag{13.14}$$

$$\begin{pmatrix}-a_u P & *\\ Pe^{AT_{\text{nom}}} & -P\end{pmatrix} < 0, \tag{13.15}$$

*then system* (9.1) *under the zero-control strategy has the ACIMR* (13.9).

Both Theorem 13.1 and Corollary 13.1 are expressed in terms of nonlinear matrix inequalities, which can not be directly solved. For instance, a straightforward procedure to apply Corollary 13.1 is as follows: First fix $a_s$ and $a_u$ and then (13.14) and (13.15) are linear matrix inequalities (LMIs). If the LMIs (13.14) and (13.15) are feasible, we have an *ACIMR*. To find the optimal *ACIMR*, denoted by *ACIMR**, we have to repeat this procedure and manually set $a_s$ and $a_u$ each time. This is a two-variable optimization problem, which is computation demanding.

The optimization problem when using Corollary 13.1 can be transformed into a standard semi-definite program (*SDP*) problem. In order to reduce the space, we only transform the problem when using Corollary 13.1. Similar transformation can be proposed in the case of Theorem 13.1.

**Algorithm 13.1** [3]

Step 1: Solve the following generalized eigenvalue minimization problem

$$\min a_s$$
$$\begin{pmatrix} -Q & * \\ P\Phi(T_{\text{nom}}) & -P \end{pmatrix} < 0,$$
$$Q \le a_s P,$$

and find the minimum value of $a_s$, denoted as $\underline{a_s}$.

Step 2: Choose a small step length $\delta$. For $a_s = \underline{a_s} : \delta : 1 - \delta$, solve the following generalized eigenvalue minimization problem

$$\min a_u$$
$$\begin{pmatrix} -a_s P & * \\ P\Phi(T_{\text{nom}}) & -P \end{pmatrix} < 0,$$
$$\begin{pmatrix} -R & * \\ Pe^{AT_{\text{nom}}} & -P \end{pmatrix} < 0,$$
$$R \le a_u P.$$

For each $a_s$, and the solution, denoted as $a_u^*(a_s)$, compute $\overline{D} = -\log_{a_u^*(a_s)}(a_s)$.

Step 3: Select the maximum $\overline{D}$, denoted as $D^*$ from Step 2. The maximum *ACIMR*, *ACIMR*$^*$, is $\frac{D^*}{D^*+1}$.

Let us give an example to show the effectiveness of the proposed method.

*Example 13.1* Consider system (9.1) with matrices $A$, $B$, and $K$ given in (10.1)

$$A = \begin{pmatrix} 0 & 1 \\ 0 & -0.1 \end{pmatrix}, \qquad B = \begin{pmatrix} 0 \\ 0.1 \end{pmatrix}, \qquad K = (-3.75 \quad -11.5).$$

Assume the sampling period is not constant. We first compute the *ACIMR* for the following four cases, using Theorem 13.1 with $N = 5$. See the results in Table 13.1.

Case 1: $T_{\text{nom}} = 0.1$, $\Delta = 0.01$.
Case 2: $T_{\text{nom}} = 0.5$, $\Delta = 0.01$.
Case 3: $T_{\text{nom}} = 1.0$, $\Delta = 0.01$.
Case 4: $T_{\text{nom}} = 1.7$, $\Delta = 0.01$.

For this particular example, we observe an interesting property: if the sampling period is approaching 1.7, the *ACIMR* is approximately 1. In other words, the system appears to be stable even if almost all the samplings are ineffective.

---

[3] Algorithm 13.1 is a one-dimensional problem, which can be easily solved by the LMI toolbox in MATLAB.

**Table 13.1** *ACIMR* calculated by using Theorem 13.1

| | $D^*$ | $ACIMR^*$ |
|---|---|---|
| Case 1 | 0.3891 | 0.2801 |
| Case 2 | 0.532 | 0.3473 |
| Case 3 | 0.844 | 0.4577 |
| Case 4 | $7.2571 \times 10^6$ | $\approx 1$ |

Using *Theorem 1* in *Zhang and Yu* [265], the *ACIMR* under the constant sampling period can be computed.

Let us compare the two methods:

- If $T_{\text{nom}} = 0.1$, the calculated *ACIMR* by *Theorem 1* in *Zhang and Yu* [265] are: infeasible, 0.1782, 0.2837, when the average dwell time are $T_{\text{nom}}$, $1.5T_{\text{nom}}$, $2T_{\text{nom}}$, respectively. By Corollary 13.1 given here, the *ACIMR* is 0.5968.
- If $T_{\text{nom}} = 0.5$, the calculated *ACIMR* by *Theorem 1* in *Zhang and Yu* [265] are: 0.1001, 0.2756, 0.3634, when the average dwell time are $T_{\text{nom}}$, $1.5T_{\text{nom}}$, $2T_{\text{nom}}$, respectively. From the calculation done based on Corollary 13.1, the *ACIMR* is 0.5903.
- If $T_{\text{nom}} = 0.7$, the calculated *ACIMR* by *Theorem 1* in *Zhang and Yu* [265] are: 0.0371, 0.2104, 0.2971, when the average dwell time are $T_{\text{nom}}$, $1.5T_{\text{nom}}$, $2T_{\text{nom}}$, respectively. Corollary 13.1 allows to calculate the value of *ACIMR* equal to 0.5861.
- If $T_{\text{nom}} = 1.0$, *Theorem 1* in *Zhang and Yu* [265] is not feasible. From Corollary 13.1, the *ACIMR* calculated value is 0.5790.

Based on the computations above, it is easy to see that our approach leads to less conservative results. In addition, *Theorem 1* in *Zhang and Yu* [265] requires an additional condition on the average dwell time.

In the sequel, we will give a useful application of the *ACIMR* concept: design of a *hyper-sampling sequence or period* based on the *ACIMR*. Assume that a *DCES* is running under the standard *single-sampling mode*. In order to save system resources, the processor may "positively" discard some sample-and-control actuation and at the "discarded" sampling instants the control signal is generated by a compensation strategy.

For this example, we use the zero-control strategy. We may design a hyper-sampling period composed of $h$ ($h$ is a positive integer) sub-sampling periods. The hyper-sampling period should satisfy that $\frac{E(h)}{h} \leq ACIMR$, where $E(h)$ is the number of discarded samplings in a hyper-sampling period. It is clear that a longer hyper-sampling period (i.e., a larger $h$) may lead to a more flexible design.

For case 1, if the hyper-sampling period is composed of 10 sub-sampling periods ($h = 10$), the hyper-sampling period allows to "discard" up to 2 samplings in every hyper-sampling period. Compared with the standard single-sampling mode, $\frac{2}{10}$ sampling frequency can be reduced. A possible form of the resultant hyper-sampling pe-

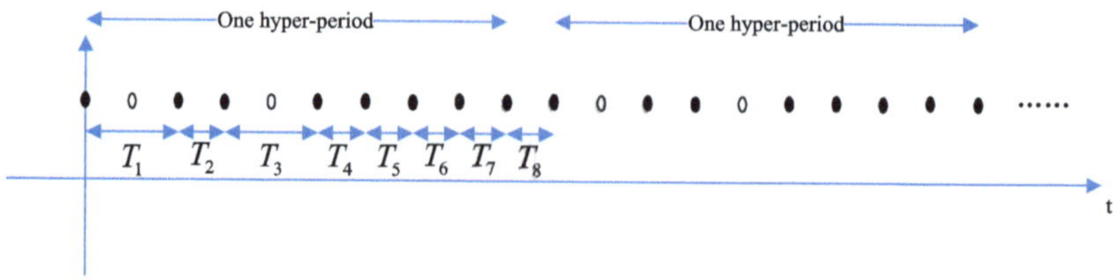


**Fig. 13.1** A hyper-sampling period for Example 13.1

riod[4] is shown in Fig. 13.1, where the white dots stand for the discarded samplings. From the processor side, this hyper-sampling period produces a new sampling sequence: $T_1, T_2, \ldots, T_8$, where $T_1$ and $T_2$ are twice of the original sampling period.

If $h = 100$, the hyper-sampling period should satisfy the condition that in every 100 samplings the system may "discard" up to 28 samplings. Compared with the single-sampling mode, $\frac{28}{100}$ sampling frequency can be reduced. One can see that a longer hyper-sampling period in general may save more calculation resources.

Using the same idea, we can design a hyper-sampling period for cases 2 and 3. If $h = 10$, $\frac{3}{10}$ and $\frac{4}{10}$ sampling frequency can be saved, respectively, for cases 2 and 3.

For case 4, we prefer to design a hyper-sampling period with a very large $h$: in every $h$ samplings, the system only samples once. *The resultant sampling frequency is near zero.* Under the single-sampling mode, this system is asymptotically stable with the minimum allowable sampling frequency 0.58, see, for instance, *Oishi and Fujioka* [188]. Using the hyper-sampling mode, the minimum allowable sampling frequency is near 0. One can see that much system calculation resources may be saved.

### 13.2.2 Dynamics Under the Hold-Control Strategy

To the best of our knowledge, no results have been reported on computing the *ACIMR* under the hold-control strategy. In the sequel, we will point out that *it is in fact impossible to compute the* ACIMR *under the hold-control strategy.*

Denoting the effective sampling instants by $t_1^e, t_2^e, \ldots$, we have the following discrete-time model of a *DCES* with the hold-control strategy

$$x\left(t_{k+1}^e\right) = \Phi\left(t_{k+1}^e - t_k^e\right)x\left(t_k^e\right). \tag{13.16}$$

The transition matrix function $\Phi(t_{k+1}^e - t_k^e)$ is time-varying and uncertain. We consider here a simple scenario with $t_{k+1} - t_k \equiv T_{\text{nom}}$. We will see that, even for this simple case, it is impossible to compute the *ACIMR* under the hold-control strategy.

---

[4]In fact, we have 45 (computed as the number of 2-combinations from a set of 10 elements) possible forms of the hyper-sampling period if we "discard" 2 samplings in every 10 samplings.

It is natural to treat (13.16) from a switched-system perspective. Formula (13.16) corresponds to a switched system consisting of possible sub-systems: $x(t_{k+1}^e) = \Phi(T_{\text{nom}})x(t_k^e)$, $x(t_{k+1}^e) = \Phi(2T_{\text{nom}})x(t_k^e)$, $x(t_{k+1}^e) = \Phi(3T_{\text{nom}})x(t_k^e)$, ... Since it involves an infinite number of sub-systems, it is impossible to analyze its stability. Thus, no *ACIMR* exits for the *DCES* with the hold-control strategy.

In order to analyze the dynamics under the hold-control strategy, we need to set an upper bound on the execution time of a hold control. It is important to point out that such an upper bound is not needed when computing the *ACIMR* under the zero-control strategy. Introduce now the following:

**Assumption 13.1** We will assume that the hold-control (13.3) can be used for at most $\lambda$ consecutive ineffective sampling instants.

The existence of such an upper bound (as stated in the assumption above) makes the analysis tractable. In fact, such an hypothesis is reasonable since $t_{k+1}^e - t_k^e$ can not be unbounded in practice.

### 13.2.3 *Optimization of the Switched Hold-Zero (HZ) Control*

Assume the largest *ACIMR* under the zero control, denoted by *ACIMR*$^*$, is obtained by Proposition 13.1 when $a_s = a_s^*$ and $a_u = a_u^*$. In this subsection, we will study how to optimally set the switching parameter $\sigma$ of the switched *HZ* controller. In addition, due to Assumption 13.1, the choice of $\sigma$ is restrained in the set $\{0, 1, \ldots, \lambda\}$.

Our optimization problem can be formulated as follows: Given $a_s = a_s^*$ and $a_u = a_u^*$, set the switching parameter $\sigma$, within the set $\{0, 1, \ldots, \lambda\}$, such that we can obtain the largest *ACIMR*. If the obtained $\sigma$ is neither 0 nor $\lambda$, we conclude that the switched *HZ* control strategy is superior to both the zero-control and the hold-control strategies.

Before presenting the main results, we need to revisit the property of the distribution information of control input missings. Since an ineffective sampling instant $t_k$ with $\theta_k = 1$ must come after an effective sampling, it is true that $M_1(k) \leq k + 1 - M(k)$. Thus, we have the following constraint on the distribution indices:

$$DR_1 \leq 1, \tag{13.17}$$

where $D = \frac{M(k-1)}{k-M(k-1)}$.

Similarly, it follows that $M_{a+1}(k) \leq M_a(k)$ ($a$ is a positive integer) since an ineffective sampling with $\theta_k = a + 1$ must take place after an ineffective sampling with $\theta_{k-1} = a$. Thus, it follows that $R_1 \geq R_2 \geq \cdots$.

Let us first present the following proposition.

**Proposition 13.2** *For given $a_s^*$ and $a_u^*$, there exist a positive-definite matrix $P$, positive scalars $o_1, o_2, \ldots, o_\lambda$, satisfying the following conditions*

$$\Phi'(T)P\Phi(T) < a_s^* P, \quad T \in [T_{\text{nom}} - \Delta, T_{\text{nom}} + \Delta], \tag{13.18}$$

$$\left(e^{AT}\right)' P e^{AT} < a_u^* P, \quad T \in [T_{\text{nom}} - \Delta, T_{\text{nom}} + \Delta], \tag{13.19}$$

$$\Phi'(T) P \Phi(T) < o_i P, \quad T \in \left[(i+1)(T_{\text{nom}} - \Delta), (i+1)(T_{\text{nom}} + \Delta)\right], \ 1 \le i \le \lambda. \tag{13.20}$$

*Let* $\theta_1 = \frac{o_1}{a_s^*}$ *and* $\theta_i = \frac{o_i}{o_{i-1}}, i = 2, \ldots, \lambda$. *If there exists a positive integer* $l$ *such that* $\max\{\theta_1, \ldots, \theta_l\} < a_u^*$ *and* $\min\{\theta_{l+1}, \ldots, \theta_\lambda\} > a_u^*$, *we have that* $ACIMR_{\sigma=\lambda} \le \cdots \le ACIMR_{\sigma=l+1} \le ACIMR_{\sigma=l} \ge ACIMR_{\sigma=l-1} \ge \cdots \ge ACIMR_{\sigma=1} \ge ACIMR_{\sigma=0}$.

(*Here* $ACIMR_{\sigma=\kappa}$ $(\kappa = 0, \ldots, \lambda)$ *denotes the ACIMR under the switched HZ control with switching parameter* $\sigma = \kappa$.)

*Proof* We first need to prove the existence of $P, o_1, o_2, \ldots, o_\lambda$, satisfying (13.18), (13.19), and (13.20). As $a_s^*$ and $a_u^*$ are the optimal solution for the zero-control strategy, for $a_s^*$ and $a_u^*$ there exists a positive-definite $P$ satisfying (13.18) and (13.19). For this $P$, (13.20) must hold for sufficiently large $o_1, o_2, \ldots, o_\lambda$. Thus, for given $a_s^*$ and $a_u^*$, the conditions (13.18), (13.19), and (13.20) are necessarily feasible. Here, we denote the $\overline{D}$ under the switched *HZ* control with switching parameter $\sigma = \kappa$ by $\overline{D}_{\sigma=\kappa}$ $(\kappa = 0, \ldots, \lambda)$. Recall that $\overline{D}_{\sigma=0} = -\log_{a_u^*} a_s^*$. Let us compute $\overline{D}_{\sigma=1}$. When $\sigma = 1$,

$$V(t_k) < \left( \left(a_u^*\right)^{D(1-R_1)} \left(\frac{o_1}{a_s^*}\right)^{DR_1} a_s^* \right)^{k-M(k-1)} V(t_0),$$

$$D = \frac{M(k-1)}{k - M(k-1)}.$$

We have that

$$\overline{D}_{\sigma=1} = \begin{cases} -\log_{(a_u^{*(1-R_1)} (\frac{o_1}{a_s^*})^{R_1})} a_s^*, & \text{if } a_u^{*(1-R_1)} (\frac{o_1}{a_s^*})^{R_1} > 1 \text{ and } \overline{D}_{\sigma=1} \le \frac{1}{R_1}, \\ \frac{1}{R_1}, & \text{otherwise.} \end{cases}$$

It follows that if $\frac{o_1}{a_s^*} < a_u^*$, $ACIMR_{\sigma=1} \ge ACIMR_{\sigma=0}$.

We proceed to consider the case when $\sigma = 2$. If $\sigma = 2$, it follows that

$$V(t_k) < \left( \left(a_u^*\right)^{D(1-R_1-R_2)} \left(\frac{o_2}{o_1}\right)^{DR_2} \left(\frac{o_1}{a_s^*}\right)^{DR_1} a_s^* \right)^{k-M(k-1)} V(t_0).$$

We have that

$$\overline{D}_{\sigma=2} = \begin{cases} -\log_{(a_u^{*(1-R_1-R_2)} (\frac{o_2}{o_1})^{R_2} (\frac{o_1}{a_s^*})^{R_1})} a_s^*, \\ \quad \text{if } a_u^{*(1-R_1-R_2)} (\frac{o_2}{o_1})^{R_2} (\frac{o_1}{a_s^*})^{R_1} > 1 \text{ and } \overline{D}_{\sigma=2} \le \frac{1}{R_1}, \\ \frac{1}{R_1}, \quad \text{otherwise.} \end{cases}$$

It is true that if $\frac{o_2}{o_1} < a_u^*$, $ACIMR_{\sigma=2} \ge ACIMR_{\sigma=1}$. Following this way, we can prove that $ACIMR_{\sigma=l} \ge ACIMR_{\sigma=l-1} \ge \cdots \ge ACIMR_{\sigma=1} \ge ACIMR_{\sigma=0}$ under the condition of the proposition.

For $\lambda \geq \sigma > l$, we have that

$$V(t_k) < \left( \left(a_u^*\right)^{D(1-R_1-\cdots-R_\sigma)} \left(\frac{o_\sigma}{o_{\sigma-1}}\right)^{DR_\sigma} \cdots \left(\frac{o_1}{a_s^*}\right)^{DR_1} a_s^* \right)^{k-M(k-1)} V(t_0).$$

Thus, we conclude that $ACIMR_{\sigma=\lambda} \leq \cdots \leq ACIMR_{\sigma=l+1} \leq ACIMR_{\sigma=l}$, and the proof is complete. □

We will set the optimal switching parameter as $\sigma = l$ obtained by Proposition 13.2 and the corresponding *ACIMR* is given by:

$$ACIMR_{\sigma=l} = \begin{cases} 1 - \frac{1}{1-\log_{\Upsilon_l} a_s^*}, & \text{if } \Upsilon_l > 1 \text{ and } -\log_{\Upsilon_l} a_s^* \leq \frac{1}{R_1}, \\ 1 - \frac{1}{1+\frac{1}{R_1}}, & \text{otherwise}, \end{cases} \tag{13.21}$$

where

$$\Upsilon_l = \left(a_u^*\right)^{(1-\sum_{i=1}^{l} R_i)} \left(\frac{o_l}{o_{l-1}}\right)^{R_l} \cdots \left(\frac{o_1}{a_s^*}\right)^{R_1}.$$

Similar to Proposition 13.1, Proposition 13.2 has also an infinite-dimensional character. We give the following numerically tractable condition, based on the discretization technique as we used it in Sect. 13.2.1.

Divide intervals $[T_{\text{nom}} - \Delta, T_{\text{nom}} + \Delta]$, $[2(T_{\text{nom}} - \Delta), 2(T_{\text{nom}} + \Delta)]$, $[3(T_{\text{nom}} - \Delta), 3(T_{\text{nom}} + \Delta)]$, …, into $N_1, N_2, N_3, \ldots$, sub-intervals, $[T_{\text{nom},1,j} - \frac{\Delta}{N_1}, T_{\text{nom},1,j} + \frac{\Delta}{N_1}]$ ($T_{\text{nom},1,j} = T_{\text{nom}} - \Delta + \frac{(2j-1)\Delta}{N_1}$, $j = 1, \ldots, N_1$), $[T_{\text{nom},2,j} - \frac{2\Delta}{N_2}, T_{\text{nom},2,j} + \frac{2\Delta}{N_2}]$ ($T_{\text{nom},2,j} = 2(T_{\text{nom}} - \Delta) + \frac{(2j-1)2\Delta}{N_2}$, $j = 1, \ldots, N_2$), $[T_{\text{nom},3,j} - \frac{3\Delta}{N_3}, T_{\text{nom},3,j} + \frac{3\Delta}{N_3}]$ ( $T_{\text{nom},3,j} = 3(T_{\text{nom}} - \Delta) + \frac{(2j-1)3\Delta}{N_3}$, $j = 1, \ldots, N_3$), …

Further, for $T \in [T_{\text{nom},i,j} - \frac{i\Delta}{N_i}, T_{\text{nom},i,j} + \frac{i\Delta}{N_i}]$ ($i = 1, 2, 3, \ldots$),

$$\begin{aligned} \Phi(T) &= \Phi(T_{\text{nom},i,j}) + \Delta_i(\upsilon)(A \quad B)F(T_{\text{nom},i,j})(I \quad K)', \\ e^{AT} &= e^{AT_{\text{nom},i,j}} + \Delta_i(\upsilon)Ae^{AT_{\text{nom},i,j}}, \end{aligned}$$

with

$$\begin{aligned} F(T_{\text{nom},i,j}) &= \exp\left(\left(\begin{smallmatrix} A & B \\ 0 & 0 \end{smallmatrix}\right) T_{\text{nom},i,j}\right), \\ \Delta_i(\upsilon) &= \int_0^{\upsilon} e^{A\theta}\, d\theta, \quad \upsilon \in \left[-\frac{i\Delta}{N_i}, \frac{i\Delta}{N_i}\right]. \end{aligned}$$

Through the discretization above, there exist $\beta_i$ such that $\|\Delta_i(\upsilon)\| \leq \beta_i$ ($i = 1, \ldots, \lambda + 1$). We have the following theorem.

**Theorem 13.2** *For given $a_s^*$ and $a_u^*$, there exist a positive-definite matrix $P$, positive scalars $\varepsilon_{1,j}, \ldots, \varepsilon_{\lambda+2,j}$, satisfying the following matrix inequalities*

$$\begin{pmatrix} -a_s^* P + \varepsilon_{1,j}(A\Phi(T_{\text{nom},1,j}) + BK)'(A\Phi(T_{\text{nom},1,j}) + BK) & * & * \\ P\Phi(T_{\text{nom},1,j}) & -P & * \\ 0 & P & -\frac{\varepsilon_{1,j}}{\beta_1^2} I \end{pmatrix} < 0, \quad j = 1, \ldots, N_1,$$

$$\begin{pmatrix} -a_u^* P + \varepsilon_{2,j}(Ae^{AT_{\text{nom},1,j}})'(Ae^{AT_{\text{nom},1,j}}) & * & * \\ Pe^{AT_{\text{nom},1,j}} & -P & * \\ 0 & P & -\frac{\varepsilon_{2,j}}{\beta_1^2} I \end{pmatrix} < 0, \quad j = 1, \ldots, N_1,$$

$$\begin{pmatrix} -o_1 P + \varepsilon_{3,j}(A\Phi(T_{\text{nom},2,j}) + BK)'(A\Phi(T_{\text{nom},2,j}) + BK) & * & * \\ P\Phi(T_{\text{nom},2,j}) & -P & * \\ 0 & P & -\frac{\varepsilon_{3,j}}{\beta_2^2} I \end{pmatrix} < 0, \quad j = 1, \ldots, N_2,$$

$$\begin{pmatrix} -o_2 P + \varepsilon_{4,j}(A\Phi(T_{\text{nom},3,j}) + BK)'(A\Phi(T_{\text{nom},3,j}) + BK) & * & * \\ P\Phi(T_{\text{nom},3,j}) & -P & * \\ 0 & P & -\frac{\varepsilon_{4,j}}{\beta_3^2} I \end{pmatrix} < 0, \quad j = 1, \ldots, N_3,$$

$$\vdots$$

$$\begin{pmatrix} -o_\lambda P + \varepsilon_{\lambda+2,j}(A\Phi(T_{\text{nom},\lambda+1,j}) + BK)'(A\Phi(T_{\text{nom},\lambda+1,j}) + BK) & * & * \\ P\Phi(T_{\text{nom},\lambda+1,j}) & -P & * \\ 0 & P & -\frac{\varepsilon_{\lambda+2,j}}{\beta_{\lambda+1}^2} I \end{pmatrix} < 0, \quad j = 1, \ldots, N_{\lambda+1}.$$

*If there exists a positive integer $l$ satisfying the conditions* $\max\{\theta_1, \ldots, \theta_l\} < a_u^*$ *and* $\min\{\theta_{l+1}, \ldots, \theta_\lambda\} > a_u^*$, *then we have that $ACIMR_{\sigma=\lambda} \leq \cdots \leq ACIMR_{\sigma=l+1} \leq ACIMR_{\sigma=l} \geq ACIMR_{\sigma=l-1} \geq \cdots \geq ACIMR_{\sigma=1} \geq ACIMR_{\sigma=0}$. The* ACIMR *is given by* (13.21).

The proof follows the same ideas to the ones proposed for Theorem 13.1, and hence it is omitted.

If the sampling period is constant, we can also derive a condition similar to Corollary 13.1. Suppose that the constant sampling period of system (9.1) is $T_{\text{nom}}$ and that the largest *ACIMR* under the zero control, denoted by *ACIMR*$^*$, is obtained by Corollary 13.1, when $a_s = a_s^*$ and $a_u = a_u^*$.

**Corollary 13.2** *With the hypothesis proposed above, for given* $a_s^*$ *and* $a_u^*$*, there exist a positive-definite matrix* $P$*, positive scalars* $o_1, \dots, o_\lambda$*, satisfying the following conditions*

$$\begin{pmatrix} -a_s^* P & * \\ P\Phi(T_{\text{nom}}) & -P \end{pmatrix} < 0, \tag{13.22}$$

$$\begin{pmatrix} -a_u^* P & * \\ P e^{AT_{\text{nom}}} & -P \end{pmatrix} < 0, \tag{13.23}$$

$$\begin{pmatrix} -o_i P & * \\ P\Phi((i+1)T_{\text{nom}}) & -P \end{pmatrix} < 0, \quad i = 1, \dots, \lambda. \tag{13.24}$$

*If there exists a positive integer* $l$ *satisfying the conditions* $\max\{\theta_1, \dots, \theta_l\} < a_u^*$ *and* $\min\{\theta_{l+1}, \dots, \theta_\lambda\} > a_u^*$*, then we have that* $ACIMR_{\sigma=\lambda} \le \dots \le ACIMR_{\sigma=l+1} \le ACIMR_{\sigma=l} \ge ACIMR_{\sigma=l-1} \ge \dots \ge ACIMR_{\sigma=1} \ge ACIMR_{\sigma=0}$.

*Remark 13.3* The system under consideration is fundamentally a time-varying system. For a time-variant system, unlike for a time-invariant one, it is very difficult to obtain a necessary and sufficient stability condition. The method used in this chapter is a Lyapunov-based one in the time-domain framework, which leads to sufficient though not necessary conditions.

*Remark 13.4* We refer to the maximum *ACIMR* under the zero-control strategy when designing the switched *HZ* control. Thus, we set $a_s = a_s^*$ and $a_u = a_u^*$. This manipulation considerably simplifies our optimization problem. If $a_s$ and $a_u$ are also free parameters, the computation workload may increase as this is an *NP-complete problem*. As a perspective for research direction, we may consider how to optimize these two parameters, which probably further enhances the *ACIMR*.

In the following, we will give two examples to illustrate the approach given by Theorem 13.2.

*Example 13.2* Consider system (9.1) with the following matrices

$$A = \begin{pmatrix} 0.1 & 1 \\ 0 & 1 \end{pmatrix}, \qquad B = \begin{pmatrix} 0 \\ 1 \end{pmatrix}, \qquad K = (-2.31 \quad -4.1).$$

Let us suppose that $\lambda = 5$ and we choose $N_i = 2$ $(i = 1, \dots, 5)$.

Case 1: $T(k) \in [0.25 - 0.001, 0.25 + 0.001]$.
Under the zero-control strategy, from the Theorem 13.1, we have that $a_s^* = 0.67, a_u^* = 2.75, ACIMR^* = 0.2836$. Let us compute the *ACIMR* under the switched *HZ* strategy. Using Theorem 13.2, we have that $o_1 = 1.00$, $o_2 = 8.57$, $o_3 = 33.51$, $o_4 = 97.41$, $o_5 = 241.47$. As $\frac{o_1}{a_s^*} < a_u^*$ and $\min\{\frac{o_2}{o_1}, \frac{o_3}{o_2}, \frac{o_4}{o_3}, \frac{o_5}{o_4}\} > a_u^*$, we set $\sigma = 1$. The relation between the *ACIMR* and $R_1$ is shown in Fig. 13.2. When $\sigma = 1$, the switched *HZ* control is superior to both the zero control and the hold control.

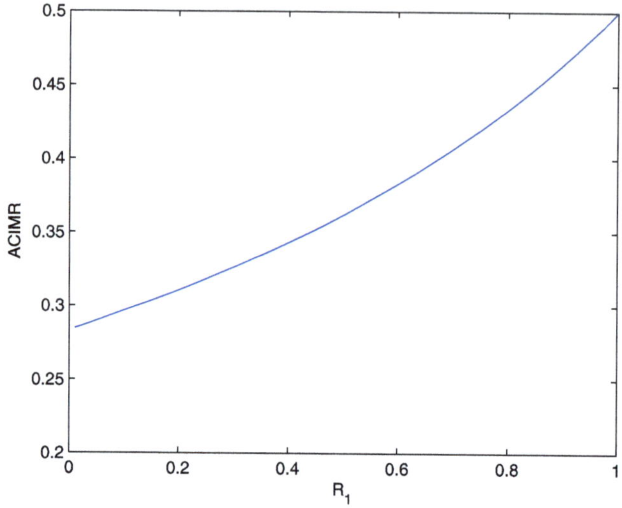


**Fig. 13.2** $R_1$ vs *ACIMR* for Example 13.2

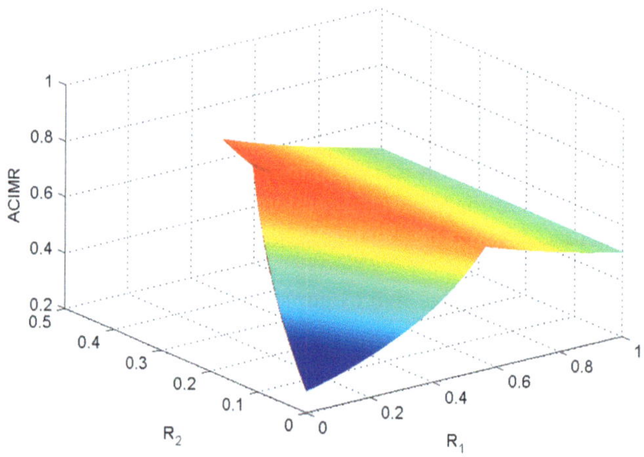


**Fig. 13.3** $R_1, R_2$ vs *ACIMR* for Example 13.2

Case 2: $T(k) \in [0.15 - 0.001, 0.15 + 0.001]$.

Under the zero-control strategy, we have that $a_s^* = 0.79, a_u^* = 1.86, ACIMR^* = 0.2753$ by using Theorem 13.1. Let us compute also the *ACIMR* under the switched *HZ* strategy. Using Theorem 13.2, we have that $o_1 = 0.62, o_2 = 0.50, o_3 = 2.80, o_4 = 8.50, o_5 = 20.10$. As $\max\{\frac{o_1}{a_s^*}, \frac{o_2}{o_1}\} < a_u^*$ and $\min\{\frac{o_3}{o_2}, \frac{o_4}{o_3}, \frac{o_5}{o_4}\} > a_u^*$, we set $\sigma = 2$. The relation among *ACIMR*, $R_1$, and $R_2$ is shown in Fig. 13.3. It is seen that a much larger *ACIMR* is obtained.

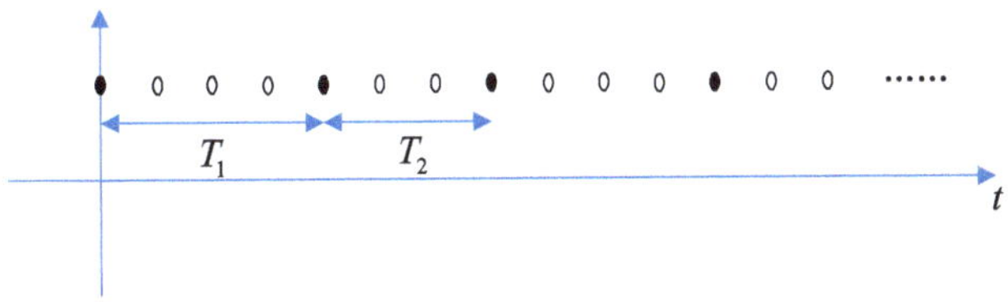


**Fig. 13.4** A hyper-sampling period for Example 13.2

Through this example, we see that the *ACIMR* depends on the distribution indices $R_1, R_2, R_3, \ldots$, if hold control is used.

We can use the obtained *ACIMR* in order to design a hyper-sampling sequence. For this example, we use the switched *HZ* control strategy. Assume we allocate a sampling period with $T_{\text{nom}} = 0.15$. In order to save system resources in situations of processor overload (e.g., when new aperiodic tasks are released), we may positively discard some sample-and-control executions, as we did in Example 13.1. According to the obtained *ACIMR*, we could define a new sampling sequence: sample once, discard three samplings, sample once and discard two samplings. This is a simple hyper-period case, with two sub-samplings: $T_1$ (three times of the original sampling period) and $T_2$ (twice of the original sampling period), as shown in Fig. 13.4. As a consequence, the average sampling frequency under the hyper-sampling mode is $\frac{2}{7}$ times of that under the simple sampling mode.

*Example 13.3* We now apply the proposed method to a benchmark example. We consider a batch reactor system, which is controlled by a sampled state-feedback control, see, for instance, *Gommans et al.* in [97].

A linearized model of the batch reactor process can be found in *Green and Limebeer* [103] and is given by:

$$\dot{x}(t) = \begin{pmatrix} 1.38 & -0.2077 & 6.715 & -5.676 \\ -0.5814 & -4.29 & 0 & 0.675 \\ 1.067 & 4.273 & -6.654 & 5.893 \\ 0.048 & 4.273 & 1.343 & -2.104 \end{pmatrix} x(t)$$
$$+ \begin{pmatrix} 0 & 0 \\ 5.679 & 0 \\ 1.136 & -3.146 \\ 1.136 & 0 \end{pmatrix} u(t).$$

Similar to *Gommans et al.* [97], we assume that the sampling period is constant and employ the sampled-data control (9.8)

$$u(t) = Kx(kT), \quad kT \le t < (k+1)T, \ k \in \mathbb{N}.$$

Let us choose the sampling period as $T = 0.1$ and the feedback gain as

$$K = \begin{pmatrix} 0.3262 & -0.6890 & 0.2002 & -0.6832 \\ 2.0250 & -0.1612 & 1.5572 & -1.0585 \end{pmatrix}.$$

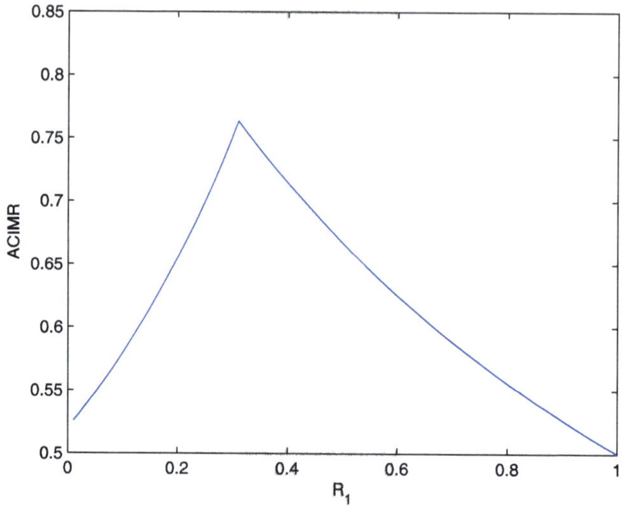


**Fig. 13.5** $R_1$ vs *ACIMR* for Example 13.3

We first use the results of Corollary 13.1 in order to compute the *ACIMR* under the zero-control strategy. We have that $ACIMR^* = 0.51$ with the associated $a_s^* = 0.65$, $a_u^* = 1.51$. We next use Corollary 13.2 to calculate the *ACIMR* under the switched *HZ* control strategy. Suppose $\lambda = 5$. We have that $o_1 = 0.40, o_2 = 1.06, o_3 = 3.27, o_4 = 7.32, o_5 = 16.59$. As $\frac{o_1}{a_s^*} < a_u^*$ and $\min\{\frac{o_2}{o_1}, \frac{o_3}{o_2}, \frac{o_4}{o_3}, \frac{o_5}{o_4}\} > a_u^*$, we set $\sigma = 1$.

The *ACIMR* under the switched *HZ* control is illustrated in Fig. 13.5, where we see that the switched *HZ* control leads to a larger *ACIMR*. From the figure, we see that *ACIMR* first increases with respect to $R_1$; then after reaching its maximum point, *ACIMR* reduces with respect to $R_1$. This reducing curve is due to the distribution constraint (13.17). For example, as $R_1 = 1$, it means that all the ineffective samplings are with $\theta_k = 1$. That is, the system samples once and then a control input missing must follow. Such a process repeats. One can see that the corresponding control input missing rate is 0.5 and, we can see from the figure, that the related *ACIMR* is also 0.5.

*Remark 13.5* For simplicity, we have chosen the value of $\lambda = 5$ (a small value) in all the examples of this chapter. When $\lambda$ increases, we have a larger range of choices, but the computation time increases accordingly. As this computation is done "offline", we are allowed to make such a choice (choose a large $\lambda$) in practice.

## 13.3 Notes and Comments

In this chapter, we proposed a new switched hold-zero (*HZ*) control strategy. When a control input missing occurs, the switched *HZ* control strategy has two candidate control laws: the hold control and the zero control. We study how to make the optimal switching between the two control laws, by setting a switching parameter, such that the switched *HZ* control admits the maximum admissible control input missing rate (*ACIMR*). With the optimal switching parameter, the switched *HZ* control strategy is shown to work better than both the zero-control and the hold-control ones. In addition, we proposed an application of the *ACIMR*: designing a hyper-sampling period based on the obtained *ACIMR*. According to the obtained hyper-sampling period, the system may positively discard some sample-and-control executions during the run-time. In this way, we may reduce the system resource utilization.

In the literature, there are two commonly used compensation strategies for control input missings: the zero-control and the hold-control ones. These two strategies as well as the one we are proposing are simple to implement in practice, as basically they do not require extra dedicated hardware. The actuators in the *DCES* architecture typically work in the event-driven fashion. Thus, if a control input missing takes place (i.e., the actuator is not triggered to update its control input to the controlled plant), the actuator generally holds its current control signal. That is, the hold-control appears as a default compensation strategy for an event-driven actuator. The zero-control is also very simple to implement since the actuator only needs to set its control signal to zero. There are other classes of compensation strategies reported in the literature for the control input missings. To name only a few, a model-based compensator was studied in *Gommans et al.* [97] and a predictive compensation was investigated in *Henriksson et al.* [115]. A common feature of the above mentioned compensation strategies is that they require an extra observer or predictor. The effectiveness and advantages are seen in the above papers. When a control input missing takes place, the actuator will use the estimated state from the observer or predictor to generate the control input to the controlled plant. When a *DCES* is controlled with the hold-control strategy, the dynamics can be analyzed from a delay-system point of view (see also Chaps. 9 and 12). If consecutive control input missings happen, the (artificial) input delay is time-varying and uncertain. To the best of our knowledge, there exists no analytical approach giving a complete characterization of the such time-varying delay systems.

In comparison with the previous mentioned methods, our approach leads to (sufficient though not necessary) conditions relatively simple and natural to apply for *DCESs* but, unfortunately, still conservative. We believe that there exists much room for further reduction of the conservatism. For instance, we find through simulations that the *ACIMR* index, in fact, approaches 1 for all the cases studied in Example 13.1. Finally, throughout this chapter, we have adopted a simple common Lyapunov function to develop the stability conditions. A potential way to further reduce the conservatism is to use multiple Lyapunov functions.

# Resumé

Distributed control and embedded systems (*DCESs*) incorporating real-time control and communication functions implemented on a distributed Hardware/Software dedicated architectures are ubiquitous in various engineering fields. Their major industrial challenge is the increasing systems complexity. In all the fields, customer needs are translated to "functions" that all require control and information. For that reason, connectivity is increasing. There is a need from system engineers to understand and master the behavior of the global system.

What we propose in this monograph is a natural extension of control theory towards this direction. Classical control modeling and design models are refined with computing and networking aspects, towards this goal. Several results on joint design of control laws and scheduling algorithms as well as stability analysis of some special cases of DCESs with respect to induced delays are presented. This study is addressed by considering different aspects of the limitations imposed by the use of communication channels as well as embedded node processors composing the HW/SW architecture of DCESs. We specially focused on limitations in terms of network communication bandwidth and processor calculation power inducing sampling and period jitter, communication delay and signal quantization limitation.

This monograph will be of great interest to advanced students and researchers in cyber-physical systems, as well as to engineers that need to design and develop DCESs with computing power or bandwidth limitations or delays.

A. Çela et al., *Optimal Design of Distributed Control and Embedded Systems*,
Communications and Control Engineering, DOI 10.1007/978-3-319-02729-6,

# Appendix A
# Four-Wheels Active Suspension System Model

## A.1 Model Description

The active suspension model (Fig. A.1) considered in this book was adopted from [60]. It consists of a seven degree-of-freedom system. In this model, the car body, or sprung mass, is free to heave, roll and pitch. In order to obtain a linear model, roll and pitch angles are assumed to be small. The suspension system connects the sprung mass to the four unsprung masses (front–left, front–right, rear–left and rear–right wheels), which are free to bounce vertically with respect to the sprung mass. The suspension element consists of a spring, a shock absorber and a hydraulic actuator at each corner. The shock absorbers are modeled as linear viscous dampers, and the tires are modeled as linear springs in parallel to linear dampers.

In order to describe this system, fifteen main variables need to be considered:

$z$ heave position of the center of gravity of the sprung mass
$\varphi$ pitch angular position of the sprung mass
$\theta$ roll angular position of the sprung mass
$d_{fl}$ front–left suspension deflection
$d_{rl}$ rear–left suspension deflection
$d_{rr}$ rear–right suspension deflection
$d_{fr}$ front–right suspension deflection
$V_{fl}$ front–left unsprung mass velocity
$V_{rl}$ rear–left unsprung mass velocity
$V_{rr}$ rear–right unsprung mass velocity
$V_{fr}$ front–right unsprung mass velocity
$d'_{fl}$ front–left tire deflection
$d'_{rl}$ rear–left tire deflection
$d'_{rr}$ rear–right tire deflection
$d'_{fr}$ front–right tire deflection

Road disturbances acting on the four vehicle wheels consist of height displacement inputs $(d_{\xi_{fl}}, d_{\xi_{rl}}, d_{\xi_{rr}}, d_{\xi_{fr}})$ and height velocity inputs $(V_{\xi_{fl}}, V_{\xi_{rl}}, V_{\xi_{rr}}, V_{\xi_{fr}})$ defined with respect to an inertial reference frame.

A. Çela et al., *Optimal Design of Distributed Control and Embedded Systems*, Communications and Control Engineering, DOI 10.1007/978-3-319-02729-6,

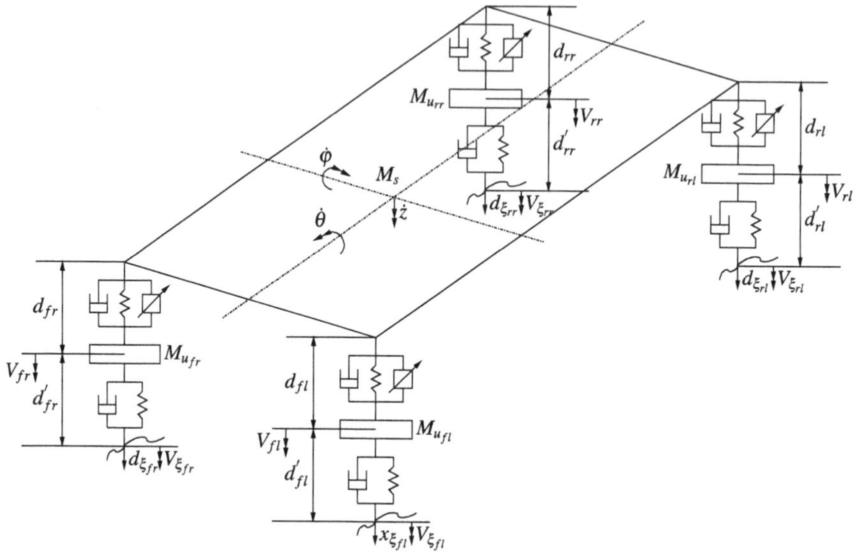


**Fig. A.1** Full vehicle model

**Table A.1** Sprung and unsprung mass

| Notation | Description | Value | Unit |
|---|---|---|---|
| $M_S$ | Sprung mass | 1460 | kg |
| $M_{u_{fl}}$, $M_{u_{fr}}$ | Front–left and front–right unsprung mass | 40 | kg |
| $M_{u_{rl}}$, $M_{u_{rr}}$ | Rear–left and rear–right unsprung mass | 35.5 | kg |
| $I_{xx}$ | Roll moment of inertia | 460 | kg m$^2$ |
| $I_{yy}$ | Pitch moment of inertia | 2460 | kg m$^2$ |

The suspension model has seven degrees of freedom. Consequently, only fourteen state variables are needed to describe it. The extra variable may be eliminated if the tire deflections are expressed as a function of three state variables $d'_{fl}$, $d'_{rl}$ and $d'_{rr}$ and of the road disturbances $d_{\xi_{fl}}$, $d_{\xi_{rl}}$, $d_{\xi_{rr}}$, $d_{\xi_{fr}}$ as illustrated in [60].

## A.2 Vehicle Parameters

The following tables, adopted from [60], summarizes the involved vehicle parameters and their values. Table A.1 is related to the sprung and unsprung masses.

Table A.2 summarizes suspension elements characteristics. Tire stifness $K_{u_{fl}}$, $K_{u_{fr}}$, $K_{u_{rl}}$ and $K_{u_{rr}}$ are equal to 175500 N/m. Front track $t_f$ and rear track $t_r$ are respectively, equal to 1.522 m and 1.510 m. Finally, longitudinal distance from sprung

**Table A.2** Suspension elements characteristics

| Notation | Description | Value | Unit |
|---|---|---|---|
| $K_{S_{fl}}, K_{S_{fr}}$ | Front–left and front–right suspension stiffness | 19960 | N/m |
| $K_{S_{rl}}, K_{S_{rr}}$ | Rear–left and rear–right suspension stiffness | 17500 | N/m |
| $C_{S_{fl}}, C_{S_{fr}}$ | Front–left and front–right suspension damping | 17500 | N/m |
| $C_{S_{rl}}, C_{S_{rr}}$ | Rear–left and rear–right suspension damping | 17500 | N/m |

mass center of gravity to front axle is $l_f = 1.011$ m whereas longitudinal distance from rear axle to sprung mass center of gravity is $l_r = 1.803$ m.

## A.3 State Equations

Applying a force-balance analysis to the model in Fig. A.1, the state equations may be derived from the following equations of motion.

First, the equations of motion, along the vertical direction, of the sprung mass:

$$\begin{aligned} M_S\ddot{z} &+ K_{S_{fl}}d_{fl} + K_{S_{rl}}d_{rl} + K_{S_{rr}}d_{rr} + K_{S_{fr}}d_{rr} \\ &+ C_{S_{fl}}\dot{d}_{fl} + C_{S_{rl}}\dot{d}_{rl} + C_{S_{rr}}\dot{d}_{rr} + C_{S_{fr}}\dot{d}_{fr} \\ &+ u_{fl} + u_{rl} + u_{rr} + u_{fr} = 0. \end{aligned} \tag{A.1}$$

The pitch dynamics of the sprung mass:

$$\begin{aligned} I_{yy}\ddot{\varphi} &+ l_r(K_{S_{rl}}d_{rl} + K_{S_{rr}}d_{rr}) - l_f(K_{S_{fl}}d_{fl} + K_{S_{fr}}d_{fr}) \\ &+ l_r(C_{S_{rl}}\dot{d}_{rl} + C_{S_{rr}}\dot{d}_{rr}) - l_f(K_{S_{fl}}\dot{d}_{fl} + K_{S_{fr}}\dot{d}_{fr}) \\ &+ l_r(u_{rl} + u_{rr}) - l_f(u_{fl} + u_{fr}) = 0. \end{aligned} \tag{A.2}$$

The roll dynamics of the sprung mass:

$$\begin{aligned} I_{xx}\ddot{\theta} &+ \frac{t_f}{2}K_{S_{fr}}d_{fr} + \frac{t_r}{2}K_{S_{rr}}d_{rr} - \frac{t_f}{2}K_{S_{fl}}d_{fl} - \frac{t_r}{2}K_{S_{rl}}d_{rl} \\ &+ \frac{t_f}{2}C_{S_{fr}}\dot{d}_{fr} + \frac{t_r}{2}C_{S_{rr}}\dot{d}_{rr} - \frac{t_f}{2}C_{S_{fl}}\dot{d}_{fl} - \frac{t_r}{2}C_{S_{rl}}\dot{d}_{rl} \\ &- \frac{K_r}{t_r}(d_{rl} - d_{rr}) - \frac{K_f}{t_f}(d_{fl} - d_{fr}) \\ &+ (u_{rr} - u_{rl})\frac{t_r}{2} + (u_{fr} - u_{fl})\frac{t_f}{2} = 0. \end{aligned} \tag{A.3}$$

Then, equations of motion of sprung masses can be written. The Vertical dynamics of the front–left sprung mass:

$$M_{u_{fl}}\dot{V}_{fl} + K_{u_{fl}}d'_{fl} - K_{S_{fl}}d_{fl} - C_{S_{fl}}\dot{d}_{fl} - \frac{K_f}{t_f^2}(d_{fl} - d_{fr})$$
$$-\frac{t_f}{4}\frac{(d_{\xi_{fl}} - d_{\xi_{fr}})}{t_f} - \frac{(d_{\xi_{rl}} - d_{\xi_{rr}})}{t_r} + C_{u_{fl}}(V_{fl} - V_{\xi_{fl}}) - u_{fl} = 0. \tag{A.4}$$

The vertical dynamics of the rear–left sprung mass:

$$M_{u_{rl}}\dot{V}_{rl} + K_{u_{rl}}d'_{rl} - K_{S_{rl}}d_{rl} - C_{S_{rl}}\dot{d}_{rl} - \frac{K_r}{t_r^2}(d_{rl} - d_{rr})$$
$$-\frac{t_r}{4}\frac{(d_{\xi_{fl}} - d_{\xi_{fr}})}{t_f} - \frac{(d_{\xi_{rl}} - d_{\xi_{rr}})}{t_r} + C_{u_{rl}}(V_{rl} - V_{\xi_{rl}}) - u_{rl} = 0. \tag{A.5}$$

The vertical dynamics of the rear–right sprung mass:

$$M_{u_{rr}}\dot{V}_{rr} + K_{u_{rr}}d'_{rr} - K_{S_{rr}}d_{rr} - C_{S_{rr}}\dot{d}_{rr} - \frac{K_r}{t_r^2}(d_{rl} - d_{rr})$$
$$-\frac{t_r}{4}\frac{(d_{\xi_{fl}} - d_{\xi_{fr}})}{t_f} - \frac{(d_{\xi_{rl}} - d_{\xi_{rr}})}{t_r} + C_{u_{rr}}(V_{rr} - V_{\xi_{rr}}) - u_{rr} = 0. \tag{A.6}$$

The vertical dynamics of the front–right sprung mass, based on the geometric considerations described previously:

$$M_{u_{fr}}\dot{V}_{fr} + K_{u_{fr}}\left(d_{fl} + d'_{fl} - d_{fr}\right) - \left(d_{rl} + d'_{rl}\right) - \frac{t_f}{t_r}\left(d_{rr} + d'_{rr}\right)$$
$$- K_{S_{fr}}d_{fr} - C_{S_{fr}}\dot{d}_{fr} - \frac{K_f}{t_f^2}(d_{fl} - d_{fr})$$
$$+\frac{t_f}{4}\frac{(d_{\xi_{fl}} - d_{\xi_{fr}})}{t_f} - \frac{(d_{\xi_{rl}} - d_{\xi_{rr}})}{t_r} + C_{u_{fr}}(V_{fr} - V_{\xi_{fr}}) - u_{fr} = 0. \tag{A.7}$$

The state equations are completed by the following kinematic relations.
Front–left suspension defection derivatives:

$$\dot{d}_{fl} = \dot{z} - l_f\dot{\varphi} - \frac{t_f}{2}\dot{\theta} - V_{fl}. \tag{A.8}$$

Rear–left suspension defection derivatives:

$$\dot{d}_{rl} = \dot{z} + l_r\dot{\varphi} - \frac{t_r}{2}\dot{\theta} - V_{rl}. \tag{A.9}$$

Rear–right suspension defection derivatives:

$$\dot{d}_{rr} = \dot{z} + l_r\dot{\varphi} + \frac{t_r}{2}\dot{\theta} - V_{rr}. \tag{A.10}$$

Front–right suspension defection derivatives:

$$\dot{d}_{fr} = \dot{z} - l_f\dot{\varphi} + \frac{t_f}{2}\dot{\theta} - V_{fr}. \tag{A.11}$$

Front–left tire defection derivatives:

$$\dot{d}'_{fl} = V_{fl} - \frac{1}{4}3V_{\xi_{fl}} + V_{\xi_{fr}} + \frac{(V_{\xi_{rl}} - V_{\xi_{rr}})t_f}{t_f}. \tag{A.12}$$

Rear–left tire defection derivatives:

$$\dot{d}'_{rl} = V_{rl} - \frac{1}{4}3V_{\xi_{rl}} + V_{\xi_{rr}} + \frac{(V_{\xi_{fl}} - V_{\xi_{fr}})t_f}{t_f}. \tag{A.13}$$

Rear–right tire defection derivatives:

$$\dot{d}'_{rr} = V_{rr} - \frac{1}{4}3V_{\xi_{rr}} + V_{\xi_{rl}} - \frac{(V_{\xi_{fl}} - V_{\xi_{fr}})t_f}{t_r}. \tag{A.14}$$

# Appendix B
# Quadrotor

## B.1 Introduction to Quaternions

Euler's rotation theorem states that, in three-dimensional space, any displacement of a rigid body such that a point on the rigid body remains fixed, is equivalent to a single rotation about some axis $\mathbf{e}$ that runs through the fixed point. Therefore, a rotation of a rigid body can be represented by a unit vector, $\mathbf{e}$, known as the Euler axis, and a rotation angle $\beta$ around this axis.

The quaternion $q$ is then defined as

$$q = \begin{pmatrix} \cos\frac{\beta}{2} \\ \mathbf{e}\sin\frac{\beta}{2} \end{pmatrix} = \begin{pmatrix} q_0 \\ \mathbf{q} \end{pmatrix} \in \mathbb{H}, \tag{B.1}$$

where

$$\mathbb{H} = \left\{ q \text{ such that } q = [q_0\ \mathbf{q}]^T, q_0 \in \mathbb{R},\ \mathbf{q} \in \mathbb{R}^3 \text{ and } q_0^2 + \mathbf{q}^T\mathbf{q} = 1 \right\}. \tag{B.2}$$

$\mathbf{q} = [q_1\ q_2\ q_3]^T$ and $q_0$ are respectively, known as the vector and scalar parts of the quaternion.

In attitude modeling, the unitary quaternion can be used to represent the rotation from an inertial coordinate frame $\mathbf{R}$ located at some point in the space (for instance, the earth North–East–Down frame, NED), to the body coordinate frame $\mathbf{B}$ attached to the quadrotor body. A coordinate change from $r \in \mathbb{R}^3$ in the reference frame to $b \in \mathbb{R}^3$ in the body frame is expressed with the rotation matrix $C(q)$ by

$$b = C(q)r, \tag{B.3}$$

where

$$C(q) = \left(q_0^2 - \mathbf{q}^T\mathbf{q}\right)I_3 + 2\left(\mathbf{q}\mathbf{q}^T - q_0\left[\mathbf{q}^\times\right]\right),$$

$$C(q) = \begin{pmatrix} 2(q_0^2 + q_1^2) - 1 & 2(q_1q_2 + q_0q_3) & 2(q_1q_3 - q_0q_2) \\ 2(q_1q_2 - q_0q_3) & 2(q_0^2 + q_2^2) - 1 & 2(q_0q_1 + q_2q_3) \\ 2(q_0q_2 + q_1q_3) & 2(q_2q_3 - q_0q_1) & 2(q_0^2 + q_3^2) - 1 \end{pmatrix}$$

A. Çela et al., *Optimal Design of Distributed Control and Embedded Systems*, Communications and Control Engineering, DOI 10.1007/978-3-319-02729-6,

where

$$[\mathbf{q}^\times] = \begin{bmatrix} 0 & -q_3 & q_2 \\ q_3 & 0 & -q_1 \\ -q_2 & q_1 & 0 \end{bmatrix} \tag{B.4}$$

is the skew symmetric tensor associated to $\mathbf{q}$.

$q$ is an efficient attitude representation from a computational point of view, which allows avoiding singularity problems that can occur with other methods like Euler angles or Cardan angles. However, quaternions are more difficult to be physically interpreted. Changing $q$ into the angles is a simple nonlinear transformation

$$\psi = \arctan\left(\frac{2(q_0q_3 + q_1q_2)}{1 - 2(q_2^2 + q_3^2)}\right), \tag{B.5}$$

$$\theta = \arcsin\big(2(q_0q_2 - q_1q_3)\big), \tag{B.6}$$

$$\phi = \arctan\left(\frac{2(q_0q_1 + q_2q_3)}{1 - 2(q_1^2 + q_2^2)}\right). \tag{B.7}$$

## B.2 Quadrotor Attidue Modeling

The movement of the quadrotor Fig. B.1 allows for six degrees of freedom. Two frames are considered: the inertial frame **R** and the body frame **B**, which is attached to the quadrotor with its origin at the quadrotor's center of mass. The quadrotor's orientation (also named attitude) is represented by three angles: yaw $\psi$, pitch $\theta$, and roll $\phi$. Given that the front and rear motors rotate counter-clockwise while the other ones rotate clockwise, gyroscopic effects and aerodynamic torques tend to be canceled. The throttle input is the sum of the thrusts of each rotor ($F_1 + F_2 + F_3 + F_4$). The pitch movement $\theta$ is obtained by increasing (or reducing) the velocity of the rear motor while reducing (or increasing) the velocity of the front motor. The roll movement $\phi$ is obtained similarly using the lateral motors. The yaw movement $\psi$ is obtained by increasing (or decreasing) the velocity of the front and rear motors while decreasing (or increasing) the velocity of the lateral motors. This can be done while keeping the total thrust $\mathscr{T}$ constant, which must satisfy $\mathscr{T} \geq mg$, with $g$ representing the earth's gravity.

The attitude of the quadrotor is completely described using a unitary quaternion $q$. The unit quaternion represents the rotation from an inertial frame to the body frame attached to the quadrotor. The dynamic evolution of the attitude quaternion is given by

$$\begin{pmatrix} \dot{q}_0 \\ \dot{\mathbf{q}} \end{pmatrix} = \frac{1}{2} \begin{pmatrix} -\mathbf{q}^T \\ I_3 q_0 + [\mathbf{q}^\times] \end{pmatrix} \omega, \tag{B.8}$$

**Fig. B.1** Quadrotor AscTec Pelican and its body frame

where $\omega \in \mathbb{R}^3$ is the angular velocity of the quadrotor in the body frame and $[\mathbf{q}^\times]$ is the skew symmetric tensor associated to $\mathbf{q}$ as defined in (B.4). The rotational motion of the quadrotor, neglecting the gyroscopic torques, is given by

$$I_f \dot{\omega} = -[\omega^\times] I_f \omega + \tau_a + \tau_{\text{dist}}, \tag{B.9}$$

where $I_f \in \mathbb{R}^{3\times3}$ is the inertia matrix of the quadrotor with respect to the body frame. $I_f$ is constant. $\tau_a \in \mathbb{R}^3$ represents the torques resulting from the differences of the relative speeds of the four rotors, and may be written as

$$\tau_a = \begin{bmatrix} \tau_{\text{roll}} \\ \tau_{\text{pitch}} \\ \tau_{\text{yaw}} \end{bmatrix},$$

and $\tau_{\text{dist}} \in \mathbb{R}^3$ describe the aerodynamic disturbances acting on the quadrotor.

The gyroscopic torques have been neglected. $\omega_{Mi}$ $(i = 1, \dots, 4)$ are the four motors' velocities. The components of the torque $\tau_a \in \mathbb{R}^3$ generated by the four rotors, can be approximated by

$$\begin{aligned} \tau_{\text{roll}} &= d \cdot b \cdot \left(\omega_{M2}^2 - \omega_{M4}^2\right), \\ \tau_{\text{pitch}} &= d \cdot b \cdot \left(\omega_{M1}^2 - \omega_{M3}^2\right), \\ \tau_{\text{yaw}} &= k \cdot \left(\omega_{M1}^2 + \omega_{M3}^2 - \omega_{M2}^2 - \omega_{M4}^2\right) \end{aligned} \tag{B.10}$$

where $d$ is the distance from the rotors to the quadrotor's center of mass, $b$ and $k$ are two parameters depending on the air density, the radius, the shape, the pitch of the blade and other factors.

**Table B.1** Quadrotor parameter values

| | | |
|---|---|---|
| $d$ | Distance between a rotor and the center of gravity | 0.225 m |
| $b$ | Parameter for the torque computation | $29.1 \times 10^{-5}$ |
| $k$ | Parameter for the torque computation | $1.14 \times 10^{-6}$ |
| $I_f$ | Quadrotor inertia matrix | $\mathrm{diag}\{8.28, 8.28, 5.7\} \times 10^{-3}\ \mathrm{kg\,m^2}$ |
| $m$ | Quadrotor mass | 0.520 kg |

Finally, the quadrotor parameter values, adopted from [17], are given in Table B.1.

# References


1. MIABOT BT v5.x user manual, Rev.2.3. Merlin Systems Corp. Ltd., 2005
2. S. Al-Areqi, D. Görges, S. Liu, Robust control and scheduling codesign for networked embedded control systems, in *50th IEEE Conference on Decision and Control and European Control Conference*, Orlando, FL, USA, December 2011
3. Allen-Bradley, Compact logix system (catalog numbers 1769-L20 and 1769-L30): user manual. Publication 1769-UM007C-EN-P, June 2001
4. A.D.O. Anderson, J.B. Moore, *Optimal Filtering* (Prentice-Hall, New York, 1979)
5. M. Andersson, D. Henriksson, A. Cervin, TRUETIME 1.3—reference manual. Department of Automatic Control, Lund Institute of Technology, Sweden, June 2005
6. A. Anta, P. Tabuada, Self-triggered stabilization of homogeneous control systems, in *American Control Conference*, Seattle, WA, USA, June 2008
7. A. Anta, P. Tabuada, Space-time scaling laws for self-triggered control, in *47th IEEE Conference on Decision and Control*, Cancun, Mexico, December 2008
8. A. Anta, P. Tabuada, Isochronous manifolds in self-triggered control, in *48th IEEE Conference on Decision and Control Held Jointly with the 28th Chinese Control Conference. CDC/CCC 2009*, Shanghai, China, December 2009
9. A. Anta, P. Tabuada, To sample or not to sample: self-triggered control for nonlinear systems. IEEE Transactions on Automatic Control **55**(9), 2030–2042 (2010)
10. J. Araujo, A. Anta, M. Mazo, J. Faria, A. Hernandez, P. Tabuada, K.H. Johansson, Self-triggered control for wireless sensor and actuator networks, in *Proceedings of 7th IEEE International Conference on Distributed Computing in Sensor Systems*, Barcelona, Spain, June 2011
11. K.-E. Årzén, A simple event-based PID controller, in *14th IFAC World Congress*, Beijing, China, July 1999
12. K.-E. Årzén, A. Cervin, Control and embedded computing: survey of research directions, in *Proceedings of the 16th IFAC World Congress*, Prague, Czech Republic, July 2005
13. K.-E. Årzén, A. Cervin, J. Eker, L. Sha, An introduction to control and real-time scheduling co-design, in *39th IEEE Conference on Decision and Control*, Sydney, Australia, December 2000
14. K.J. Åström, B. Wittenmark, *Computer-Controlled Systems: Theory and Design* (Prentice Hall, New York, 1997)
15. K.J. Åström, B. Bernhardsson, Comparison of Riemann and Lebesgue sampling for first order stochastic systems, in *41th IEEE Conference on Decision and Control*, Las Vegas, Nevada, USA, December 2002
16. G. Attiya, Assignment of tasks on parallel and distributed computer systems. PhD thesis, University of Marne-la-Vallée, France, October 2004



A. Çela et al., *Optimal Design of Distributed Control and Embedded Systems*, Communications and Control Engineering, DOI 10.1007/978-3-319-02729-6,

17. C. Aubrun, D. Simon, Y.Q. Song (eds.), *Co-design Approaches for Dependable Networked Control Systems* (Wiley, New York, 2010)
18. E. Bajic, B. Bouard, Réseau Profibus, in *Techniques de l'Ingénieur—Traité d'Informatique Industrielle*, vol. S 8 160 (Editions Techniques de l'Ingénieur, Paris, 2002)
19. B. Bamieh, J. Boyd Pearson, The $\mathscr{H}_2$ problem for sampled-data systems. Systems and Control Letters **19**(1), 1–12 (1992)
20. G. Belanger, J. Speyer, S. Ananyev, D. Chichka, R. Carpenter, Decentralized control of satellite clusters under limited communication, in *Proceedings of the AIAA Guidance, Navigation, and Control Conference and Exhibit*, Providence, Rhode Island, USA, August 2004
21. A. Bemporad, M. Morari, Control of systems integrating logic, dynamics, and constraints. Automatica **35**(3), 407–427 (1999)
22. M.-M. Ben Gaid, Optimal scheduling and control for distributed real-time systems. PhD thesis, Université d'Evry Val d'Essonne, France, November 2006
23. M.-M. Ben Gaid, A. Çela, Model predictive control of systems with communication constraints, in *Proceedings of the 16th IFAC World Congress on Automatic Control*, Prague, Czech Republic, July 2005
24. M.-M. Ben Gaid, A. Çela, Trading quantization precision for update rate in systems with limited communication: a model predictive approach. Automatica **46**(7), 1210–1214 (2010)
25. M.-M. Ben Gaid, A. Çela, S. Diallo, R. Kocik, R. Hamouche, A. Reama, Performance evaluation of the distributed implementation of a car suspension system, in *Proceedings of the IFAC Workshop on Programmable Devices and Embedded Systems*, Brno, Czech Republic, February 2006
26. M.-M. Ben Gaid, A. Çela, Y. Hamam, Optimal integrated control and scheduling of systems with communication constraints, in *Joint 44th IEEE Conference on Decision and Control and European Control Conference*, Seville, Spain, December 2005
27. M.-M. Ben Gaid, A. Çela, Y. Hamam, Optimal integrated control and scheduling of networked control systems with communication constraints: application to a car suspension system. IEEE Trans. Control Syst. Technol. **14**(4), 776–787 (2006)
28. M.-M. Ben Gaid, A. Çela, Y. Hamam, Optimal real-time scheduling of control tasks with state feedback resource allocation. IEEE Trans. Control Syst. Technol. **17**(2), 309–326 (2009)
29. M.-M. Ben Gaid, A. Çela, Y. Hamam, C. Ionete, Optimal scheduling of control tasks with state feedback resource allocation, in *American Control Conference*, Minneapolis, Minnesota, USA, June 2006
30. M.-M. Ben Gaid, A. Çela, R. Kocik, Distributed control of a car suspension system, in *Proceedings of the 5th EUROSIM Congress on Modeling and Simulation*, Paris, France, September 2004
31. M.-M. Ben Gaid, R. Kocik, Y. Sorel, R. Hamouche, A methodology for improving software design lifecycle in embedded control systems, in *Proceeding of Design, Automation and Test in Europe, DATE '08*, Munich, Germany, 10–14 March 2008
32. M.-M. Ben Gaid, D. Simon, O. Sename, A convex optimization approach to feedback scheduling, in *16th Mediterranean Conference on Control and Automation, MED'08*, Ajaccio, France, 2008
33. A. Berlin, K. Gabriel, Distributed MEMS: new challenges for computation. IEEE Comput. Sci. Eng. Mag. **4**(1), 12–16 (1997)
34. E. Bini, A. Cervin, Delay-aware period assignment in control systems, in *Proceedings of the 2008 Real-Time Systems Symposium*, Barcelona, Spain, November–December 2008
35. S. Bittanti, A.J. Laub, J.C. Willems (eds.), *The Riccati Equation*. The Periodic Riccati Equation (Springer, Berlin, 1991)
36. F. Blanchini, Set invariance in control. Automatica **35**(11), 1747–1767 (1999)
37. V. Blondel, J.N. Tsitsiklis, Complexity of stability and controllability of elementary hybrid systems. Automatica **35**(3), 479–489 (1999)
38. F. Borrelli, M. Baotić, A. Bemporad, M. Morari, Dynamic programming for constrained optimal control of discrete-time linear hybrid systems. Automatica **41**(10), 1709–1721 (2005)

39. B. Bouard, Ethernet en tant que réseau de terrain: standard Profinet, in *Techniques de l'Ingénieur—Traité d'Informatique Industrielle*, vol. S 8 162 (Editions Techniques de l'Ingénieur, Paris, 2005)
40. S. Boyd, L. El Ghaoui, E. Feron, V. Balakrishnan, *Linear Matrix Inequalities in System and Control Theory* (SIAM, Philadelphia, 1994)
41. M.S. Branicky, S.M. Phillips, W. Zhang, Scheduling and feedback co-design for networked control systems, in *41st IEEE Conference on Decision and Control*, Las Vegas, NV, USA, December 2002
42. R.W. Brockett, D. Liberzon, Quantized feedback stabilization of linear systems. IEEE Transactions on Automatic Control **45**(7), 1279–1289 (2000)
43. R.W. Brockett, Stabilization of motor networks, in *34th IEEE Conference on Decision and Control*, New Orleans, LA, USA, December 1995
44. R.W. Brockett, Minimum attention control, in *36th IEEE Conference on Decision and Control*, San Diego, California, USA, December 1997
45. S. Bttanti, P. Claneri, *Periodic Control Systems* (Pergamon, Elmsford, 2002)
46. A. Burns, A. Wellings, *Real-Time Systems and Programming Languages* (Addison-Wesley, Reading, 2001)
47. G. Buttazzo, M. Velasco, P. Martí, G. Fohler, Managing quality-of-control performance under overload conditions, in *Proceedings of the 16th Euromicro Conference on Real-Time Systems*, Catania, Italy, July 2004
48. G.C. Buttazzo, *Hard Real-Time Computing Systems: Predictable Scheduling Algorithms and Applications* (Prentice Hall, New York, 1997)
49. G.C. Buttazzo, G. Lipari, M. Caccamo, L. Abeni, Elastic scheduling for flexible workload management. IEEE Trans. Comput. **51**(3), 289–302 (2002)
50. E.F. Camacho, C. Bordons, *Model Predictive Control* (Springer, Berlin, 2004)
51. C. Canudas, K.J. Åström, K. Braun, Adaptive friction compensation in dc-motor drives. IEEE J. Robot. Autom. **3**(6), 681–685 (1987)
52. R. Castane, P. Martí, M. Velasco, A. Cervin, Resource management for control tasks based on the transient dynamics of closed-loop systems, in *Proceedings of the 18th Euromicro Conference on Real-Time Systems*, Dresden, Germany, July 2006, pp. 171–182
53. A. Cervin, Integrated control and real-time scheduling. PhD thesis, Department of Automatic Control, Lund Institute of Technology, Sweden, April 2003
54. A. Cervin, P. Alriksson, Optimal on-line scheduling of multiple control tasks: a case study, in *Proceedings of the 18th Euromicro Conference on Real-Time Systems*, Dresden, Germany, July 2006
55. A. Cervin, J. Eker, B. Bernhardsson, K.-E. Årzén, Feedback-feedforward scheduling of control tasks. Real-Time Syst. **23**(1–2), 25–53 (2002)
56. A. Cervin, D. Henriksson, B. Lincoln, J. Eker, K.-E. Årzén, How does control timing affect performance? IEEE Control Syst. Mag. **23**(3), 16–30 (2003)
57. A. Cervin, M. Velasco, P. Martí, A. Camacho, Optimal online sampling period assignment: theory and experiments. IEEE Trans. Control Syst. Technol. **19**(4), 902–910 (2011)
58. C.F. Mendez-Barrios, Low-order controllers for time-delay systems. An analytical approach. PhD thesis, Université Paris-Sud, Laboratory of Signals and Systems (L2S), July 2011
59. A. Chaillet, A. Bicchi, Delay compensation in packet-switching networked controlled systems, in *47th IEEE Conference on Decision and Control*, Cancun, Mexico, December 2008
60. R.M. Chalasani, Ride performance potential of active suspension systems—Part II: comprehensive analysis based on a full-car model, in *Proceedings of the Symposium on Simulation and Control of Ground Vehicles and Transportation Systems, ASME AMD*, Anaheim, CA, USA, December 1986
61. J. Chen, H.A. Latchman, Frequency sweeping tests for stability independent of delay. IEEE Transactions on Automatic Control **40**(9), 1640–1645 (1995)
62. T. Chen, B.A. Francis, $\mathscr{H}_2$-optimal sampled-data control. IEEE Trans. Autom. Control **36**(4), 387–397 (1991)
63. T. Chen, B.A. Francis, *Optimal Sampled-Data Control Systems* (Springer, Berlin, 1995)

64. M.B.G. Cloosterman, L. Hetel, N. Van de Wouw, W.P.M.H. Heemels, J. Daafouz, H. Nijmeijer, Controller synthesis for networked control systems. Automatica **46**(10), 1584–1594 (2010)
65. M.B.G. Cloosterman, N. Van de Wouw, W.P.M.H. Heemels, H. Nijmeijer, Robust stability of networked control systems with time-varying network-induced delays, in *Proceedings of the 45th Conference on Decision and Control*, San Diego, CA, December 2006
66. ControlNet International, ControlNet specification, Release 2.0, including Errata 2, December 1999
67. F. Cottet, J. Delacroix, C. Kaiser, Z. Mammeri, *Scheduling in Real-Time Systems* (Wiley, New York, 2003)
68. R.J. Dakin, A tree search algorithm for mixed integer programming problems. Comput. J. **8**(3), 250–255 (1965)
69. C. De Paersis, F. Mazenc, Stability of quantized time-delay nonlinear systems: a Lyapunov-Krasowskii-functional approach. Math. Control Signals Syst. **21**(4), 337–370 (2010)
70. C. De Persis, R. Sailer, F. Fabian, On a small-gain approach to distributed event-triggered control, in *18th IFAC World Congress*, Milan, Italy, 2011
71. D. Decotigny, Une infrastructure de simulation modulaire pour l'évaluation de performances de systèmes temps-réel. PhD thesis, Université de Rennes 1, April 2003
72. D.F. Delchamps, Stabilizing a linear system with quantized state feedback. IEEE Transactions on Automatic Control **35**(8), 916–924 (1990)
73. S.K. Dhall, C.L. Liu, On a real-time scheduling problem. Oper. Res. **26**(1), 127–140 (1978)
74. M. Di Natale, Scheduling the CAN bus with earliest deadline techniques, in *Proceedings of the 21st IEEE Real-Time Systems Symposium*, Orlando, Floria, USA, September 2000
75. M.C.F. Donkers, W.P.M.H. Heemels, Output-based event-triggered control with guaranteed $L_\infty$-gain and improved and decentralised event-triggering. IEEE Transactions on Automatic Control **57**(6), 1362–1376 (2012)
76. M.C.F. Donkers, W.P.M.H. Heemels, N. Van de Wouw, L. Hetel, Stability analysis of networked control systems using a switched linear systems approach. Automatic Control, IEEE Transactions on **56**(9), 2101–2115 (2011)
77. R. Dorf, M. Farren, C. Phillips, Adaptive sampling frequency for sampled-data control systems. IEEE Transactions on Automatic Control **7**(1), 38–47 (1962)
78. J.C. Doyle, K. Glover, P.P. Khargonekar, B. Francis, State-space solutions to the standard $H_2$ and $H_\infty$ control problems. IEEE Transactions on Automatic Control **34**(8), 831–847 (1989)
79. J. Eker, Flexible Embedded Control Systems. Design and Implementation. PhD thesis, Department of Automatic Control, Lund Institute of Technology, Sweden, December 1999
80. N. Elia, S.K. Mitter, Stabilization of linear systems with limited information. IEEE Transactions on Automatic Control **46**(9), 1384–1400 (2001)
81. T. Estrada, P.J. Antsaklis, Stability of model-based networked control systems with intermittent feedback, in *17th IFAC World Congress*, Seoul, Korea, 2008
82. F. Fagnani, S. Zampieri, Quantized stabilization of linear systems: complexity versus performance. IEEE Transactions on Automatic Control **49**(9), 1534–1548 (2004)
83. H. Fazelinia, R. Sipahi, N. Olgac, Stability robustness analysis of multiple time-delayed systems using building block concept. IEEE Transactions on Automatic Control **52**(5), 799–810 (2007)
84. C. Fiter, Contribution to the control of systems with time-varying and state-dependent sampling. PhD thesis, LAGIS, Ecole Centrale de Lille, Cité Scientifique, BP 48, 59651 Villeneuve d'Ascq Cedex France, November 2012
85. R. Fletcher, S. Leyffer, Numerical experience with lower bounds for MIQP branch-and-bound. SIAM J. Optim. **8**(2), 604–616 (1998)
86. O.E. Flippo, A.H.G. Rinnoy Kan, A note on benders decomposition in mixed-integer quadratic programming. Oper. Res. Lett. **9**(2), 81–83 (1990)
87. A. Franci, A. Chaillet, Quantized control of nonlinear systems: a robust approach. Int. J. Control **83**(12), 2453–2462 (2010)

88. E. Fridman, A. Seuret, J.-P. Richard, Robust sampled-data stabilization of linear systems: an input delay approach. Automatica **40**(8), 1441–1446 (2004)
89. H. Fujioka, A discrete-time approach to stability analysis of systems with aperiodic sample-and-hold devices. IEEE Transactions on Automatic Control **54**(10), 2440–2445 (2009)
90. H. Fujioka, Stability analysis of systems with aperiodic sample-and-hold devices. Automatica **45**(3), 771–775 (2009)
91. H. Gao, T. Chen, New results on stability of discrete-time systems with time-varying state delay. IEEE Transactions on Automatic Control **52**(2), 328–334 (2007)
92. H. Gao, T. Chen, J. Lam, A new delay system approach to network-based control. Automatica **44**(1), 39–52 (2008)
93. M.R. Garey, D.S. Johnson, *Computers and Intractability: A Guide to the Theory of NP-Completeness* (Freeman, New York, 1979)
94. R. Gielen, S. Olaru, M. Lazăr, On polytopic embeddings as a modeling framework for networked control systems, in *Proceedings of the 3rd International Workshop on Assessment and Future Directions of Nonlinear Model Predictive Control*, Pavia, Italy, 2008
95. R.H. Gielen, S. Olaru, M. Lazăr, W.P.M.H. Heemels, N. van de Wouw, S.-I. Niculescu, On polytopic inclusions as a modelling framework for systems with time-varying delays. Automatica **46**(3), 615–619 (2010)
96. R. Goebel, R.G. Sanfelice, A.R. Teel, Hybrid dynamical systems. IEEE Control Syst. **29**(2), 28–93 (2009)
97. T.M.P. Gommans, W.P.M.H. Heemels, N.W. Bauer, N. van de Wouw, Compensation-based control for lossy communication networks, in *American Control Conference*, Montreal, Canada, June 2012
98. T.M.P. Gommans, W.P.M.H. Heemels, N.W. Bauer, N. van de Wouw, Compensation-based control for lossy communication networks. Int. J. Control **86**(10), 1880–1897 (2013)
99. G.C. Goodwin, H. Haimovich, D.E. Quevedo, J.S. Welsh, A moving horizon approach to networked control systems design. IEEE Transactions on Automatic Control **49**(9), 1427–1445 (2004)
100. D. Görges, *Optimal Control of Switched Systems with Application to Networked Embedded Control Systems* (Logos, Berlin, 2012)
101. D. Görges, M. Izák, S. Liu, Optimal control and scheduling of switched systems. IEEE Transactions on Automatic Control **56**(1), 135–140 (2011)
102. T. Grandpierre, C. Lavarenne, Y. Sorel, Optimized rapid prototyping for real-time embedded heterogeneous multiprocessors, in *Proceedings of 7th International Workshop on Hardware/Software Co-design*, Rome, Italy, May 1999
103. M. Green, D.J.N. Limebeer, *Linear Robust Control* (Prentice-Hall, New York, 1995)
104. K. Gu, V.L. Kharitonov, J. Chen, *Stability of Time-Delay Systems* (Birkhäuser, Basel, 2003)
105. K. Gu, S.-I. Niculescu, J. Chen, On stability of crossing curves for general systems with two delays. J. Math. Anal. Appl. **311**(1), 231–253 (2005)
106. Z.-H. Guan, J. Huang, G. Chen, M. Jian, On the stability of networked impulsive control systems. Int. J. Robust Nonlinear Control **22**(17), 1952–1968 (2012)
107. G. Guo, H. Jin, A switching system approach to actuator assignment with limited channels. Int. J. Robust Nonlinear Control **20**(12), 1407–1426 (2010)
108. S.C. Gupta, Adaptive gain and adaptive sampling sampled-data systems, in *Proceedings of the IEEE Winter General Meeting*, New York, USA, January 1963
109. S.C. Gupta, Increasing the sampling efficiency for a control system. IEEE Transactions on Automatic Control **8**(3), 263–264 (1963)
110. W.P.M.H. Heemels, R.J.A. Gorter, A. van Zijl, P.P.J. van den Bosch, S. Weiland, W.H.A. Hendrix, M.R. Vonder, Asynchronous measurement and control: a case study on motor synchronization. Control Eng. Pract. **7**(12), 1467–1482 (1999)
111. W.P.M.H. Heemels, J.H. Sandee, P. Bosch, Analysis of event-driven controllers for linear systems. Int. J. Control **81**(4), 571–590 (2008)
112. W.P.M.H. Heemels, N. Van de Wouw, R.H. Gielen, M.C.F. Donkers, L. Hetel, S. Olaru, M. Lazar, J. Daafouz, S. Niculescu, Comparison of overapproximation methods for stabil-

ity analysis of networked control systems, in *Proceedings of the 13th ACM International Conference on Hybrid Systems: Computation and Control, HSCC '10*, New York, NY, USA, 2010, pp. 181–190
113. T. Henningsson, E. Johannesson, A. Cervin, Sporadic event-based control of first-order linear stochastic systems. Automatica **44**(11), 2890–2895 (2007)
114. D. Henriksson, A. Cervin, Optimal on-line sampling period assignment for real-time control tasks based on plant state information, in *Joint 44th IEEE Conference on Decision and Control and European Control Conference*, Seville, Spain, December 2005
115. E. Henriksson, H. Sandberg, K.H. Johansson, Predictive compensation for communication outages in networked control systems, in *47th IEEE Conference on Decision and Control*, Cancun, Mexico, December 2008
116. T. Henzinger, J. Sifakis, The embedded systems design challenge, in *Proceedings of the 14th International Symposium on Formal Methods (FM)*. Lecture Notes in Computer Science (Springer, Berlin, 2006)
117. J.P. Hespanha, Y. Xu, Communication logics for networked control systems, in *Proceedings of the 2004 American Control Conference*, Boston, Massachusetts, USA, June 2004
118. J.P. Hespanha, Y. Xu, Optimal communication logics for networked control systems, in *43rd IEEE Conference on Decision and Control*, Paradise Island, Bahamas, December 2004
119. J.P. Hespanha, P. Naghshtabrizi, Y. Xu, A survey of recent results in networked control systems. Proc. IEEE **95**(1), 138–162 (2007)
120. L. Hetel, J. Daafouz, C. Iung, LMI control design for a class of exponential uncertain systems with application to network controlled switched systems, in *American Control Conference, New York*, New York, USA, 2007, pp. 1401–1406
121. L. Hetel, A. Kruszewski, W. Perruquetti, J.-P. Richard, Discrete and intersample analysis of systems with aperiodic sampling. Automatic Control, IEEE Transactions on **56**(7), 1696–1701 (2011)
122. D. Hristu-Varsakelis, Optimal control with limited communication. PhD thesis, Division of Engineering and Applied Sciences, Harvard University, June 1999
123. D. Hristu-Varsakelis, Short-period communication and the role of zero-order holding in networked control systems. IEEE Transactions on Automatic Control **53**(5), 1285–1290 (2008)
124. D. Hristu-Varsakelis, W.S. Levine, *Handbook of Networked and Embedded Control Systems* (Birkhäuser, Boston, 2005)
125. T.C. Hsia, Comparisons of adaptive sampling control laws. IEEE Transactions on Automatic Control **17**(6), 830–831 (1972)
126. T.C. Hsia, Analytic design of adaptive sampling control law in sampled-data systems. IEEE Transactions on Automatic Control **19**(1), 39–42 (1974)
127. Verimag. http://www-verimag.imag.fr/DIST-TOOLS/TEMPO/kronos/. Kronos
128. Esterel-Technologies. http://www.esterel-technologies.com/. SCADE
129. INRIA. http://www.syndex.org/. SynDEx
130. HYCON2. Prospective en automatique. Technical report, GDR MACS, December 2011
131. C. Ionete, A. Çela, Structural properties and stabilization of NCS with medium access constraints, in *Proceedings of the 45th IEEE Conference on Decision and Control*, San Diego, California, USA, December 2006
132. ISO, Road vehicles: interchange of digital information—controller area network (CAN) for high-speed communication. ISO standard-11898, November 1993
133. X. Jiang, Q.-L. Han, S. Liu, A. Xue, A new $H_\infty$ stabilization criterion for networked control systems. IEEE Transactions on Automatic Control **53**(4), 1025–1032 (2008)
134. T.A. Johansen, A. Grancharova, Approximate explicit model predictive control implemented via orthogonal search tree partitioning, in *15th IFAC World Congress*, Barcelona, Spain, July 2002
135. M. Joseph, P. Pandya, Finding response times in a real-time system. Comput. J. **29**(5), 390–395 (1986)
136. R.E. Kalman, B. Ho, K. Narendra, Controllability of linear dynamical systems. Contrib. Differ. Equ. **1**, 188–213 (1963)

137. P. Khargonekar, N. Sivashankar, $\mathscr{H}_2$ optimal control for sampled-data systems. Systems and Control Letters **17**(6), 425–436 (1991)
138. R. Kocik, M.-M. Ben Gaid, R. Hamouche, Software implementation simulation to improve control laws design, in *Proceedings of the First European Congress: Sensors & Actuators for Advanced Automotive Applications*, Paris, France, December 2005
139. R. Kocik, Y. Sorel, A methodology to design and prototype optimized embedded robotic systems, in *Proceedings of 2nd IMACS International Multiconference, CESA'98*, Hammamet, Tunisia, April 1998
140. I. Kolmanovsky, E.G. Gilbert, Theory and computation of disturbance invariant sets for discrete-time linear systems. Math. Probl. Eng. **4**(4), 317–367 (1998)
141. H. Kopetz, W. Ochsenreiter, Clock synchronization in distributed real-time systems. IEEE Trans. Comput. **36**(8), 933–940 (1987)
142. C.M. Krishna, K.G. Shin, *Real-Time Systems* (McGraw-Hill, New York, 1997)
143. M. Kuschel, P. Kremer, S. Hirche, M. Buss, Lossy data reduction methods for haptic telepresence systems, in *Proceedings of the IEEE International Conference on Robotics and Automation*, Orlando, Florida, USA, May 2006
144. Y.-K. Kwok, I. Ahmad, Benchmarking and comparison of the task graph scheduling algorithms. J. Parallel Distrib. Comput. **59**(3), 381–422 (1999)
145. A.H. Land, A.G. Doig, An automatic method for solving discrete programming problems. Econometrica **28**(3), 497–520 (1960)
146. R. Lazimy, Improved algorithm for mixed-integer quadratic programs and a computational study. Math. Program. **32**(1), 100–113 (1985)
147. M.S. Lee, C.S. Hsu, On the $\tau$-decomposition method of stability analysis for retarded dynamical systems. SIAM J. Control **7**, 242–259 (1969)
148. X.-G. Li, A. Çela, S.-I. Niculescu, A. Reama, Some remarks on robust stability of networked control systems, in *18th IEEE International Conference on Control Applications*, Saint Petersburg, Russia, July 2009, pp. 19–24
149. X.-G. Li, A. Çela, S.-I. Niculescu, A. Reama, Stability analysis of networked control systems based on a switched control, in *Proceedings of the 1st IFAC Workshop on Estimation and Control of Networked Systems*, Venice, Italy, September 2009
150. X.-G. Li, A. Çela, S.-I. Niculescu, A. Reama, A switched control method for networked control systems, in *8th IFAC Workshop on Time Delay Systems*, Sinaia, Romania, June 2009
151. X.-G. Li, A. Çela, S.-I. Niculescu, A. Reama, Some problems in the stability of networked-control systems with periodic scheduling. Int. J. Control **83**(5), 996–1008 (2010)
152. F.L. Lian, Analysis, design, modeling, and control of networked control systems. PhD thesis, Univ. Michigan, 2001
153. F.L. Lian, J.R. Moyne, D.M. Tilbury, Performance evaluation of control networks: Ethernet, ControlNet, and DeviceNet. IEEE Control Syst. Mag. **21**(1), 66–83 (2001)
154. D. Liberzon, *Switching in Systems and Control* (Birkhäuser, Boston, 2003)
155. B. Lincoln, B. Bernhardsson, LQR optimization of linear system switching. IEEE Transactions on Automatic Control **47**(10), 1701–1705 (2002)
156. C.L. Liu, J.W. Layland, Scheduling algorithms for multiprogramming in a hard-real-time environment. J. ACM **20**(1), 46–61 (1973)
157. J.W.S. Liu, *Real-Time Systems* (Prentice Hall, New York, 2000)
158. X. Liu, L. Sha, M. Caccamo, G. Buttazzo, Online control optimization using load driven scheduling, in *Proceedings of the 39th IEEE Conference on Decision and Control*, Sydney, Australia, December 2000
159. S. Longo, G. Herrmann, P. Barber, Optimization approaches for controller and schedule codesign in networked control, in *6th IFAC Symposium on Robust Control Design*, Haifa, Israel, June 2009
160. C. Lu, J. Stankovic, G. Tao, S. Son, Feedback control real-time scheduling: framework, modeling and algorithms. Special issue on control-theoretic approaches to real-time computing. J. Real-Time Syst. **23**(1/2), 85–126 (2002)

161. L. Lu, L. Xie, M. Fu, Optimal control of networked systems with limited communication: a combined heuristic and convex optimization approach, in *Proceedings of the 42nd IEEE Conference on Decision and Control*, Hawaii, USA, December 2003
162. R. Luck, A. Ray, An observer-based compensator for distributed delays. Automatica **26**(5), 903–908 (1990)
163. J. Lunze, D. Lehmann, A state-feedback approach to event-based control. Automatica **46**(1), 211–215 (2010)
164. MAP/TOP Users Group, Manufacturing automation protocol specification—version 3.0, August 1988
165. N. Marchand, S. Durand, J.F. Guerrero-Castellanos, A general formula for event-based stabilization of nonlinear systems. IEEE Transactions on Automatic Control **58**(5), 1332–1337 (2013)
166. P. Martí, Analysis and design of real-time control systems with varying control timing constraints. PhD thesis, Technical University of Catalonia, 2002
167. P. Martí, J.M. Fuertes, G. Fohler, K. Ramamritham, Improving quality-of-control using flexible timing constraint: metric and scheduling issues, in *Proceedings of the 23rd IEEE Real-Time Systems Symposium*, Austin, Texas, USA, December 2002
168. P. Martí, J.M. Fuertes, K. Ramamritham, G. Fohler, Jitter compensation for real-time control systems, in *Proceedings of the 22nd IEEE Real-Time Systems Symposium*, London, England, December 2001
169. P. Martì, C. Lin, S.A. Brandt, M. Velasco, J.M. Fuertes, Draco: efficient resource management for resource-constrained control tasks. IEEE Trans. Comput. **58**(1), 90–105 (2009)
170. M. Mazo, A. Anta, P. Tabuada, On self-triggered control for linear systems: guarantees and complexity, in *10th European Control Conference* (2009)
171. M. Mazo, M. Cao, Decentralized event-triggered control with asynchronous updates, in *IEEE Conference on Decision and Control*, Orlando, USA, December 2011
172. M. Mazo, P. Tabuada, Input-to-state stability of self-triggered control systems, in *48th IEEE Conference on Decision and Control, Held Jointly with the 28th Chinese Control Conference. CDC/CCC 2009*, Shanghai, China, December 2009
173. C.F. Mendez-Barrios, S.-I. Niculescu, J. Chen, M. Maya-Mendez, Output feedback stabilisation of single-input single-output linear systems with I/O network-induced delays. An eigenvalue-based approach. Int. J. Control (accepted). doi:10.1080/00207179.2013.834075
174. W. Michiels, S.-I. Niculescu, *Stability and Stabilization of Time-Delay Systems. An Eigenvalue-Based Approach* (SIAM, Philadelphia, 2007)
175. L. Mirkin, Some remarks on the use of time-varying delay to model sample-and-hold circuits. IEEE Transactions on Automatic Control **52**(6), 1109–1112 (2007)
176. J.R. Mitchell, W.L. McDaniel Jr., Sensitivity of discrete systems to variation of sampling interval. IEEE Transactions on Automatic Control **14**(2), 200–201 (1969)
177. L.A. Montestruque, P.J. Antsaklis, On the model-based control of networked systems. Automatica **39**(10), 1837–1843 (2003)
178. L.A. Montestruque, P.J. Antsaklis, Stability of model-based networked control systems with time-varying transmission times. IEEE Transactions on Automatic Control **49**(9), 1562–1572 (2004)
179. H. Mounier, J.-P. Richard, Special issue on time-delay systems for communication networks. Int. J. Syst. Sci. **34**(10–11), 561–571 (2003)
180. H. Mounier, J. Rudolph, *Time Delay Systems*. Encyclopedia of Life and Support Systems, vol. 6.43.19.4 (2003)
181. R.M. Murray, K.J. Åström, S. Boyd, R.W. Brockett, G. Stein, Future directions in control in an information-rich world. IEEE Control Syst. Mag. **23**(2), 20–33 (2003)
182. G.N. Nair, R.J. Evans, Stabilization with data-rate-limited feedback: tightest attainable bounds. Systems and Control Letters **41**(1), 49–56 (2000)
183. N. Navet, Y. Song, F. Simonot-Lion, C. Wilwert, Trends in automotive communication systems. Proc. IEEE **93**(6), 1204–1223 (2005)

184. D. Nešić, D. Liberzon, A united framework for design and analysis of networked and quantized control systems. IEEE Transactions on Automatic Control **54**(4), 732–747 (2009)
185. D. Nešić, A.R. Teel, Input-output stability properties of networked control systems. IEEE Trans. Autom. Control **49**(10), 1650–1667 (2004)
186. S.-I. Niculescu, *Delay Effects on Stability. A Robust Control Approach*. LNCIS, vol. 269 (Springer, Heidelberg, 2001)
187. J. Nilsson, Real-time control systems with delays. PhD thesis, Department of Automatic Control, Lund Institute of Technology, Sweden, January 1998
188. Y. Oishi, H. Fujioka, Stability and stabilization of aperiodic sampled-data control systems using robust linear matrix inequalities. Automatica **46**(8), 1327–1333 (2010)
189. S. Olaru, S.-I. Niculescu, Predictive control for linear systems with delayed input subject to constraints, in *17th IFAC World Congress*, Seoul, Korea, July 2008
190. N. Olgac, R. Sipahi, An exact method for the stability analysis of time-delayed linear time-invariant (LTI) systems. IEEE Transactions on Automatic Control **47**(5), 793–797 (2002)
191. L. Palopoli, A. Bicchi, A. Sangiovanni-Vincentelli, Numerically efficient control of systems with communication constraints, in *Proceedings of the 41st IEEE Conference on Decision and Control*, Las Vegas, USA, December 2002
192. L. Palopoli, C. Pinello, A. Bicchi, A. Sangiovanni-Vincentelli, Maximizing the stability radius of a set of systems under real-time scheduling constraints. IEEE Transactions on Automatic Control **50**(11), 1790–1795 (2005)
193. L. Palopoli, C. Pinello, A.L. Sangiovanni-Vincentelli, L. El-Ghaoui, A. Bicchi, Synthesis of robust control systems under resource constraints, in *Hybrid Systems: Computation and Control*, ed. by M. Greenstreet, C. Tomlin. Lecture Notes in Computer Science, vol. 2289 (Springer, Heidelberg, 2002), pp. 337–350
194. A.S. Poznyak, *Advanced Mathematical Tools for Automatic Control Engineers*. Deterministic Techniques, vol. 1 (Elsevier, Amsterdam, 2008)
195. A. Quazi, W. Konrad, Underwater acoustic communications. IEEE Commun. Mag. **20**(2), 24–30 (1982)
196. A. Rachid, F. Collet, Bus CAN, in *Techniques de l'Ingénieur—Traité Informatique Industrielle*, vol. S 8 140 (Editions Techniques de l'Ingéieur, Paris, 2000)
197. S.V. Rakovic, E.C. Kerrigan, K.I. Kouramas, D.Q. Mayne, Invariant approximations of the minimal robust positively invariant set. IEEE Transactions on Automatic Control **50**(3), 406–410 (2005)
198. A. Ramirez, R. Garrido, S. Mondie, Integral retarded velocity control of DC servomotors, in *11th IFAC Workshop on Time-Delay Systems*, Grenoble, France, February 2013
199. H. Rehbinder, M. Sanfridson, Integration of off-line scheduling and optimal control, in *Proceedings of the 12th Euromicro Conference on Real-Time Systems*, Stockholm, Sweden, June 2000
200. H. Rehbinder, M. Sanfridson, Scheduling of a limited communication channel for optimal control. Automatica **40**(3), 491–500 (2004)
201. U. Rettig, O. von Stryk, Numerical optimal control strategies for semi-active suspension with electrorheological fluid dampers, in *Fast Solution of Discretized Optimization Problems* (Springer, Berlin, 2001), pp. 221–241
202. J.-P. Richard, Time-delay systems: an overview of some recent advances and open problems. Automatica **39**(10), 1667–1694 (2003)
203. D. Robert, O. Sename, D. Simon, Sampling period dependent RST controller used in control/scheduling co-design, in *16th IFAC World Congress*, Prague, Czech Republic, July 2005
204. W.J. Rugh, *Linear System Theory* (Prentice Hall, New York, 1996)
205. S. Samii, A. Cervin, P. Eles, Z. Peng, Integrated scheduling and synthesis of control applications on distributed embedded systems, in *Proceedings of the Conference on Design, Automation and Test in Europe*, Dresden, Germany, March 2009, pp. 57–62
206. S. Samii, P. Eles, Z. Peng, A. Cervin, Quality-driven synthesis of embedded multi-mode control systems, in *Proceedings of the Conference on Design, Automation and Test in Europe*, Dresden, Germany, July 2009, pp. 864–869

207. L. Schenato, To zero or to hold control inputs with lossy links? IEEE Transactions on Automatic Control **54**(5), 1093–1099 (2009)
208. D. Seto, J.P. Lehoczky, L. Sha, K.G. Shin, On task schedulability in real-time control systems, in *Proceedings of the 17th IEEE Real-Time Systems Symposium*, Los Alamitos, CA, USA, December 1996
209. A. Seuret, J.-P. Richard, Control of a remote system over network including delays and packet dropout, in *17th IFAC World Congress*, Seoul, Korea, 2008, pp. 6336–6341
210. G. Seyboth, D.V. Dimarogonas, K.H. Johansson, Control of multi-agent systems via event-based communication, in *18th IFAC World Congress*, Milano, Italy, 2011
211. L. Sha, T. Abdelzaher, K.-E. Årzén, A. Cervin, T. Baker, A. Burns, G. Buttazzo, M. Caccamo, J. Lehoczky, A.K. Mok, Real-time scheduling theory: a historical perspective. Real-Time Syst. **28**(2–3), 101–155 (2004)
212. D. Simon, D. Robert, O. Sename, Robust control/scheduling co-design: application to robot control, in *Proceedings of the 11th IEEE Real-Time and Embedded Technology and Applications Symposium*, San Francisco, USA, March 2005
213. B. Sinopoli, C. Sharp, L. Schenato, S. Schaffert, S.S. Sastry, Distributed control applications within sensor networks. Proc. IEEE **91**(8), 1235–1246 (2003)
214. J. Skaf, S. Boyd, Analysis and synthesis of state-feedback controllers with timing jitters. IEEE Transactions on Automatic Control **54**(3), 652–657 (2009)
215. S. Skogestad, I. Postlethwaite, *Multivariable Feedback Control: Analysis and Design* (Wiley, New York, 1996)
216. M. Smith Jr., An evaluation of adaptive sampling. IEEE Transactions on Automatic Control **16**(3), 282–284 (1971)
217. Y. Sorel, Syndex: system-level cad software for optimizing distributed real-time embedded systems. J. ERCIM News **59**, 68–69 (2004)
218. E. Sozer, M. Stojanovic, J. Proakis, Underwater acoustic networks. IEEE J. Ocean. Eng. **25**(1), 72–83 (2000)
219. A. Speranzon, J. Silva, J. de Sousa, K.H. Johansson, On collaborative optimization and communication for a team of autonomous underwater vehicles, in *Proceedings of Reglermöte*, Stockholm, Sweden, May 2004
220. M. Stojanovic, Recent advances in high-speed underwater acoustic communications. IEEE J. Ocean. Eng. **21**(2), 125–136 (1996)
221. T. Su, S. Longo, G. Herrmann, P. Barber, Computation of an optimal communication schedule in a nonlinear networked control system using sum-of-squares. Systems and Control Letters **61**(3), 387–396 (2012)
222. Y.S. Suh, Stability and stabilization of nonuniform sampling systems. Automatica **44**(12), 3222–3226 (2008)
223. Y.S. Suh, V.H. Nguyen, Y.S. Ro, Modified Kalman filter for networked monitoring systems employing a send-on-delta method. Automatica **43**(2), 332–338 (2007)
224. X.-M. Sun, G.-P. Liu, D. Rees, W. Wang, A novel method of stability analysis for networked control systems, in *17th IFAC World Congress*, Seoul, Korea, July 2008, pp. 4852–4856
225. P. Tabuada, Event-triggered real-time scheduling of stabilizing control tasks. IEEE Transactions on Automatic Control **52**(9), 1680–1685 (2007)
226. P. Tabuada, X. Wang, Preliminary results on state-trigered scheduling of stabilizing control tasks, in *45th IEEE Conference on Decision and Control*, San Diego, USA, December 2011
227. S. Tatikonda, S. Mitter, Control under communication constraints. IEEE Transactions on Automatic Control **49**(7), 1056–1068 (2004)
228. Automatic Control Laboratory. http://control.ee.ethz.ch/~hybrid/hysdel/. ETH. HYSDEL
229. UC Berkeley EECS Department. http://ptolemy.eecs.berkeley.edu/ptolemyII/. Ptolemy
230. UC EECS Department. http://embedded.eecs.berkeley.edu/research/hytech/. HyTech
231. J.-P. Thomesse, A review of the fieldbuses. Annu. Rev. Control **22**, 35–45 (1998)
232. K. Tindell, A. Burns, Guaranteeing message latencies on control area network (CAN), in *Proceedings of the 1st International CAN Conference*, Mainz, Germany, September 1994

233. K. Tindell, A. Burns, A.J. Wellings, Calculating controller area network (CAN) message response times. Control Eng. Pract. **3**(8), 1163–1169 (1995)
234. R. Tomovic, G. Bekey, Adaptive sampling based on amplitude sensitivity. IEEE Transactions on Automatic Control **11**(2), 282–284 (1966)
235. R. Tomovic, G. Bekey, Sensitivity of discrete systems to variation of sampling interval. IEEE Transactions on Automatic Control **11**(2), 284–287 (1966)
236. E. Tovar, F. Vasques, Cycle time properties of the Profibus timed-token protocol. Comput. Commun. **22**(13), 1206–1216 (1999)
237. K. Tsumura, H. Ishii, H. Hoshina, Tradeoffs between quantization and packet loss in networked control of linear systems. Automatica **45**(12), 2963–2970 (2009)
238. TTTech, Time-triggered protocol TTP/C—high-level specification document—protocol version 1.1, November 2003
239. UPPAAL, Department of Information Technology, Aalborg University. http://www.uppaal.org/
240. N. Van de Wouw, P. Naghshtabrizi, M.B.G. Cloosterman, J.P. Hespanha, Tracking control for sampled-data systems with uncertain time-varying sampling intervals and delays. Int. J. Robust Nonlinear Control **20**(4), 387–411 (2010)
241. M. Velasco, P. Martí, E. Bini, On Lyapunov sampling for event-driven controllers, in *48th IEEE Conference on Decision and Control*, Shanghai, China, December 2009
242. P. Velasco, P. Martí, J. Fuertes, The self triggered task model for real-time control systems, in *Proceedings of the Work-in-Progress Session 24th IEEE Real-Time Syst. Symp. (RTSSÓ3)*, 2003
243. G.C. Walsh, O. Beldiman, L.G. Bushnell, Error encoding algorithms for networked control systems. Automatica **38**(2), 261–267 (2002)
244. G.C. Walsh, H. Ye, Scheduling of networked control systems. IEEE Control Syst. Mag. **21**(1), 57–65 (2001)
245. G.C. Walsh, H. Ye, L.G. Bushnell, Asymptotic behavior of nonlinear networked control systems. IEEE Trans. Autom. Control **46**(7), 1093–1097 (2001)
246. J. Wang, H. Mounier, A. Çela, S.-I. Niculescu, Event driven intelligent PID controllers with applications to motion control, in *18th IFAC World Congress*, Milano, Italy, August 2011
247. X. Wang, M.D. Lemmon, Event design in event-triggered feedback control systems, in *47th IEEE Conference on Decision and Control 2008*, Cancun, Mexico, December 2008
248. X. Wang, M.D. Lemmon, State based self-triggered feedback control systems with $L_2$ stability, in *17th IFAC World Congress*, Seoul, Korea, July 2008
249. X. Wang, M.D. Lemmon, Self-triggered feedback control systems with finite-gain $l_2$ stability. IEEE Transactions on Automatic Control **54**(3), 452–467 (2009)
250. X. Wang, M.D. Lemmon, Self-triggering under state-independent disturbances. IEEE Transactions on Automatic Control **55**(6), 1494–1500 (2010)
251. X. Wang, M.D. Lemmon, Event-triggering in distributed networked control systems. IEEE Transactions on Automatic Control **56**(3), 586–601 (2011)
252. B. Wittenmark, J. Nilson, M. Torngren, Timing problems in real-time control systems, in *Proceedings of the American Control Conference*, Seattle, Washington, USA, 1995
253. W.S. Wong, R.W. Brockett, Systems with finite communication bandwidth constraints—part I: state estimation problems. IEEE Transactions on Automatic Control **42**(9), 1294–1299 (1997)
254. W.S. Wong, R.W. Brockett, Systems with finite communication bandwidth constraints—part II: stabilization with limited information feedback. IEEE Transactions on Automatic Control **44**(5), 1049–1053 (1999)
255. F. Xia, Y. Sun, Control-scheduling codesign: a perspective on integrating control and computing. J. Dyn. Contin. Discret. Impuls. Syst., Ser. B (2006)
256. L. Xie, H. Zhou, C. Zhang, $H_2$ optimal deconvolution of periodic IIR channels: an LMI approach, in *Proceedings of the 6th International Symposium on Signal Processing and Its Applications*, Kuala-Lumpur, Malaysia, August 2001

257. T. Yang, A. Gerasoulis, List scheduling with and without communication delays. Parallel Comput. **19**(12), 1321–1344 (1993)
258. Y. Ye, A fully polynomial-time approximation algorithm for computing a stationary point of the general linear complementarity problem. Math. Oper. Res. **18**(2), 334–345 (1993)
259. J.K. Yook, D.M. Tilbury, N.R. Soparkar, Trading computation for bandwidth: reducing communication in distributed control systems using state estimators. IEEE Trans. Control Syst. Technol. **10**(4), 503–518 (2002)
260. D. Yue, Q.-L. Han, C. Peng, State feedback controller design of networked control systems. IEEE Transactions on Circuits and Systems II **51**(11), 640–644 (2004)
261. S. Zampieri, Trends in networked control systems, in *17th IFAC World Congress*, Seoul, Korea, July 2008
262. L. Zhang, D. Hristu-Varsakelis, Stabilization of networked control systems: communication and controller co-design, in *16th IFAC World Congress*, Prague, Czech Republic, September 2005
263. W. Zhang, M.S. Branicky, Stability of networked control systems with time-varying transmission period, in *Allerton Conference on Communication Control and Computing*, Allerton, 2001
264. W. Zhang, M.S. Branicky, S.M. Phillips, Stability of networked control systems. IEEE Control Syst. Mag. **21**(6), 84–99 (2001)
265. W.-A. Zhang, L. Yu, Stabilization of sampled-data control systems with control inputs missing. IEEE Transactions on Automatic Control **55**(2), 447–452 (2010)
266. H. Zhou, L. Xie, C. Zhang, A direct approach to $H_2$ optimal deconvolution of periodic digital channels. IEEE Trans. Signal Process. **50**(7), 1685–1698 (2002)
267. H. Zimmermann, OSI reference model—the ISO model of architecture for open system interconnection. IEEE Trans. Commun. **28**(4), 425–432 (1980)

# Index



A. Çela et al., *Optimal Design of Distributed Control and Embedded Systems*, Communications and Control Engineering, DOI 10.1007/978-3-319-02729-6,

MIX
Papier aus verantwortungsvollen Quellen
Paper from responsible sources
FSC® C105338

If you have any concerns about our products,
you can contact us on
**ProductSafety@springernature.com**

In case Publisher is established outside the EU,
the EU authorized representative is:
**Springer Nature Customer Service Center GmbH**
**Europaplatz 3, 69115 Heidelberg, Germany**

Printed by Libri Plureos GmbH
in Hamburg, Germany